MODERN CARPENTRY

MODERN CARPENTRY

Robert F. Baudendistel

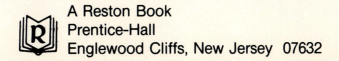
A Reston Book
Prentice-Hall
Englewood Cliffs, New Jersey 07632

© 1986 by **Prentice-Hall**
    A Division of Simon & Schuster, Inc.
    Englewood Cliffs, NJ 07632

10 9 8 7 6 5 4 3 2 1

PRINTED IN THE UNITED STATES OF AMERICA

I bestow my thanks to the following for all their help and encouragement;
My wife Marty,
My mother Ida and my aunt Freda,
My five wonderful children, David, Betsy, Stacy, Jason and Jessica,
My friend Rennie Richmond Schwartz
and most of all, to my greatest joy:
My grand-daughter Jennifer Nicole

CONTENTS

PREFACE

Modern Carpentry presents the fundamentals of carpentry in an easy to read, logical presentation to fulfill the needs of various segments of our population. Students in high school, vocational school, building construction programs as well as those taking college level carpentry courses will all benefit from the wealth of information presented within, as will homeowners and those enrolled in various apprenticeship programs.

The sequence of chapters has been designed to direct its readers through the various steps that are involved in the actual construction of a residence. Diagrams and tables are used to complement and reinforce the text so that its readers will have no difficulties in their comprehension of the written material.

The text has several unique features that elevate it far above its competition. The chapter entitled *Techniques, Tips and Time Savers* is designed to acquaint the novice with information usually only associated with seasoned veterans.

The *Alternate Building Methods* chapter offers its readers up to date information on several modern day building methods to enable them to intelligently converse with anyone intent on employing any of these methods.

Even though the chapters dealing with home heating, plumbing and electricity are not directly related to the carpentry profession, they have been included to help broaden the basic knowledge and understanding of every carpenter. The information presented should enable the carpenter to more competently cooperate with any subcontractor specializing in any of these fields during the construction of a new home, or to attempt the actual installation without the aid of the subcontractor.

The chapter entitled *Alterations and Additions* is timely with respect to a large segment of our nation's homeowners. Because of high interest rates, many homeowners have decided to "stay put" and make alterations or additions to their present dwelling rather than to move into a larger, more expensive residence.

The feasibility of harnessing the sun's energy to reduce the consumption of fossil fuels is thoroughly discussed in specific areas throughout the text. The textbook also acquaints the reader with information concerning the other alternate "free" sources of energy that may be harnessed to heat or light our homes.

The concerns for energy conservation are emphasized in pertinent areas throughout the text. Important features such as house placement, landscaping effects, thermal insulation and adequate ventilation, must all be thoroughly understood before they can be treated properly to achieve the optimum in energy conservation.

The final chapter, concerned with the various methods used to estimate quantities of materials, should prove invaluable to any carpenter who wants to convert from employee to employer. A skilled artisan that remains dependent upon others to estimate materials, as well as labor costs, will always remain an employee.

The carpentry profession encompasses all those individuals who build with wood. As a profession, it has enormous appeal because it provides both the young apprentice and the seasoned artisan with a daily sense of accomplishment. The object created, ranging from the simplest concrete form to a complex roofing design, provides its creator with a deep sense of satisfaction.

Every carpenter can personally appreciate "the fruits of his or her labors," a living testimony to their abilities as a carpenter, and have the creation acknowledged by society as well.

Because of the diversity which exists within the carpentry profession, the employment possibilities are endless. For example, one individual can specialize in assembling concrete forms, another in cabinetmaking, while a third individual may prefer to concentrate on applying the finish roofing material to the roof frame.

Hopefully, the impact of this textbook on its readers will encourage them to pursue carpentry as their life's chosen profession.

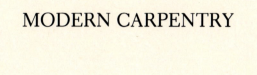

MODERN CARPENTRY

1

TOOLS

As a carpenter, you must possess the ability and knowledge to

1. Select quality tools that are durable, efficient, well-constructed, and easily maintained
2. Use your tools properly, effectively, and safely to produce a quality product
3. Keep your tools in top shape by performing the necessary periodic maintenance

SELECTION OF TOOLS

Your first concern, when purchasing any tool, is that it will provide you with many years of trouble-free service and enable you to produce a quality product. This need will often prevent you from selecting a "bargain" tool. On the other hand, many of the larger tool manufacturers guarantee their hand tools against defects. So, even though the initial price of a quality tool may be higher than another "questionable brand," it assures life-time service that will save you money in the long run.

For some tools, the intended purpose may restrict your selection. For example, a 12-ounce claw hammer, as opposed to the customary 16-, 20-, or 22-ounce hammer, would not be suitable for you to use to rough frame a house, but would work admirably to nail the inside trim in place.

Every craftsperson should realize that the purchase of quality tools is a never-ending process designed to eliminate or reduce some laborious procedure as well as to improve the quality of the finished project.

BASIC HAND TOOLS

The basic hand tools needed by every aspiring carpenter are discussed in the following categories: measuring, marking, cutting, fastening (and pulling apart), drilling (boring) and finishing.

Measuring Tools

Since measurement plays such a vital role in every carpentry project, every carpenter should possess the following:

1. A flexible, retractable steel tape (Figure 1:1) that has a hooked end to hold on to the edge of a piece of

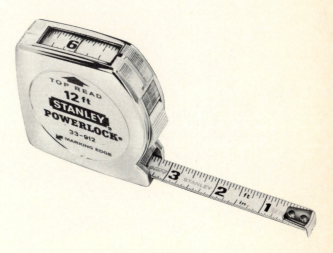

FIGURE 1:1

wood. The most popular lengths are 12′ and 16′. The base of its case is also marked (either 2″ or 3″) to allow for inside measurements (Figure 1:2). The various graduations printed on carpenter's rules are presented in Figure 1:3.

2. A 25′, 50′, or 100′ steel tape for longer measurements such as house foundations (Figure 1:4).

3. A 6′ or 8′ folding (zig-zag) wooden rule with a sliding extension to make precise inside measurements (Figure 1:5).

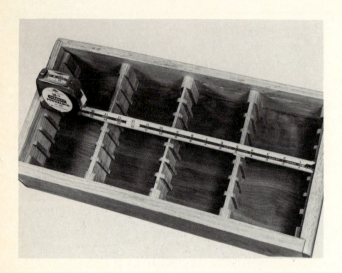

FIGURE 1:2

Sixteenth inch graduations

Tenth inch graduations

Eighth inch graduations

Twelfth inch graduations

FIGURE 1:4
Steel tape.

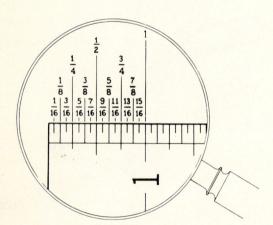

FIGURE 1:3
The various graduations used on carpenter's rules are shown here.

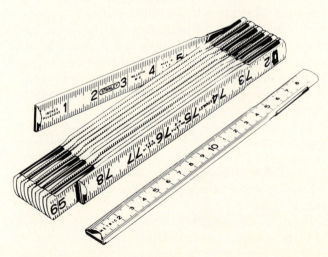

FIGURE 1:5
Folding rules are available in inch and metric calibrations.

Calipers are used to make small accurate measurements. They may be designed for inside (Figure 1:6), outside (Figure 1:7), or both (Figure 1:8) kinds of measurement. A compass or *wing divider* can be used for various purposes. It is especially effective for duplicating irregularities of one surface and transferring them to another surface (Figure 1:9).

A carpenter may use several different squares throughout the daily routine. The features common to all squares are their ability to measure short lengths, to

FIGURE 1:9
Wing dividers or compass duplicates irregularities of one surface on another.

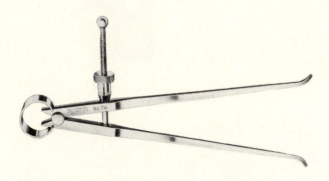

FIGURE 1:6

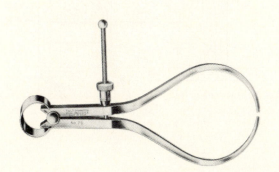

FIGURE 1:7

FIGURE 1:8

check for angles (mainly right angles) and to test for square. Each type, however, has one or more unique features.

The *try square* consists of a 6″ to 12″ long steel blade that is attached to a squared-off wooden handle (Figure 1:10). Its design allows it to be used to both check for

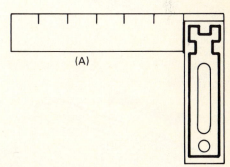

(A)

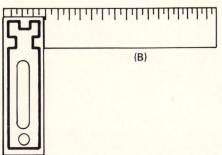

(B)

FIGURE 1:10
Typical try square. (A) inch scale. (B) metric scale.

square and measure the inside and the outside of a joint. It is also used to check the flatness of a board's edge (Figure 1:11).

A *combination square* is used to square small pieces of wood and to mark a precise right angle (90°), a miter (45°), or a straight line (Figure 1:12). The various ways that a combination square can be used are presented in Figure 1:13. In tight quarters where it is impossible to use a carpenter's level, the spirit level on the combination square can be used (Figure 1:14).

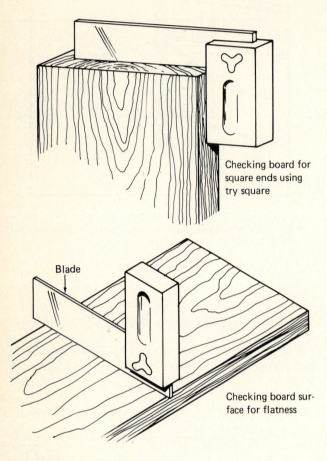

Checking board for square ends using try square

Blade

Checking board surface for flatness

Blade

Testing edge of board for square-ness by moving try square up and down

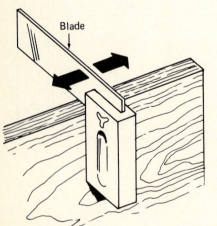

FIGURE 1:11
(A) Checking if the end of a board is square. (B) To test the surface of the work for flatness place the try square as shown, and see if the blade sits evenly or rocks. (C) To test if the edge of a board is square, hold the handle of the try square against the surface of the board and move it up and down to see if the blade constantly touches the edge of the board.

FIGURE 1:12
Combination square.

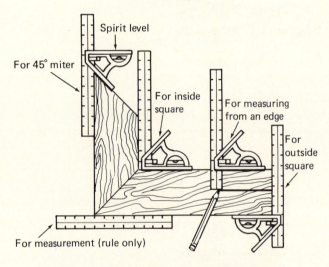

Spirit level

For 45° miter

For inside square

For measuring from an edge

For outside square

For measurement (rule only)

FIGURE 1:13
Ways to use a combination square.

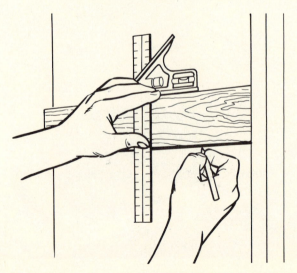

FIGURE 1:14

A *T-bevel*, or adjustable square, enables the carpenter to duplicate any angle between 0° and 180° by simply loosening the wing nut and moving the adjustable blade (Figure 1:15) to create the desired angle.

A *carpenter's framing square* has a short arm called the "tongue" that is 16″ long and 1½″ wide. The long arm of the square is called the "body." It is 24″ long and 2″ wide (Figure 1:16).

The "heel" is the outside corner where tongue and body meet. The same designation is sometimes given to the inside corner. The "face" is the front of the square and is determined by the surface having the manufacturer's name or trademark stamped on it. It is also distinguished by holding the tongue with your right hand and placing the body in front of you (Figure 1:17).

This square, also called a *rafter square*, is calibrated so that you can lay out and mark roof rafters and stair stringers using the most appropriate angle for the task at hand (Figure 1:18). This square is also used to check building corners and door and window frames for square (Figure 1:19).

FIGURE 1:16

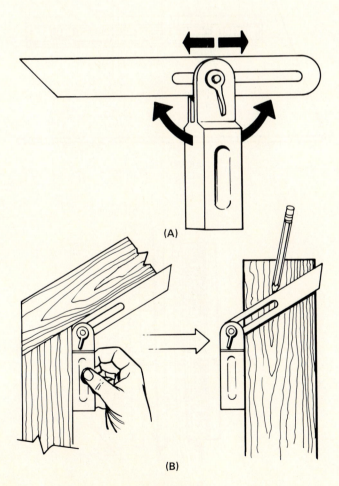

FIGURE 1:15
(A) T-bevel. (B) Sliding T-bevel transfers angles.

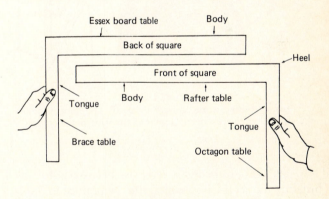

FIGURE 1:17
Parts of the framing square.

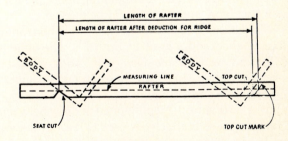

FIGURE 1:18
Rafter layout (bird's mouth).

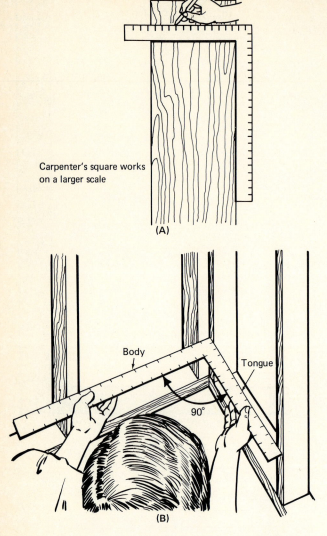

Carpenter's square works
on a larger scale

(A)

Body

Tongue

90°

(B)

FIGURE 1:19
(A) Carpenter's square works on a larger scale. (B) The corner is square
if body and tongue rest neatly against all the structural members.

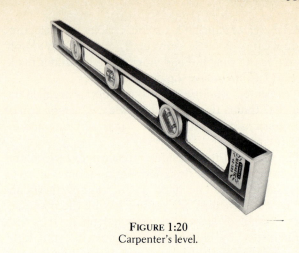

FIGURE 1:20
Carpenter's level.

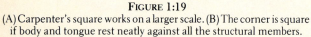

FIGURE 1:21
Level both directions.

A *24″–28″ carpenter's level* is necessary for all
building projects. It enables the carpenter to both
plumb (vertical) and level (horizontal) structural framing
members and foundations (Figure 1:20). Without its
use, walls would lean, shelves would tilt, doors and windows would bind, and floors would be uneven.

It is imperative that both directions of a horizontal
surface be checked for level (Figure 1:21). The same is
true for any vertical member; both sides must be
checked for plumb, especially those framing members
that comprise a corner (Figure 1:22).

A line level (Figure 1:23) positioned midway between
two remote points is effectively used to level extra-long
spans, to anticipate how to construct the foundation, as
well as to calculate existing grades.

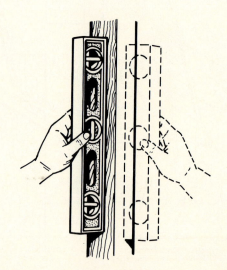

FIGURE 1:22
Plumb both sides.

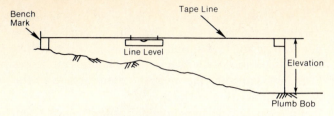

FIGURE 1:23
Line level.

To establish true vertical lines, a *plumb bob* is used. It is excellent for locating the exact location directly below a fixed point (Figure 1:24).

A roll of nylon *string* has no equal when a level line is needed between two fixed points. When stretched taut, it does not sag nor is it affected by adverse weather conditions. It is used for both the line level and the plumb bob.

Marking Tools

Most building materials are marked with either an ordinary #2 pencil or a *carpenter's pencil* (Figure 1:25). The carpenter's pencil is a thick, flat pencil containing a thick core of graphite. It is ideal for rough framing work because it makes a coarse line that can be easily seen, and pieces so labeled are easily recognizable long after they have been marked.

FIGURE 1:24
Locating a point with a plumb bob.

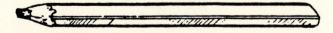

FIGURE 1:25
Carpenter's pencil.

Figure 1:26 shows a cut-off mark being made on a board and the resultant "X" label indicating the waste portion of the board.

A *chalk line* (Figure 1:27) is used to "snap" long, straight lines on paneling, wallboard, and lumber. Once both ends of the line are positioned over the predetermined marks, the line is drawn taut and then is snapped in the center. This procedure should be done only once (Figure 1:28).

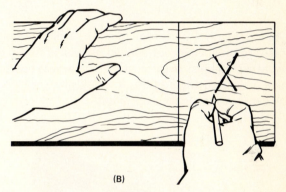

(A)

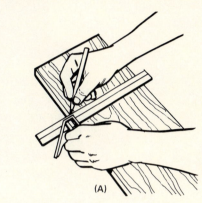

(B)

FIGURE 1:26

FIGURE 1:27

A *marking gauge* (Figure 1:31) is used to make lines parallel to the edge of a board. The desired width is determined and transferred to the gauge, which is then locked in position with its setscrew. The steel pin scores a straight, knife-like line over the entire length of the wood's surface (Figure 1:32).

A *rule* can be used like a marking gauge to mark a continuous line parallel to the side or edge of a piece of wood (Figure 1:33).

FIGURE 1:28

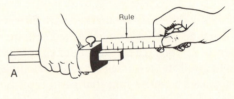

FIGURE 1:31

An *awl* (Figure 1:29) is used to make small pilot holes for screws and to mark cut-off lines on lumber. It can be used to punch holes in soft materials, as well as to score plastic laminates (Figure 1:30). It should always be used with care because of its sharp point.

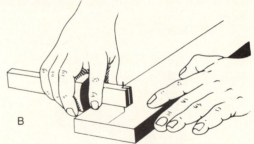

FIGURE 1:32

FIGURE 1:29
Awl.

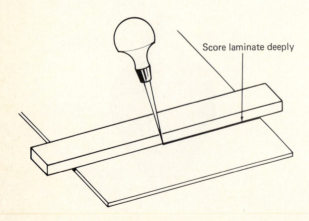

FIGURE 1:30

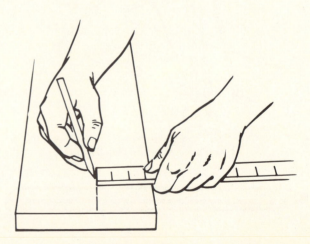

FIGURE 1:33

A *contour gauge* (Figure 1:34) has the ability to transfer an odd shape from one material, such as a molding, to another. The one side of the gauge reproduces the same shape as the object being copied while the other side gives the complementary shape.

A *drywall square* (Figure 1:35), resembling a large, 4' T-square, is used to cut drywall panels along straight lines. The square is placed over the panel so that a razor knife can follow its edge while making the initial cut through the top surface of the panel (Figure 1:36).

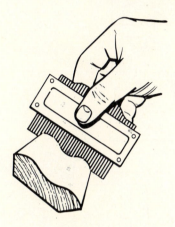

FIGURE 1:34

dry wall squares

FIGURE 1:35

FIGURE 1:36

Cutting Tools

Every good carpenter should own at least six different types of hand saws. The *backsaw* (miter saw) with its complementing miter box is needed to cut perfect miter joints. It is easily identified as the saw having a reinforced metal back for rigidity (Figure 1:37).

To trim or "cope" these miter joints, a *coping saw* is used (Figure 1:38). It is also used to make intricate, unusually shaped cuts in materials as varied as wood, plastic, and metal.

The *crosscut* and *rip saws* appear to be identical. The purpose intended for each, with respect to how each is expected to cut lumber, requires that each saw type have a different tooth layout. The *crosscut saw* cuts lumber across the grain, has eight to eleven cutting points per inch of the cutting edge, and is 26" long (Figure 1:39).

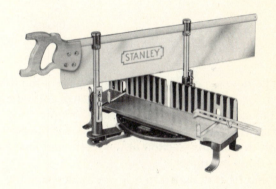

FIGURE 1:37
Backsaw and miter box.

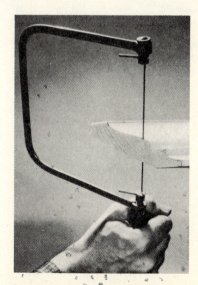

FIGURE 1:38

TOP VIEW OF CROSS-CUT TEETH

By sighting along the top of the teeth you will notice a "V" groove down which a needle will slide when the saw is properly set and filed.

Look down on the teeth and notice that the teeth are set evenly about ¼ the thickness of the blade.

FIGURE 1:39
Crosscut hand saw. (Courtesy of Stanley Tools).

The *ripsaw* is used to cut with the grain of the wood, has five to seven cutting points per inch of cutting edge, and ranges in length from 26″ to 30″ (Figure 1:40).

Figure 1:41 compares the two saws with respect to tooth shape, cross section of the teeth, and how each cuts through lumber.

Table 1:1 is designed to help you understand the in-

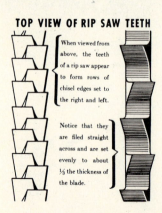

TOP VIEW OF RIP SAW TEETH

When viewed from above, the teeth of a rip saw appear to form rows of chisel edges set to the right and left.

Notice that they are filed straight across and are set evenly to about ⅓ the thickness of the blade.

FIGURE 1:40
Ripsaw.

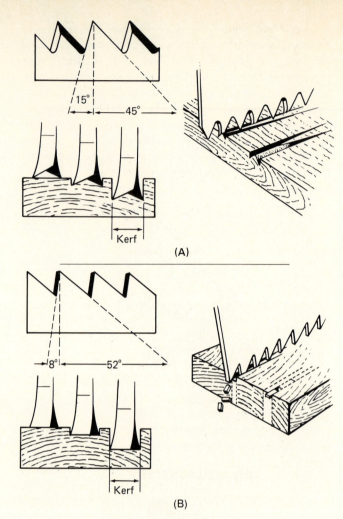

FIGURE 1:41
(A) Top: Shape of crosscut teeth. Bottom: Cross section of crosscut teeth. Right: How crosscut saw cuts. (B) Top: Shape of ripsaw teeth. Bottom: Cross section of rip teeth. Right: How ripsaw cuts.

herent differences between a six-point and an eleven-point saw.

TABLE 1:1

Number of points	Size of teeth	Speed of cut	Quality of cut
6-point saw	Large	Fast	Rough
11-point saw	Small	Slow	Fine

Thus, for fine cuts, a saw having a large number of points per inch along its cutting edge (blade) should be selected. Usually, the number of points is clearly stamped on the blade of a saw.

Figure 1:42 shows how the number of points for a hand saw is determined. Note that a saw having eight points per inch has only seven teeth spanning the same distance. This measurement must be taken from the tip of the two points.

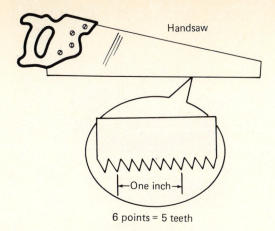

FIGURE 1:42
Hand saw.

A crosscut saw should be used at a 45° angle with the face of the work (Figure 1:43). Its teeth are set (bent) alternately right and left, so they make a saw cut, or "kerf," wider than the thickness of the saw blade (Figure 1:44). This design prevents the saw from binding on the cut surface. It performs a dual-action cutting motion as it travels through the wood, since it cuts both as a knife and as a chisel.

A ripsaw has chisel-like teeth which are used to cut wood fibers along the grain (Figure 1:45). It should be held at a 60° angle with the face of the work in order to be effective (Figure 1:46).

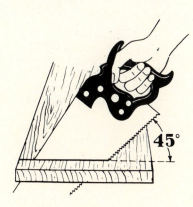

FIGURE 1:43

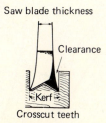

Saw blade thickness

Clearance

Kerf

Crosscut teeth

FIGURE 1:44

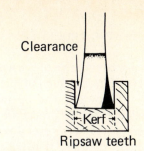

Clearance

Kerf

Ripsaw teeth

FIGURE 1:45

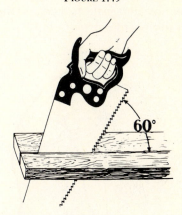

FIGURE 1:46

With any hand saw, every attempt should be made to keep the saw's handle directly above the guide line to ensure a straight cut (Figure 1:47).

The *hacksaw* is used to cut metals. Most are designed with an adjustable frame, or bow, to accept various blade lengths. It is ideal for cutting through nails, bolts, screws, and metal pipes (Figure 1:48).

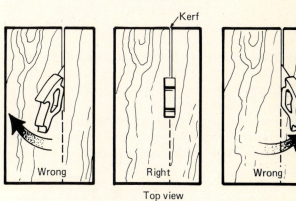

Kerf

Wrong   Right   Wrong

Top view

FIGURE 1:47

FIGURE 1:48
Hacksaw.

SAWING PROCEDURE

To avoid damage to the edge of the wood, the saw is first placed over the marked cut line before it is drawn backwards towards the operator (Figure 1:49). The balance of the procedure involves a push-pull operation until the excess is cut off. It is advisable to hold the piece that is being cut off when the saw nears the completion of its cut, so that it does not damage the rear edge as the cut-off piece falls to the ground.

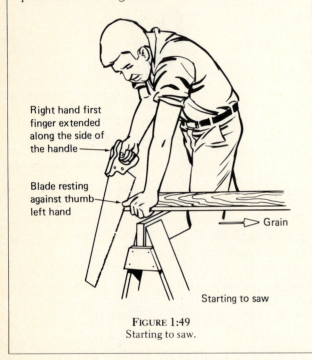

FIGURE 1:49
Starting to saw.

Both the *keyhole* and the *compass saws* are used to cut curves, to make irregular cuts once a starting hole has been made, to be used in tight locations and to cut out electrical boxes (Figure 1:50).

A *retractable utility knife* or a *pocket knife* is used to cut a variety of building materials. Because of their razor-sharp blades, both tools must be used with care (Figure 1:51).

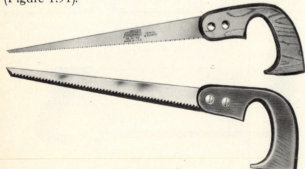

FIGURE 1:50
(A) Compass saw. (B) Keyhole saw.

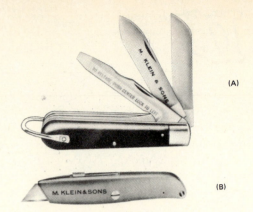

FIGURE 1:51
(A) Pocket knife with two folding blades. (B) Utility knife.

Fastening Tools

Claw hammers are used to drive nails into wood. In rough construction work, either a 16-, 20-, or 22-ounce hammer can be used; for finish work, a 12- or 16-ounce hammer is best (Figure 1:52). The number of ounces refers to the weight of the hammer's drop-forged steel head. A hammer having a curved head is better suited to the removal of nails than the straight or rip claw type which is better suited to rip apart pieces already fastened together (Figure 1:53).

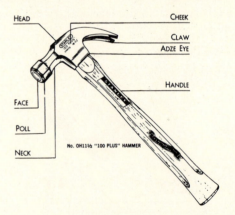

FIGURE 1:52

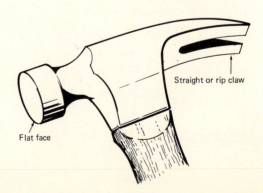

FIGURE 1:53

The face of a hammer can be either flat or bell-shaped. The bell-shaped face is used in finishing work because it is less likely to display hammer marks on the wood (Figure 1:54).

Make certain that you purchase a drop-forged hammer head that is permanently bonded to a steel or fiberglass handle. Wood handles and cast iron hammer heads are weak and very dangerous.

Other hammers that a carpenter should own include the ball-pein, the hatchet, and the soft face (Figure 1:55).

A *stapler* is used to apply building paper over roof and exterior sheathing. The hammer type is faster, holds twice as many staples, and is less tedious to use than the squeeze type, which is better suited for the installation of ceiling tile (Figure 1:56).

Every carpenter should have a complete set of *screwdrivers,* both Phillips and slotted, for installing and removing wood screws (Figure 1:57). A screwdriver must be matched perfectly to both the size and type of screw being driven or removed. Using an improper screwdriver will damage screw heads and make the screws difficult to remove later.

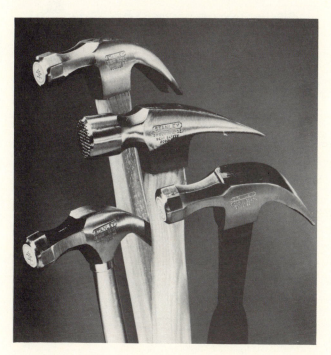

FIGURE 1:54

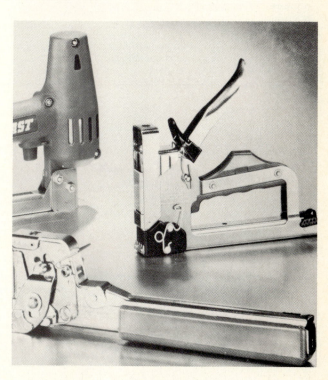

FIGURE 1:56

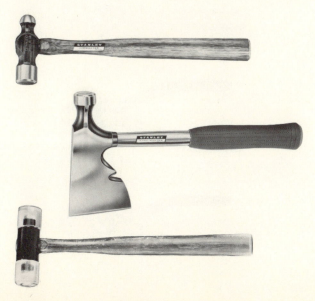

FIGURE 1:55

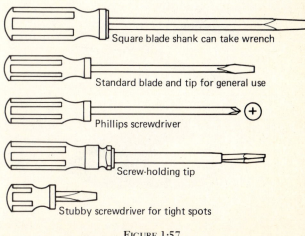

Square blade shank can take wrench

Standard blade and tip for general use

Phillips screwdriver

Screw-holding tip

Stubby screwdriver for tight spots

FIGURE 1:57
Screwdrivers.

A spiral ratchet screwdriver is ideal to use whenever a large number of screws must be driven by hand (Figure 1:58).

An offset screwdriver is often the only type that can be used to drive or remove screws that are in locations where a normal screwdriver could not be used or turned (Figure 1:59).

Various *wrenches* are needed to fasten nuts, bolts, and lag screws whenever they are used (Figure 1:60).

Clamps are used by carpenters to hold glued pieces together while the glue sets and dries, and also to tem-

porarily hold pieces together while another function is performed. C, bar, parallel, pipe, and spring are the types of clamps that can be used (Figure 1:61).

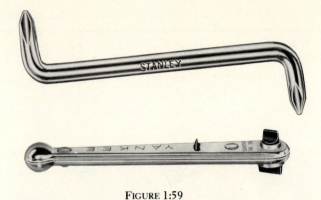

FIGURE 1:59

FIGURE 1:58
Spiral rachet screwdriver.

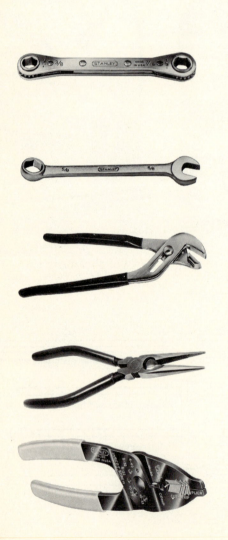
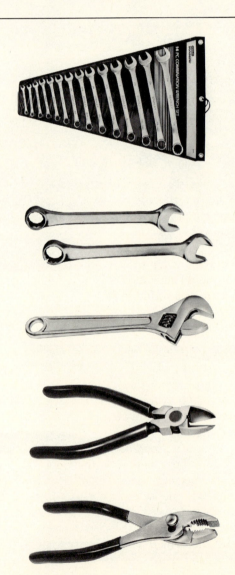

FIGURE 1:60
Pliers and wrenches.

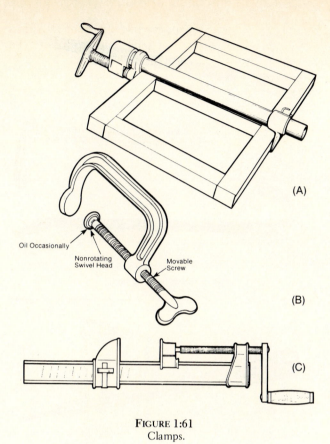

FIGURE 1:61
Clamps.

It must be used with care when prying apart materials that will be reused so they are not marred or damaged.

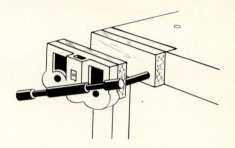

FIGURE 1:62
Bench vise.

FIGURE 1:63
Nail Sets.

FIGURE 1:64
Nail puller.

FIGURE 1:65

A *bench vise* (Figure 1:62) can also be used to hold wood. Note that the vise's jaws are lined with wood to avoid damaging the wood being held by the vise.

To recess the head of a finishing nail beneath the wood's surface, a nail set (punch) is used (Figure 1:63). Since there are differences in the size of nail heads, a carpenter should have several sizes and may prefer a set with a square or a round head.

Tools for Prying and Pulling Lumber Apart

A *nail puller* (Figure 1:64) is used to remove bent or imbedded nails by placing its pointed end beneath the nail head and driving the other end with a hammer (Figure 1:65). To avoid marring the wood's surface, a small piece of lumber should be placed between the nail puller and the wood.

A *pry bar* or *offset ripping chisel* (Figure 1:66) can be used to remove bent nails or to pry structural members apart. It is especially useful for alteration work. The longer the bar, the more leverage you have to pry materials apart.

The *ripping chisel* (Figure 1:67) resembles the ripping bar on one end, which has two slots for pulling nails.

The *crow bar* (Figure 1:68) has a curved end with a slot to pull nails and a tapered end for prying. It resembles a ripping bar except that both ends lie in the same plane.

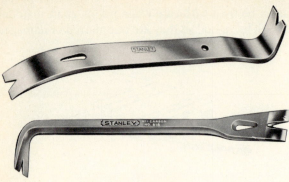

FIGURE 1:66
(A) Pry bar. (B) Offset ripping chisel.

FIGURE 1:67
The ripping chisel.

FIGURE 1:68
The crow bar (ripping bar).

Drilling Tools

There are three hand tools that a carpenter uses to make holes in wood. For small holes, up to $\frac{11}{64}$", a carpenter can use an *automatic push drill* (Figure 1:69) or a *hand drill* (Figure 1:70). The *bit brace* (Figure 1:71) is used for holes $\frac{3}{16}$" in diameter and larger. Its moveable jaws are designed to hold bits having square tapered shanks, called tangs (Figure 1:72). Most braces have a ratchet mechanism that enables the carpenter to drill a hole even when the working quarters are cramped.

FIGURE 1:69
A push drill. (Courtesy of Stanley Tools)

DRILLING TIPS

Two procedures can be used to avoid splitting out the back of the wood during a drilling operation. One method requires that you clamp a scrap piece of wood below the desired piece and drill through until the bit penetrates the scrap surface (Figure 1:73).

The other method requires that you proceed cautiously with the drilling operation until you can see the tip of the drill bit protruding through the wood. At this point you stop, remove the bit from the hole, and then direct the drill bit from the opposite side (Figure 1:74). This is especially important when drilling the holes in doors for the lock set.

To join two pieces of wood with wood screws, you must drill a hole in each board. The top board must receive a hole equal in diameter to the screw's shank. This is called the shank or anchor hole. The lower board must receive a pilot hole with a diameter similar to that of the screw's core (Figure 1:75).

Table 1-2 presents a comparison of the various drill bit types and specialty items.

FIGURE 1:70
A hand drill. (Courtesy of Stanley Tools)

Table 1:2
Drill Bits and Some Specialty Attachments

Comparison of the various drill bit types and specialty items

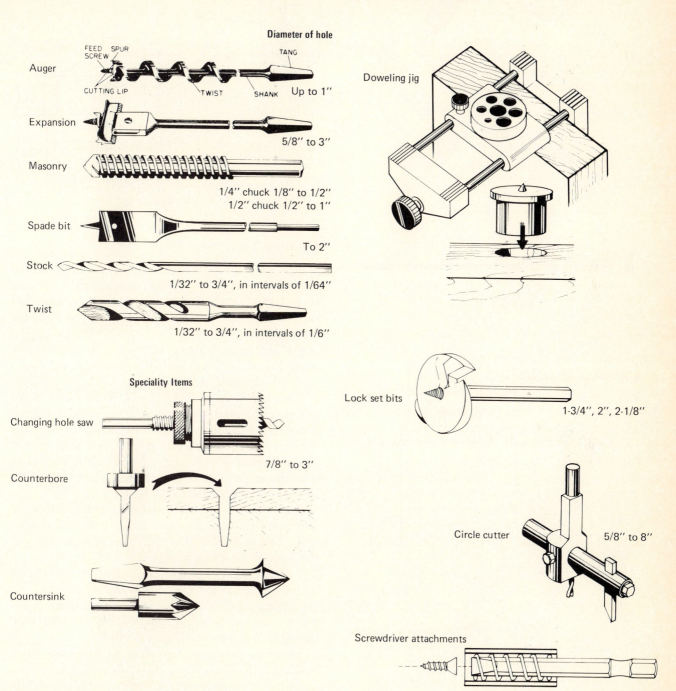

Diameter of hole

Auger
FEED SCREW
SPUR
CUTTING LIP
TWIST
SHANK
TANG
Up to 1''

Expansion
5/8'' to 3''

Masonry
1/4'' chuck 1/8'' to 1/2''
1/2'' chuck 1/2'' to 1''

Spade bit
To 2''

Stock
1/32'' to 3/4'', in intervals of 1/64''

Twist
1/32'' to 3/4'', in intervals of 1/6''

Speciality Items

Changing hole saw
7/8'' to 3''

Counterbore

Countersink

Doweling jig

Lock set bits
1-3/4'', 2'', 2-1/8''

Circle cutter
5/8'' to 8''

Screwdriver attachments

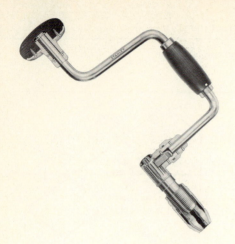

FIGURE 1:71
Parts of a brace. (Courtesy of Stanley Tools)

FIGURE 1:72

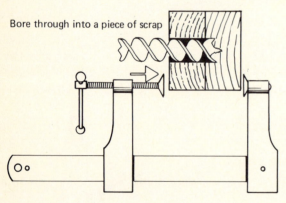

Bore through into a piece of scrap

FIGURE 1:73

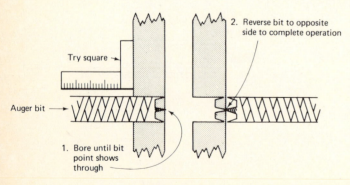

2. Reverse bit to opposite side to complete operation

Try square

Auger bit

1. Bore until bit point shows through

FIGURE 1:74

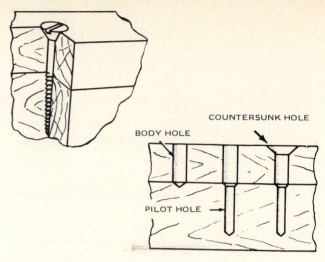

COUNTERSUNK HOLE
BODY HOLE
PILOT HOLE

FIGURE 1:75

Finishing Tools

Planes are used by the carpenter to cut, smooth, and dress the wood to its finished dimensions. There are several different types and sizes of planes available, but the two used most commonly are the jack plane and the block plane.

The *jack plane* is always used along the grain of the wood (Figure 1:76). Any item to be planed should always be firmly clamped in a vise (Figure 1:77).

The *block plane* is used to plane across the lumber's end grain (Figure 1:78). To avoid damage to the edge, it is advisable to plane in one direction up to the edge and then reverse the direction (Figure 1:79). The plane iron of a block plane is set at a much more acute angle (12°) than the angle of most conventional planes (20°–30°), as shown in Figure 1:80.

Wood chisels are used to cut, notch, and smooth wood joints and to shape mortises for door hinges. (Figure 1:81). The beveled edge of the chisel must always face the wood to be removed (Figure 1:82).

To avoid a deeper gouge than desired, make your cuts with the chisel in the direction of the cut and with the grain (Figure 1:83). When chiseling across the grain and the area is enclosed, the beveled edge of the chisel should face downward (Figure 1:84).

There are two important rules to remember when working with chisels. They are (1) use only sharp chisels and for the use they were intended and (2) keep both hands well away from the cutting edges.

The two handles of the spokeshave (Figure 1:85) enable the carpenter to do the finished shaping of most curved surfaces having small diameters.

Surform tools (Figure 1:86) are used for various finishing procedures on wood, plastics, and even soft metals. They are handled like a small plane.

FIGURE 1:76

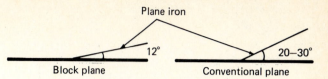

Plane iron

Block plane 12°   Conventional plane 20—30°

FIGURE 1:80

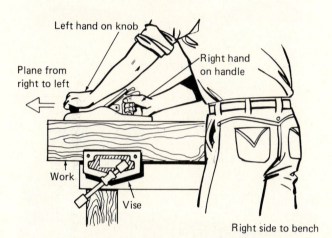

Left hand on knob

Right hand on handle

Plane from right to left

Work

Vise

Right side to bench

FIGURE 1:77

FIGURE 1:81

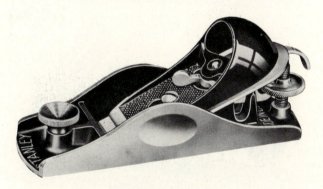

FIGURE 1:78

FIGURE 1:79

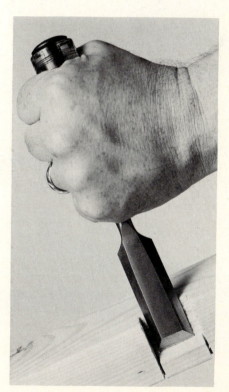

FIGURE 1:82

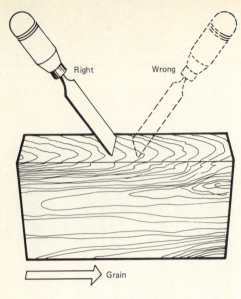

Right　Wrong

Grain

FIGURE 1:83

FIGURE 1:84

FIGURE 1:85
Spokeshave.

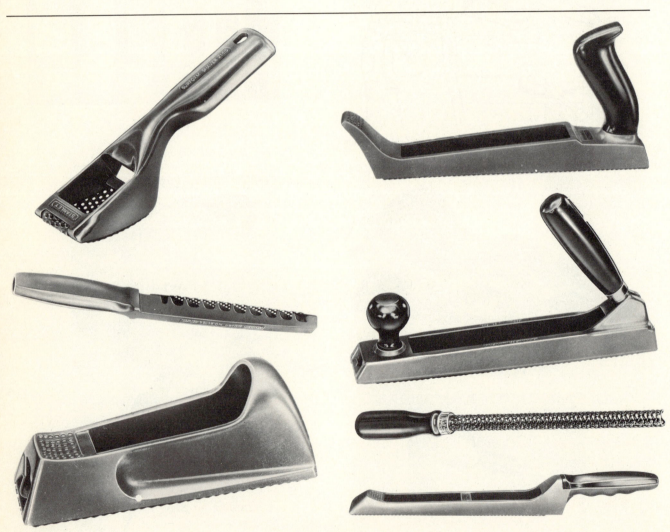

FIGURE 1:86
"Surform" tool.

A putty knife, 1″ to 2″ wide, is used to glaze glass, to apply wood fillers, as well as to remove old paint and varnish from wood (Figure 1:87).

A joint compound knife, 3″ to 10″ wide, can be used to apply joint compound to wallboard so that its joints and seams are concealed. Use joint compound, then tape, then more joint compound (Figure 1:88).

There are three criteria for classifying files. The first is cross-sectional shape. A file may be flat, round, half round, square, pillar, or triangular (Figure 1:89). The second criterion is length, which affects the file's coarseness; the longer the file is, the larger its teeth are.

The coarseness of a file's teeth is the third criterion. The range includes coarse, bastard, second cut, and, finally, dead smooth. The common cuts of a file's teeth can be single cut, double cut, rasp, or curved (Figure 1:90).

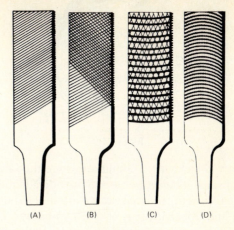

FIGURE 1:90

FIGURE 1:91
File brush clears metal particles from teeth.

A single-cut file will produce a smooth surface under light pressure. A double-cut file, using heavier pressure, will remove more material faster but the surface will not be as smooth.

Materials to be filed should be clamped in a vise. A file brush should be used to remove any build-up of metal filings in the teeth (Figure 1:91).

CORRECT POWER TOOL USAGE FOR OPTIMUM SUCCESS

Circular Saw

Start by checking the saw blade to make sure it is secure and the electrical cord for wear. Use a scrap of wood to make test cuts and precision cuts before making the actual cut.

Make certain that the base or shoe of the saw rests firmly on the piece to be cut and that it is properly supported. Depressing the trigger switch starts the motor and spins the saw blade. Slowly push the saw along the marked line until the cut has been completed.

When trimming an edge, make sure that the saw's housing passes over the material to be kept. The material to be cut off should always be to the right of the saw blade (Figure 1:92).

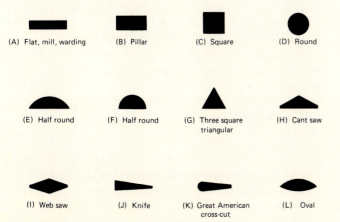

FIGURE 1:87

FIGURE 1:88

(A) Flat, mill, warding   (B) Pillar   (C) Square   (D) Round

(E) Half round   (F) Half round   (G) Three square triangular   (H) Cant saw

(I) Web saw   (J) Knife   (K) Great American cross-cut   (L) Oval

FIGURE 1:89

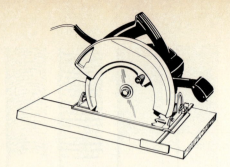

FIGURE 1:92

To avoid kickback, a condition where the saw jumps back towards its user, be sure to provide adequate support near the cut, and make every attempt to cut on the drawn line. Kickback may also be caused by a dull saw blade. Place the "good side" down, especially when cutting wood panels, so that any splintering will occur on the top surface.

When the saw is used to make angle cuts, its blade must be adjusted to the desired setting and then used to cut the material. For a successful cut, the saw's base or shoe must lie flat on the material (Figure 1:93).

A pocket cut requires the following steps

1. Adjust the blade to the desired depth.
2. Tilt the blade guard so that the saw blade nearly touches the drawn line. Then release the guard so that it rests on the material to prevent the saw blade from missing its cut.
3. As you begin to cut, gradually lower the saw blade until the base or shoe of the saw lies flat on the material; then complete the desired cut.
4. Allow the blade to come to a complete halt before withdrawing the blade from the material (Figure 1:94).

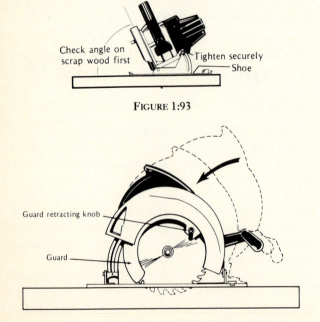

FIGURE 1:93

FIGURE 1:94
Making pocket cuts with a circular saw.

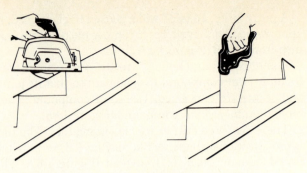

FIGURE 1:95

A hand saw may be needed to complete notched cuts that require two perpendicular cuts, such as the formation of stair stringers at a job site (Figure 1:95).

Saber Saw

Before using a *saber saw*, it is advisable to clamp down thin material such as sheet metal, formica, plastics, and composition tile to guarantee straight cuts and to minimize vibration. Excessive vibration can damage the material. Unlike circular saws, most saber saws are able to cut intricate designs in a variety of materials. The procedure for making pocket cuts and bevel angle cuts with a saber saw is similar to that explained in the circular saw discussion. See Figure 1:96.

Reciprocating Saw

Operation of a *reciprocating saw* is very similar to that of a saber saw. It has the advantage of being able to make deeper cuts than the saber saw because it can hold blades up to 10″ in length (Figure 1:97). It is especially handy for doing alterations on existing structures.

Electric Drill

Because of its adaptability to a variety of attachments, the *electric drill* is an absolute necessity for every carpenter. With just one tool—and the appropriate attachments (Figure 1:98)—you can drill holes, sand, grind, buff, and insert and remove screws (provided the drill is equipped with a reversible chuck rotation).

Other Power Tools

The *power miter saw* (Figure 1:99) provides the carpenter with a tool that is fast, accurate and light enough to be carried to and used at the building site. It can cut and duplicate any angle between 0° and 180°.

Two other saws that are used in woodworking, but which are usually too bulky to be taken to the building site, are the *bench saw* (Figure 1:100) and the *radial arm saw* (Figure 1:101). The bench saw blade is hidden

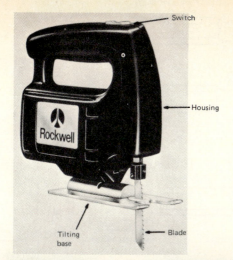

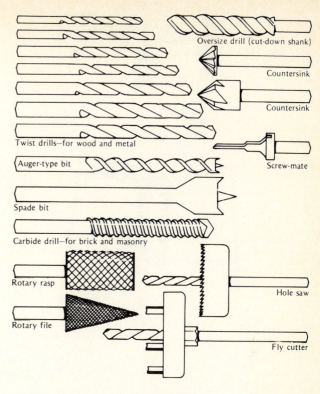

FIGURE 1:98

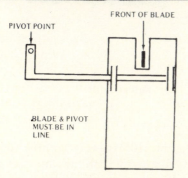

A

B

FIGURE 1:96
(A) Parts of a saber saw. (Courtesy of Rockwell) (B) Crosscutting with a saber saw.

FIGURE 1:99
Parts of a power miter saw. (Courtesy of DeWalt)

FIGURE 1:97
A reciprocating saw. (Courtesy of Rockwell)

FIGURE 1:100

FIGURE 1:101

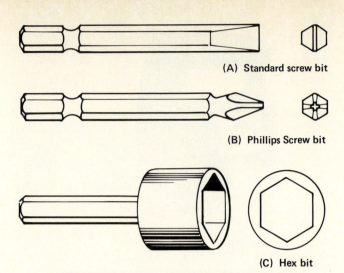

(A) Standard screw bit

(B) Phillips Screw bit

(C) Hex bit

FIGURE 1:103
(A) Standard screw bit. (B) Phillips screw bit. (C) Hex bit.

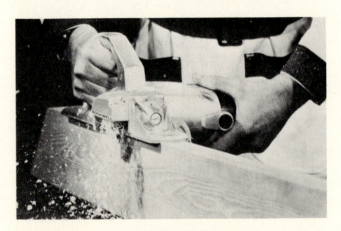

FIGURE 1:104

beneath the worktable, while the radial arm saw blade lies above the table and cuts from above.

A *power screwdriver* (Figure 1:102) is equipped with a chuck that is designed to hold three different types of bits (Figure 1:103). The chuck can be set to drive screws flush with the material's surface or countersink them below the surface. The setting works in tandem with the tool's clutch, which disengages once the screw has reached its desired destination. This tool is ideal for installing wallboard and other sheathing to wood or metal studs.

An *electric jack plane* (Figure 1:104) is much faster, produces better results, and is less fatiguing to use than a hand plane.

Screw gun

FIGURE 1:102
Screw gun.

The *router* (Figure 1:105) is a very versatile power tool that has a variety of uses for both the carpenter and the cabinet maker. By selecting the correct bit (Figure 1:106), the router can make groves and notches, cut mortises, trim laminates, round corners, and make dovetail, rabbet, and dado joints.

It requires a great deal of experimentation to become proficient with a router. It is important to realize from the start that you are to make no adjustment unless the electrical plug has been pulled from its socket, and you must always test the bit setting and depth on a scrap of wood before you turn it loose on the desired material.

Because of their versatility, pneumatic staplers/nailers are becoming increasingly more popular with builders who realize that the initial expense to purchase one of these tools and its supporting air compressor is soon recovered because of the vast savings in

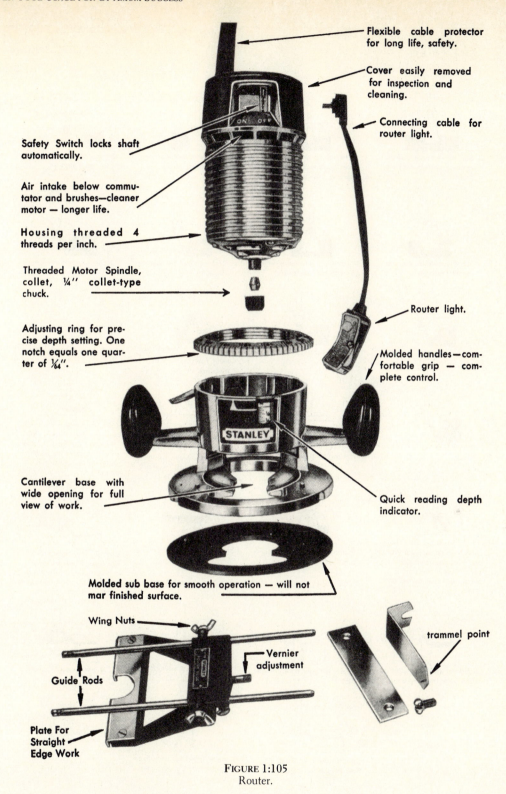

Flexible cable protector
for long life, safety.

Cover easily removed
for inspection and
cleaning.

Connecting cable for
router light.

Safety Switch locks shaft
automatically.

Air intake below commu-
tator and brushes—cleaner
motor — longer life.

Housing threaded 4
threads per inch.

Threaded Motor Spindle,
collet, ¼" collet-type
chuck.

Router light.

Adjusting ring for pre-
cise depth setting. One
notch equals one quar-
ter of 1/64".

Molded handles—com-
fortable grip — com-
plete control.

Cantilever base with
wide opening for full
view of work.

Quick reading depth
indicator.

Molded sub base for smooth operation — will not
mar finished surface.

Wing Nuts

trammel point

Vernier
adjustment

Guide Rods

Plate For
Straight
Edge Work

FIGURE 1:105
Router.

manhours. They are used to apply sheathing, subfloor-ing and roofing in much less time than it takes by conventional nailing methods (Figure 1:107).

A *ramset* (Figure 1:108) resembles a long-barreled pistol and sounds like one when the trigger is pulled thus

activating the powder charge. It is designed to drive a variety of fastener types into wood, concrete or metal, provided the correct charge has been selected. Extreme care must be exercised when using this tool, since goggles are a must and ear plugs are helpful.

Straight—for general stock removal, slotting, grooving, rabbeting

Veining—for decorative free-hand routing such as carving, inlay work

Sash bead—for beading inner side of window frames

Sash cope—for coping window rails to match bead cut

Core box—for fluting and general ornamentation

Dovetail—for dovetailing joints. Use with dovetail templet

Corner round—for edge rounding

Bead—for decorative edging

Cove—for cutting coves

45° bevel chamfer—for bevel cutting

Mortise—for stock removal, dados, rabbets, hinge butt mortising

Rabbeting—for rabbeting or step-cutting edges

Roman ogee—for decorative edging

Panel pilot—for cutting openings and for through-cutting

Pilot spiral (down)—for operations where plunge cutting is required in conjunction with templet routing, using the pilot guide

Straight spiral (down)—for through cutting plastics and non-ferrous metals; also for deep slotting operations in wood

Straight spiral (up)—for slotting and mortising operations particularly in non-ferrous metals such as aluminum door jambs

V-groove—for simulating plank construction

Spiral—for outside and inside curve cutting

Bits for trimming plastic laminates. Solid carbide or carbide-tipped bits for flush and bevel trimming operations Solid carbide self-pivoting flush and bevel trimming bits

Solid-carbide combination flush/bevel trimming bit

Carbide-tipped bevel trimmer bit

Carbide-tipped ball bearing flush trimming bit

Carbide-tipped 25° bevel trimmer kits

Carbide-tipped combination flush/bevel trimming bit

Carbide-tipped 15° backsplash trimmer—with $\frac{5}{16}$" diameter hole

FIGURE 1:106
Router bits.

TOOL SAFETY

Each hand or power tool used by a carpenter has been developed to reduce the time needed to perform a specific task as well as to improve the accuracy of the end product. Even though each tool has been designed to be relatively safe, its user must thoroughly understand how to use it properly and how to care for and maintain it properly. Performing any task with a broken or dull tool can be dangerous and is definitely not advisable.

In all phases of house construction, electric power tools have replaced hand tools for many tasks. The carpenter must thoroughly understand the proper uses and operation of each power tool in order to prevent injury to himself or herself and damage to the tool.

The information on "Personal Power Tool Safety" presented here is intended to provide general guidelines on the correct use and handling of power tools for safety and peak performance.

FIGURE 1:107
(Courtesy of Bostitch Fastening Systems)

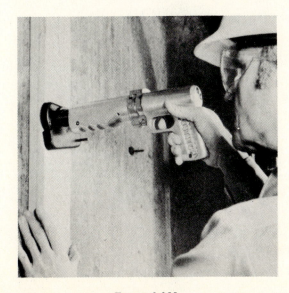

FIGURE 1:108
Ramset.

PERSONAL POWER TOOL SAFETY

1. It is imperative that the work area be safe.
   a) It should not be cluttered with debris.
   b) It should be well lighted.
   c) Bystanders must be kept at a safe distance away.
   d) Materials should be well supported and clamped down when necessary.
2. Always be careful when using electricity.
   a) No electric tool should be used in damp locations or be exposed to rain.
   b) An electric cord should never come in contact with sharp objects or heat and should never by yanked from a receptacle.
   c) Before any electric tool is serviced, it should be disconnected from its power source.
   d) An extension cord should be of adequate size and should be labeled "for outdoor use" if so used. It is essential that the cord be inserted into a "ground fault" receptacle.
   e) Never allow yourself or any material to become entangled by an electric cord.
   f) Never operate any tool if the line voltage is questionable (e.g. lights dim when tool is turned on).
   g) Always be sure that the tool is either properly grounded or purchase only those tools that are double-insulated.

(Continued)

3. Wear the proper clothing.
   a) Loose fitting clothing or jewelry should not be worn because they can get caught in the moving parts.
   b) Safety glasses should be worn whenever there is a danger of flying debris.
   c) Safety shoes should be worn at all times. Sneakers and street shoes are improper footwear at the building site.
4. You must always keep safety in mind when handling tools.
   a) When idle, a tool should be stored in a dry location and out of reach of small children.
   b) Monthly maintenance should be performed when the tool is not in use. This includes checking for sharpness of attachments and proper functioning of all moveable external parts, and following a definite lubrication schedule.
   c) When in use, a tool must never be forced beyond its capabilities. Most electric tools will characteristically sound very different from their normal performance sounds when they are pushed too far.
   d) Never use the wrong-sized attachment on any power tool; this could cause injury to yourself and the tool.
   e) Make certain that you are never off balance and that you do not overextend your reach when using a power tool.
   f) Avoid operating any electric tool in an explosive environment where fumes might be ignited by sparks generated by the tool. Clear the working area of gasoline, oil, paint thinner, or other volatile liquids.
   g) When an electric tool appears to be working at less than peak efficiency, disconnect the cord before examining the tool or its components.
   h) Never place your hands beneath the working area for any reason.
   i) Never remove any of the manufacturer's safety-designed components from any power tool.

QUESTIONS

1. Explain why a carpenter purchases a specific tool.

2. Which face of a chisel, flat or beveled, should face the wood to be removed?

3. List three (3) different types of knives and tell when and where they are used by a carpenter.

4. Explain the difference between the terms *plumb* and *level*.

5. Explain the basic difference in cutting angle between the crosscut and rip saws.

6. Should the material to be cut off be positioned to the right or to the left of the circular saw's blade?

7. Explain which power saw is best to select whenever intricate design cuts must be made.

8. Compare the speed and quality of the cut made with a 6 point saw to that made by an 11 point saw.

9. Explain the difference in grain cuts made with a crosscut saw and a rip saw. Is it possible for one to be substituted for the other?

10. List four (4) hand tools used by a carpenter.

11. Explain how a plumb bob is used.

12. Discuss one function of the T bevel.

13. Which type of plane is used to smooth end-grain wood?

14. Explain how a circular saw should be used to minimize splintering when cutting a plywood panel.

15. Name and describe the uses of a widely used portable power tool.

2

TECHNIQUES, TIPS, AND TIME SAVERS

This chapter explains some basic carpentry techniques and provides some helpful tips. It also suggests effective methods that you can use to decrease the time spent performing some tasks throughout the entire construction process.

NAILING METHODS

Nails are used to join various structural parts together during the entire house framing process. In addition to knowing the correct sequence of steps that must be followed during the framing of a house, the carpenter must know

1. The size and type of nail to use in each specific operation
2. The quantity required for each operation so that structural soundness is guaranteed
3. Which nailing method is the most appropriate

First, you should know some of the terminology used in the various nailing methods. The term *face* is used to denote the width of the board, *edge* is the thickness, and the term *end* refers to the two cut-off butt ends (Figure 2:1).

Face-nailing (Figure 2:2) is the most common method of nailing. It is performed by driving a nail through the face of one piece of lumber into a second piece.

Toe nailing (Figure 2:3) is used to join two pieces of lumber that are positioned at right angles to each other. Nails must be spaced on opposite sides so that they are able to pass each other without interference.

End-nailing (Figure 2:4) is actually a specialized form of face-nailing. With this method, nails are driven through the face of the top board into the cut-off butt

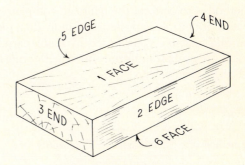

FIGURE 2:1
Key to numbered surfaces
1,6 face (width)
3,4 end (butt ends)
2,5 edge (thickness)

FIGURE 2:2
Face nailing lapped panelling.

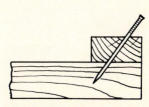

FIGURE 2:3
Toe nailing.
Note: When finished, the nail should be flush with the wood.

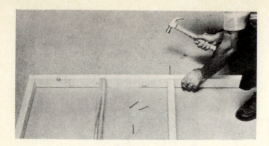

FIGURE 2:4
End nailing.

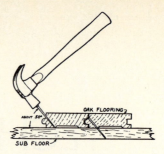

FIGURE 2:5
Edge or blind nailing. Note the angle of the nail and its position just
above the shoulder of the tongue.

end of the underlying piece. The nail should be long
enough so that it penetrates deeper into the underlying
piece than the distance it travels through the top piece.
This is done to ensure proper holding power.

Edge-nailing (Figure 2:5), also called blind nailing, is
most commonly used to attach tongue-and-groove lumber to the underlying structure. The dual purpose of
this method is to conceal the nailhead completely and
to prevent noticeable gaps between individual pieces.

A *nailing machine* is preferred by many carpenters
because it drives nails uniformly into the flooring at the
correct angle (Figure 2:6).

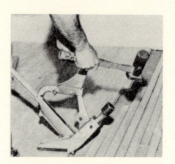

FIGURE 2:6

TABLE 2:1
Common Nailing Methods and Their Uses

Nailing method	Nails used	Quantity	Uses for joining
Face-nail	20d	Staggered 12″	Built up beams, girders
	16d	Every 16″	Sole plate to joists
	16d	Two every 12″–16″	Upper top plate to lower top plate
	10d, 16d	16 o.c.	Double studs, and at intersecting walls
	10d	5	Rafter to ceiling joist
	10d, 16d	2 each side	Collar beam to rafter
	10d	2 each side	Corner studs (built-up)
	8d	Every 6″ on edge	Wall and roof sheathing $\frac{1}{2}$″ or thicker
	6d	and 12″ intermediate	Sheathing $\frac{3}{8}$″ or less in thickness
Toe nail	16d	2 each side	Joist to sill, to beam (girder)
	16d	2–3 each end	Solid bridging to joists
	8d	2 each end	Bridging to joists
	8d	4	Stud to soleplate—wall assembled vertically
	8d	2 each side	Rafter to top plate
	8d	2 each side	Ceiling joist to top plate
	8d	4	Rafter to rafter through ridge board
End-nail	16d	3	Header to joists
	16d	2	Soleplate to stud—wall assembled horizontally
	16d	2	Top plate to stud
	10d	3	Ridge board to rafter
Edge-nail	Flooring nails (on angle)		Hardwood floor to subfloor
	6d, 8d casing		Join roof deck boards (post and beam construction)

In addition to the common nailing methods, there are several techniques that can be employed to increase a nail's holding power. They are

1. Nails driven at a slight angle (Figure 2:7) for better holding power
2. Long nails driven through both pieces of wood, bent over, and then clinched to the back of the second piece (Figure 2:8)

Table 2–1 presents the common nailing methods and their uses.

Figure 2:9, in conjunction with Table 2:2, indicates where to use the various types of nailing methods in platform construction.

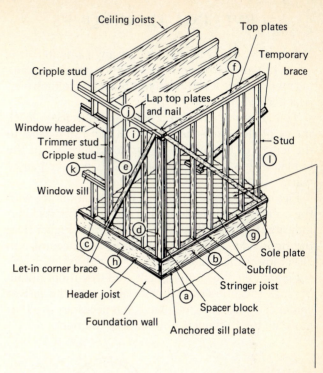

Note: subfloor should be plywood 4'x8' panels not individual boards

FIGURE 2:9

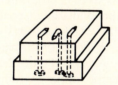

FIGURE 2:7

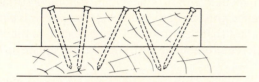

FIGURE 2:8

TABLE 2:2
Nailing Methods Used in Platform Construction

Toe nailing

(a)	Header and stringer joist to anchored sill plate
(b)	Studs to sole plate
(c)	Cripple studs, above and below window opening

Face-nailing

(d)	Corner stud to spacer block
(e)	Stud to trimmer stud
(f)	Upper top plate to lower top plate
(g)	Sole plate to subfloor, joist

End-nailing

(h)	Header joist to joist
(i)	Lower top plate to stud
(j)	Stud to window header
(k)	Window sill to cripple stud

Edge-nailing

(l)	Hardwood floor to subfloor

LEVELING TECHNIQUES

A level is an invaluable tool for any carpenter. It is constantly used to determine whether structural framing members are level (horizontal plane) and plumb (vertical plane).

To determine whether a horizontal surface is level, the carpenter must check the vials that run parallel to the length of the level. The surface is level when the bubble within the vial is centered between the pair of marks.

Figure 2:10 shows an uneven horizontal surface with the bubble in the vial to the left of the two marks. This indicates that the left side of the surface is higher than the right side. To level the surface, it is necessary to either lower the left side or raise the right side.

Figure 2:11 shows an upright stud that it not plumb because the bubbles in both vials are to the left of center. To make the stud plumb, it is necessary to either move its top to the left or its base to the right.

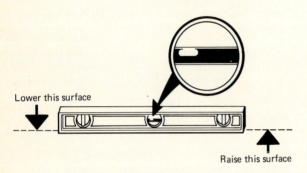

Lower this surface

Raise this surface

FIGURE 2:10

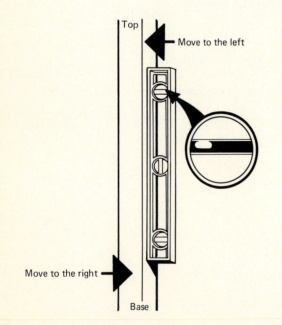

Top

Move to the left

Move to the right

Base

FIGURE 2:11

A *water level* (Figure 2:12) can be used to level footings and floors, as well as to determine the difference in grade between two isolated points. It consists of a plastic tube filled with a red colored liquid that will exhibit the same height at both ends of the tube no matter where the tube is positioned between two points.

MEASUREMENTS

Every carpenter realizes the importance accurate measurements play in each phase of house construction. The only time to use rough approximations is when estimating the lumber list.

The most common measurement made in house construction involves a determination of *length*. For a measurement of length to be accurate, its point or surface of reference must be both straight and square. It is equally important that the structural item selected to fulfill this length requirement be both straight and square on each end.

The rule or tape measure that you use to measure length is designed to emphasize the 16-inch module. The 16-inch centers have been printed in red by the manufacturer. A tape, so designed, enables you to very effectively layout the exact locations of upright wall studs, roof rafters, or floor joists that are spaced at 16-inch centers (16″ o.c.) See Figure 2:13.

To determine whether or not two adjacent walls join at exactly right angles (90°) to each other, and thus are square, measurements involving a 3-4-5 triangle or one of its multiples are used. For example, a distance of 3′ is marked off on one wall, and one of 4′ on the other wall. The diagonal distance between these two marks will be exactly 5′ only when the two walls are square with each other (Figure 2:14).

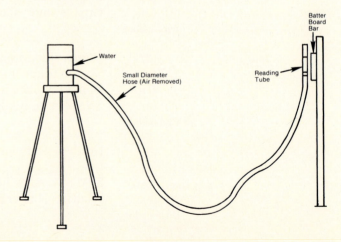

Water

Small Diameter
Hose (Air Removed)

Batter
Board
Bar

Reading
Tube

FIGURE 2:12

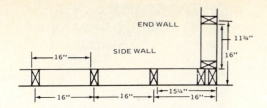

FIGURE 2:13
Stud layout on sole plate using 16″ spacing.

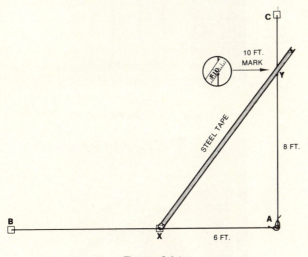

FIGURE 2:14

To check the squareness of all four walls—to achieve a perfect rectangle—the diagonal distances are measured. If they agree, the area in question is square (Figure 2:15).

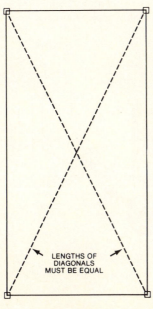

FIGURE 2:15

In Figure 2:16, the wall frame is checked with a tape measure. If the diagonal distances are equal, the wall frame section is square.

CROWNING STRUCTURAL MEMBERS

Before it is cut and nailed in place, every joist or roof rafter must be sighted along its edge to ascertain its crown. The crown, or bowed edge, must always face upward (Figure 2:17). When this simple operation is forgotten or ignored, both the floor and roof areas will be wavy.

FIGURE 2:16

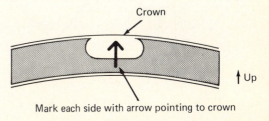

FIGURE 2:17

TIPS

Sheet Material

To add strength and rigidity to the framework of any structure, sheet materials (e.g., plywood, gypsum wallboard) should always be placed at right angles to the framing members (Figure 2:18). Note that the grain pattern of the plywood runs lengthwise along the sheet.

The joints, where two pieces meet, should always be located directly above a framing member, and should be staggered with each row of material.

Interior Coverings

Even though there are several materials that can be used as the covering for interior walls and ceilings (e.g., plaster, wallboard, tile, paneling) the room's walls should not be covered with any of these materials until the ceiling area has been completely covered.

FRAMING PLATFORM WALLS

Using doubled 2 × 12's as headers for exterior doors and windows (Figure 2:19) eliminates the need to cut and nail short cripple studs in the area between the header and the double top plate (Figure 2:20). This is necessary whenever 2 × 8's or 2 × 10's are used as headers.

Assembling corner studs long before the exterior walls are framed will reduce the actual construction time, since they have been designed to provide both the necessary exterior and interior nailing surface (Figure 2:21).

FIGURE 2:19
Solid 2 × 12 header.

FIGURE 2:20
Short cripples above header.

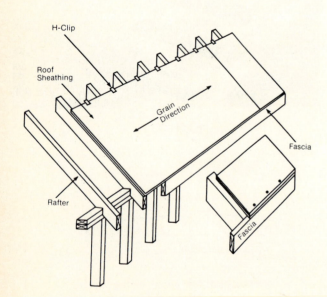

FIGURE 2:18

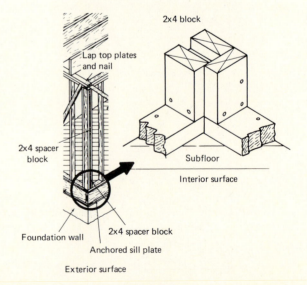

FIGURE 2:21

After the exterior walls are assembled, they are raised to their final vertical position (Figure 2:22).

This procedure can be easily and safely accomplished with a two-person crew provided the following steps are performed:

1. Nail small pieces of 2 × 4 to the outside portion of the floor frame. These serve as floor stops to help prevent the frame from sliding off the edge (Figure 2:23).

2. Use a chalk line to snap a line on the subfloor to indicate the exact desired location of the wall section, which has been previously been determined by using the 3-4-5 triangular measurement procedure (Figure 2:24).

3. Place several short pieces of 2 × 4 under the top portion of the frame to make it easy for you to grab and raise the frame in one operation (Figure 2:25).

Once the wall frames have been raised and positioned, they are then checked for plumb with a carpenter's level (Figure 2:26). Once plumb, they must be adequately braced to resist any movement in the assembled wall.

FIGURE 2:22

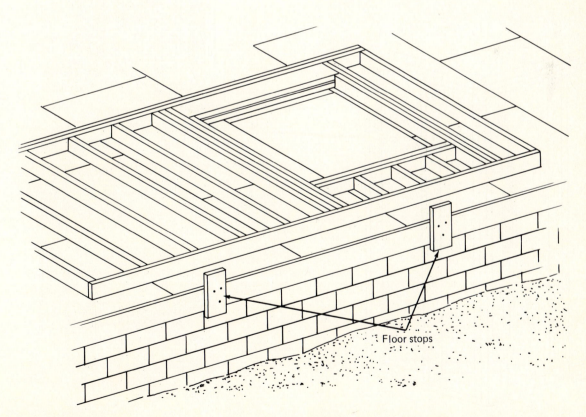

Floor stops

FIGURE 2:23

FIGURE 2:24

To prevent the top of the framed wall from either leaning inward or tilting outward, temporary 2 × 4 braces are used. One end of each brace is nailed to the top portion of one of the wall studs and the base to a wooden cleat which has been nailed to the subfloor. The braces maintain the plumbness of the wall (Figure 2:27).

To ensure that the assembled wall does not sway or lean in one direction or the other, braces are placed on a diagonal and nailed to the outer surfaces of the upright studs. These braces can be temporary and removed once the roof and ceiling joist have been installed, or they can be left permanently within the framed wall. When permanent, both the studs and the plates must be notched to conceal the brace (Figure 2:28).

Short 2x4 pieces

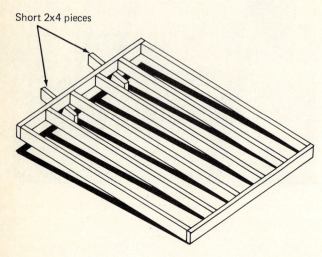

FIGURE 2:25

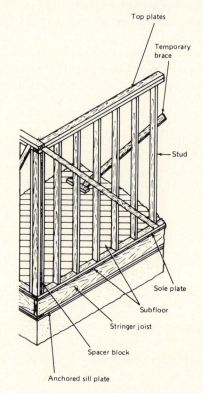

FIGURE 2:27

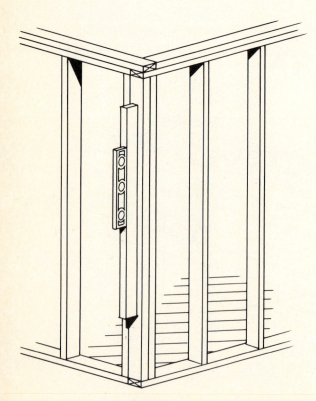

FIGURE 2:26

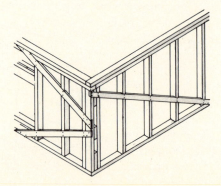

FIGURE 2:28

PANELING

When working with paneling, you must always remember to provide a slight space or gap where the panel meets the ceiling and where it meets the floor to make it easy to install the panel. Some form of molding will be used to conceal these gaps. Also, remember to cut from the finished side when using a hand or table saw and from the back side of a panel when using a circular or saber saw.

A *scriber,* or divider (Figure 2:29), is used to transfer the exact shape of a room's corner to the surface of the panel so that it can be cut to fit the corner snugly. Since most corners are somewhat irregular in shape, the divider's point will mark the panel exactly like the contour of the wall it is to abut. When performing this operation, make certain that the panel is plumb before you attempt to scribe it.

CARRYING SHEET GOODS

Full 4′ × 8′ sheets of plywood, gypsum wallboard, and paneling are difficult to carry from one location to another because of their awkward size and shape. When carrying a sheet on your right side, you should lift the sheet with your right hand and guide it with your left (Figure 2:30). The reverse is true when you carry it on your left side. When two workers carry several sheets at the same time, they both should position themselves on the same side of the panels.

TIME SAVERS

Plywood

The use of 4′ × 8′ sheets of plywood to sheath exterior walls and roof area, as well as to serve as subfloors, has become the accepted practice throughout the construction industry as one means of reducing the total man-hours required to frame and enclose any structure.

To minimize the number of saw cuts necessary for full sheets, the framing members of the structure must be positioned on either 16″ or 24″ centers. This practice can also reduce waste, since it is usually possible to use the excess from one row of plywood as the starting portion of the next row. Designing the overall structure to have dimensions which are multiples of four feet is another effective way to reduce the time spent cutting full 4′ × 8′ sheets.

Whenever possible, exterior platform walls that are assembled horizontally should be covered with sheathing before they are raised to their final vertical positions (Figure 2:31). This practice greatly reduces the time spent measuring, cutting, and nailing the various ply-

wood panels in place while working from ladders or scaffolds.

FIGURE 2:29
Scribing a panel.

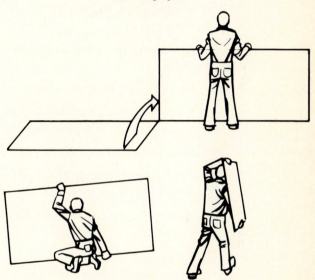

FIGURE 2:30
Lifting and carrying a large sheet.

FIGURE 2:31
A wall section is raised or "walked" into place.

Gypsum Wallboard

Most interior walls and ceilings that are covered with gypsum board (wallboard) eventually are taped to conceal joints, structural imperfections, and room corners. The length of each piece of wallboard you want to use as part of a room corner should be cut $\frac{1}{4}''$ to $\frac{1}{2}''$ shorter than the actual measurement dictates. This practice allows for any structural imperfections in the framed wall or ceiling and should prevent the need to shave or recut the edge, which is often the case when the exact measurement is used.

STANDARDIZING STRUCTURAL MEMBERS

It is important that all floor joists have the same width to ensure a level floor surface. Often, structural lumber from the same pile will vary greatly. For example, a nominal 2 × 6 may range from $5\frac{1}{4}''$ to $5\frac{3}{4}''$ in width.

Most carpenters order precut studs measuring $7'\ 8''$ or standard $8'$ studs. This practice guarantees uniformity in wall height throughout the structure, and eliminates the need to measure the length of each stud and then cut off the excess.

QUESTIONS

1. Compare face nailing with toe nailing, with respect to the position of the two boards being nailed together and the direction the nail takes as it is driven into the wood.

2. Give three (3) examples of where face nailing is used in residential construction.

3. Give three (3) examples of where toe nailing is used.

4. A level placed in a horizontal position shows the bubble in the vial to the left of center. Explain how to correct this condition.

5. State how you would determine whether an exterior outside corner is square.

6. Explain what happens to a roof surface if the carpenter neglects to place the rafters with their crowns facing upward.

7. Why is it necessary to brace all exterior walls before they are covered with sheathing?

3

ADHESIVES AND FASTENERS

Without the use of adhesives or fasteners, it would be impossible for any building material to be assembled into the desired form or to remain in its desired location. Adhesives and fasteners perform the same function; they join two building materials together. However, selection of the most appropriate item is determined by the material's composition, it's intended use, and the thorough knowledge which every craftsperson is expected to possess.

ADHESIVES

An adhesive is any substance that will bond two building materials together by surface attachment. Glue, cement, mastic, solder, and paste are examples.

Builders know that some adhesives can be used to bond only specific materials, while others can join a variety of materials.

It would be ideal if the builder could select one adhesive that would serve all bonding needs, but this is usually impossible because adhesives differ in the following characteristics:

1. The amount and kinds of purposes which they serve
2. The conditions under which they can be used (e.g., temperature)
3. The amount of surface preparation required
4. The necessity to clamp the materials
5. The physical conditions which the eventual bond will be subjected to (e.g., immersed in water)
6. Whether or not they come ready to be used or require mixing components together

7. The types of materials they can bond
   a) Porous or nonporous
   b) Rigid or flexible
   c) Rough or smooth
   d) Similar or dissimilar composition

Therefore, it is imperative for the craftsperson to choose an adhesive that is specifically designed for the task at hand and to read and follow the manufacturer's directions completely.

Table 3:1 identifies various types of adhesives and how they are marketed, and describes their strengths, weaknesses, temperature requirements and setting times, resistance to water, and clean up procedures.

FASTENERS

Nails

Common nails are the most suitable fasteners to use in rough framing any wooden structure. When they are to be exposed to the weather, they must be coated with zinc (e.g., galvanized). Common nails vary in length from 1 inch (2 penny or 2d) to 6 inches (60 penny or 60d)—Figure 3:1. Note that the wire gauge size varies with the length (penny size). Figure 3:2 compares several wire gauge sizes and clearly shows that the nail becomes thicker as the gauge number gets smaller. The penny designation also applies to box, casing and finishing nails. All may be purchased by the pound, in 5-pound boxes, or by the 50-pound box.

Box nails are thinner than their common nail counterparts and are used for light framing and toe nailing.

TABLE 3:1
Types of Adhesives

Category	How marketed	Strengths	Weaknesses	Temp./Curing time	Resistance to water	Clean-up
Wood glues Liquid hide	Ready to use	Strong joints, excellent for furniture repairs & construction	Not resistant to water or moisture	70–90°F, 8 hours	Poor	Warm water
Polyvinyl acetate (white glue)	Ready to use	Excellent household fix-it adhesive, good for porous materials	Requires clamping, corrodes metal, poor resistance to solvents	Room temp., 8 hours	Poor	Warm, soapy water
Aliphatic resin	Ready to use	Furniture repair, porous materials, resistant to heat and solvents	Needs brief clamping period	45–110°F	Poor	Warm water
Casein	Mixed with water	Nontoxic, excellent for oily woods	May stain some dark woods, dulls cutting tools when dry	32–110°F, 2–3 hours	Poor, but is water resistant	
Urea formaldehyde (plastic resin)	Mixed with water	Strong bond, does not stain, resistant to rot and mold, cheap method to glue veneers when mixed with flour	Not used on oily woods, joints must be well fitted, 12 hours of clamping is minimum	70°F (+), 12 hours	Excellent	Warm, soapy hours
Resorcinol	Resin and catalyst	Excellent for exterior use, has excellent gap-filling properties, can be used on items immersed in water	Requires clamping and eye protection, can not be removed once it dries	70°–120°F, 10	Excellent	Cool water before it sets
Acrylic resin	Resin and catalyst	Strong bond, not affected by gasoline or oil, has good gap-filling properties, fast setting	Sets too fast to bond large areas	70°–100°F, varies due to the mix	Excellent	Acetone
Contact cements Solvent	Ready to use	Fast drying, heat resistant, clamping not required	Volatile, flammable, some are toxic		Excellent	
Chlorinated-base	Ready to use	Very fast drying, nonflammable	Avoid contact with eyes			Toluene Xylene
Water based (neoprene latex)	Ready to use	Nontoxic, nonflammable, heat resistant, joins two nonporous materials	Slowest of the contacts to dry, acid destroys the bond		Excellent	Water (before dry) Toluene (once dry)

(Continued)

TABLE 3:1   (CONTINUED)

Category	How marketed	Strengths	Weaknesses	Temp./Curing time	Resistance to water	Clean-up
Construction adhesives (mastics) Used for: wood panels, wallboard, ceiling tile, ceramic tile, metal framing	Ready to use	Fast drying, applied with caulk gun or putty knife, able to fill gaps	Need slight pressure for bond to take, may attack polystyrene foams, some are toxic and flammable			Mineral spirits or lighter fluid
Specialty adhesives Cyanoacrylate	Ready to use	Dries in seconds, resistant to most chemicals, known as "super glues"	Used only on solid, nonporous materials; not effective on soft, absorbent surfaces; very expensive, may bond fingers together, not able to fill gaps		Excellent	Acetone or nail polish remover
Epoxy (2 containers)	Resin and catalyst—catalyst or hardener (polyamide resin) and epoxy resin	Bonds hard to stick materials, resistant to most solvents, no clamping required, versatile	Any materials used to mix and apply are best discarded after use	Wide temp. range, 4–6 hours	Excellent	Acetone
Polyvinyl chloride (PVC)	Ready to use	Bonds porous and nonporous materials, unaffected by oil, gasoline, or alcohol; applied to 1 surface for nonporous, and to both surfaces for porous materials; allowed to dry before second application is made		10–30 minutes	Excellent	Acetone or lacquer thinner
Cellulose nitrate (model airplane glue)	Ready to use	Quick drying	Flammable, fumes are toxic, bond unable to take stress		Good	Acetone
Acrylonitrile	Ready to use	Yields a flexible bond	Not effective for wood	175–325°F (use an iron)	Excellent	Acetone or methyl ethyl ketone
Plastic rubber cement	Ready to use	Makes permanent repairs to rubber, bonds to any clean dry surface		8–10 hours		

(Continued)

TABLE 3:1   (CONTINUED)

Category	How marketed	Strengths	Weaknesses	Temp./Curing time	Resistance to water	Clean-up
Styrene-butadiene	Ready to use	Bonds to most non-porous materials, has excellent gap-filling abilities	Cannot be used in vicinity where gasoline or oil are used		Excellent	Mineral spirits or turpentine
Silicone	Ready to use	Used as a sealant and caulking compound, flexible, used in or out	Slow setting time		Excellent	
Concrete bonders (acrylic resin)	Ready to use	Excellent outdoors, used to fortify cement, to level floors, and to seal, prime, and provide a nonslip surface over an existing masonry material			Excellent	
Hot melt glues	Solid stick inserted into a gun	Bonds most porous materials, nonflammable	Bonds too fast for large areas	60 seconds	Excellent	

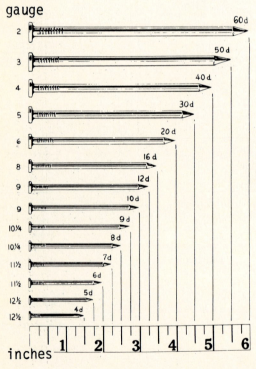

FIGURE 3:1
Length and gauge of the most common wire nails.

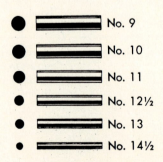

FIGURE 3:2
Wire-gauge sizes used in making nails.

Both casing and finishing nails are used to give a project a finished appearance. The nail head surface of both is much smaller than the common nail (Figure 3:3). Table 3:2 gives the number of nails per pound.

There are a variety of specialized nails, each designed to perform a specific function. A nail's function dictates the shape of its head (Figure 3:4), the shape of its shank —which determines its holding capacity (Figure 3:5)— and the style of its point (Figure 3:6).

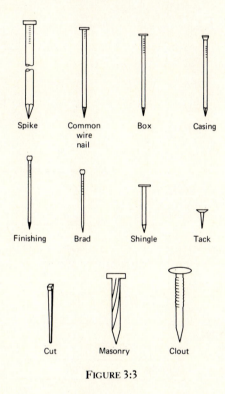

FIGURE 3:3

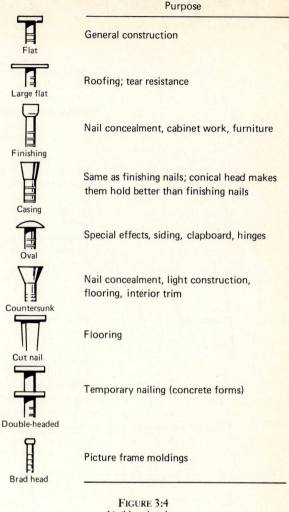

	Purpose
Flat	General construction
Large flat	Roofing; tear resistance
Finishing	Nail concealment, cabinet work, furniture
Casing	Same as finishing nails; conical head makes them hold better than finishing nails
Oval	Special effects, siding, clapboard, hinges
Countersunk	Nail concealment, light construction, flooring, interior trim
Cut nail	Flooring
Double-headed	Temporary nailing (concrete forms)
Brad head	Picture frame moldings

FIGURE 3:4
Nail head styles.

TABLE 3:2

Penny size	Length (inches)	Gauge	Number per pound
2	1	15	840
3	$1\frac{1}{4}$	14	540
4	$1\frac{1}{2}$	$12\frac{1}{2}$	300
6	2	$11\frac{1}{2}$	160
8	$2\frac{1}{2}$	$10\frac{1}{4}$	100
10	3	9	65
12	$3\frac{1}{4}$	9	65
16	$3\frac{1}{2}$	8	45
20	4	6	30
30	$4\frac{1}{2}$	5	20
40	5	4	17
50	$5\frac{1}{4}$	3	14
60	6	2	11

Brads, which have the smallest head and the thinnest cross-section, are sold by the box instead of by the pound. This is also true for many other specialized nails.

Because hand nailing is relatively slow and physically fatiguing, many builders employ pneumatic nailers to perform various nailing tasks. There are models that can be used for rough framing, for finish strip-floor installations, and for installation of roof sheathing and roof shingles (Figure 3:7).

Whenever wooden components must be fastened to either concrete or steel, a powder-actuated system can be used. This system employs a gun-like device (Figure 3:8), equipped with a small explosive charge, that when activated, drives a specially designed fastener (Figure 3:9) into the material. While using a powder-actuated tool, the operator must wear safety glasses and exercise extreme care.

Type	Abbrev.	Remarks	Illustration
		SHANKS	
Smooth	C	For normal holding power; temporary fastener.	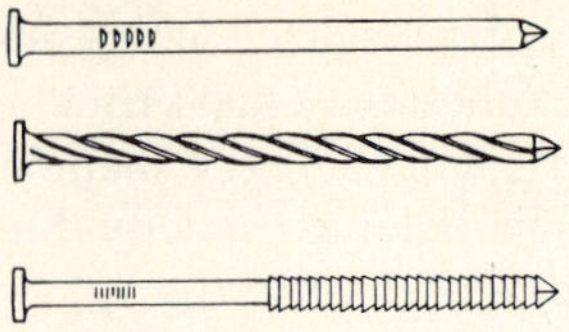
Spiral	S	For greater holding power; permanent fastener.	
Ringed	R	For highest holding power; permanent fastener.	
Flat Countersunk	Cs	For nail concealment; light construction, flooring, and interior trim.	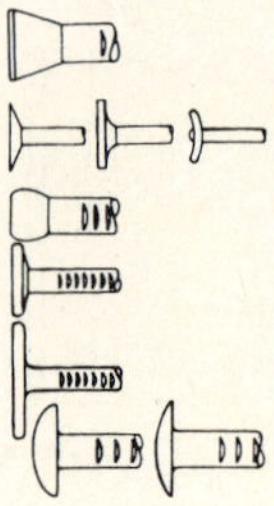
Drywall	Dw	For gypsum wallboard.	
Finishing	Bd	For nail concealment; cabinetwork, furniture.	
Flat	F	For general construction.	
Large Flat	Lf	For tear resistance; roofing paper.	
Oval	O	For special effects; siding and clapboard.	
Diamond	D	For general use, 35° angle; length about 1.5 × diameter.	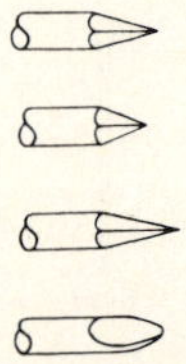
Blunt Diamond	Bt	For harder wood species to reduce splitting, 45° angle.	
Long Diamond	N	For fast driving, 25° angle; may tend to split harder species.	
Duckbill	Db	For clinching small nails.	

Reprinted with permission of Canadian Wood Council.

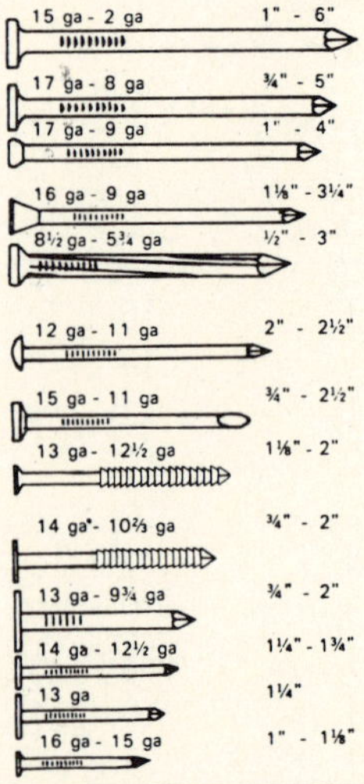

Type of Nail	Head	Shank	Point	Material	Finishes and Coatings	Size	
						Diameter	Length
Standard or Common	F	C,R,S	D	A,S	B,Ge	15 ga - 2 ga	1" - 6"
Box	F,Lf	C,R,S	D	S	B,Pt	17 ga - 8 ga	¾" - 5"
Finishing	Bd	C,S	D	S	B,Bl	17 ga - 9 ga	1" - 4"
Flooring and Casing	Cs	C,S	Bt,D	S	B,Bl	16 ga - 9 ga	1⅛" - 3¼"
Concrete	Cs	S	Bt,D	Sc	Ht	8½ ga - 5¾ ga	½" - 3"
Siding and Clapboard	F,O	C,S	D	A,S	B,Ghd	12 ga - 11 ga	2" - 2½"
Clinch	F,Lf	C,S	Db	S	B	15 ga - 11 ga	¾" - 2½"
Gypsum Wallboard	Dw,F	C,R,S	D,N	S	Bl,Ge	13 ga - 12½ ga	1⅛" - 2"
Underlay and Underlay/Subfloor	F,Cs	C,R	D	S	B,Ht	14 ga - 10⅔ ga	¾" - 2"
Roofing	Lf	C,R,S	D	A,S	B,Ghd	13 ga - 9¾ ga	¾" - 2"
Wood Shingle	F	C	D	S	B,Ghd	14 ga - 12½ ga	1¼" - 1¾"
Gypsum Lath	F	C,S	D,N	S	B,Bl,Ge	13 ga	1¼"
Wood Lath	F	C,S	D	S	Bl	16 ga - 15 ga	1" - 1⅛"

Reprinted with permission of Canadian Wood Council.

FIGURE 3:5
Types of nail shanks.

DIAMOND POINT The most common type of nail point. Nails with this type of point hold well and may be used in all but the harder woods.

CHISEL POINT Generally used on a large nail to be driven into hardwood. By cutting through the hard wood fiber, the chisel point slightly reduces the holding power of the nail, but the fibers are not forced apart causing splitting.

BLUNT POINT This type of point causes the nail to break through the fiber of hardwood eliminating splitting but decreasing holding power.

NEEDLE POINT Nails with needle points may be used on any except the harder woods, nails drive easily.

SIDE POINT Nails with side points are suited especially for clinching. They are used in hardwoods, as they drive easier than blunt points with holding power only slightly less than needle or diamond point.

DUCKBILL Facilitates the clinching of the point where it is desirable to join flat wooden members in a more permanent manner for example, wood shipping crates.

FIGURE 3:6
Types of nail points.

FIGURE 3:7
(Courtesy of Bostitch Fastening Systems)

FIGURE 3:8
(Courtesy of Ramset)

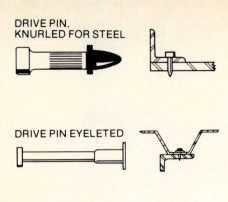

DRIVE PIN, KNURLED FOR STEEL

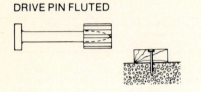

DRIVE PIN EYELETED

DRIVE PIN FLUTED

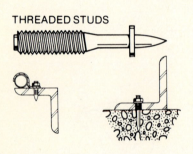

THREADED STUDS

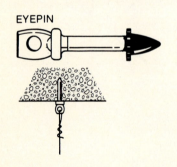

EYEPIN

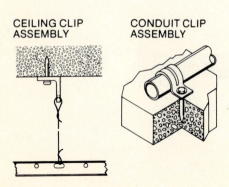

CEILING CLIP ASSEMBLY   CONDUIT CLIP ASSEMBLY

FIGURE 3:9

Screws

Because screws have greater holding capacity than nails, carpenters prefer to use them for many interior projects, especially in cabinet work. The size of a screw is related to its length and its diameter, or gauge number, which has a range of zero to 24. A screw with a large gauge number has a larger diameter and is thicker than one with a small gauge number. The difference between one gauge number to the next would be 13 thousandths (0.013) of an inch (Table 3:3).

Wood screws are tapered, and threaded on the lower end. When joining two pieces together with wood screws, drill as follows:

Top piece—*same* diameter as the screw

Bottom piece—diameter *less* than the screw (Figure 3:10)

A screw is also classified with respect to the shape of its head (round, oval, flat, standard, slotted, or Phillips head), the finish of its surface (zinc, chromate, nickel, or chromium plated), its composition (steel, brass, or aluminum) and finally its purpose. Figure 3:11 shows classifications and explains how to measure a wood screw.

A box of screws contains 100 screws and is labeled with the length, gauge number, head shape (e.g., FH for flat head), and composition or finish.

The time required to install wallboard can be reduced significantly, compared to hand-nailing, by using a properly grounded, electric screwdriver (Figure 3:12). For the best results, the tool should have a slip-clutch that will disengage when the screw head is driven firmly against the wallboard panel, plus a magnetic tip to firmly hold

TABLE 3:3

Gauge	Shank diameter (*inches*)	Pilot hole diameter *hardwoods*	(inches) *softwoods*
2	0.086	$\frac{3}{64}$	—
3	0.099	$\frac{1}{16}$	—
4	0.112	$\frac{1}{16}$	—
5	0.125 ($\frac{1}{8}$)	$\frac{5}{64}$	$\frac{1}{16}$
6	0.138	$\frac{5}{64}$	$\frac{1}{16}$
7	0.151	$\frac{3}{32}$	$\frac{1}{16}$
8	0.164	$\frac{3}{32}$	$\frac{5}{64}$
10	0.177	$\frac{7}{64}$	$\frac{3}{32}$
12	0.216	$\frac{1}{8}$	$\frac{7}{64}$
14	0.242	$\frac{9}{64}$	$\frac{7}{64}$
16	0.268	$\frac{5}{32}$	$\frac{9}{64}$
18	0.294	$\frac{3}{16}$	$\frac{9}{64}$
20	0.320	$\frac{13}{64}$	$\frac{11}{64}$
24	0.372	$\frac{7}{32}$	$\frac{3}{16}$

the correct wallboard screw to the bit. Figure 3:13 presents the various wallboard screw types, with an explanation of where each type should be used.

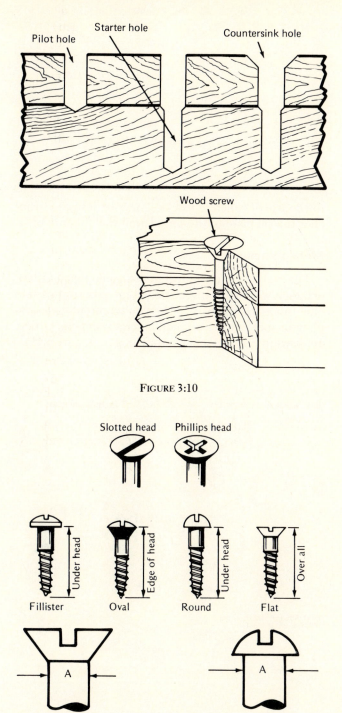

FIGURE 3:10

FIGURE 3:11

How to measure a wood screw: The diameter or gauge number is measured across the shank as indicated in the illustrations by A (*Bottom*). The length of a flat head wood screw is the overall length. The length of an oval head screw is measured from the point to the edge of the head. Round and fillister head screws are measured from the points to the under sides of the heads.

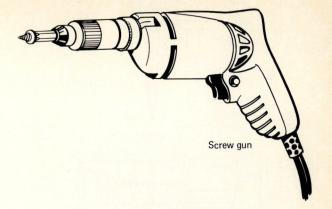

Screw gun

FIGURE 3:12

Type	Application
G	To gypsum base panels
S	To steel (25 gauge)
S–12	To steel (up to 12 gauge)
W	To wood

FIGURE 3:13
Types and applications of wallboard screws.

Lag Screws

Lag screws are used whenever nails or screws can not provide the desired structural anchorage and support. Expanding lead shields complement the lag screws whenever a wooden component must be attached to a masonary surface (Figure 3:14).

Lag screws are classified by the diameter of their square or hexagonal head and their length. Their insertion requires the drilling of a smaller pilot hole and the use of an adjustable wrench to properly anchor it. Placing a washer beneath the head helps prevent injury to the wood's surface.

Machine and Stove Bolts

Both types of bolts are used to fasten two materials together. Once inserted through a drilled hole, a bolt will effectively hold the two materials in place by simply tightening the accompanying nut (Figure 3:15).

Lag screws

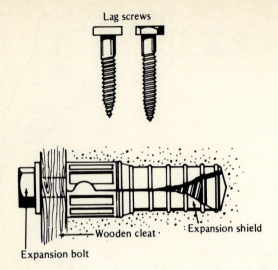

Wooden cleat · Expansion shield

Expansion bolt

FIGURE 3:14

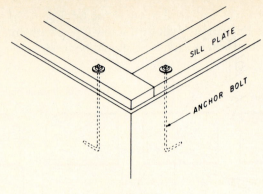

FIGURE 3:17

Machine bolts   Stove bolts

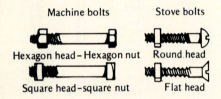

Hexagon head – Hexagon nut   Round head

Square head – square nut   Flat head

FIGURE 3:15

Carriage Bolts

A carriage bolt, instead of a lag screw, is used whenever a hole can be drilled through both the material to be anchored and the anchoring surface. Its head is round and larger than its diameter to enable it to rest against the material. Its shoulders, located directly beneath the head, prevent the bolt from turning in its hole while its nut is being tightened (Figure 3:16).

Anchor Bolts

These L-shaped bolts are inserted into either the top surface of a poured concrete foundation wall or into the

top course of concrete blocks. They are designed to anchor the sill plate to the foundation wall (Figure 3:17).

Metal Fasteners

There are a variety of different-shaped metal fasteners, each with its own specific design to perform a specific function. Figure 3:18 shows a representative sampling of some of these unique metal fasteners. Others are used in post-and-beam construction (Figure 3:19).

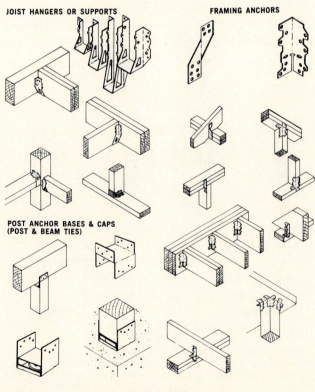

JOIST HANGERS OR SUPPORTS   FRAMING ANCHORS

POST ANCHOR BASES & CAPS
(POST & BEAM TIES)

FIGURE 3:18
(Courtesy of Teco)

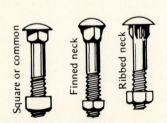

Square or common   Finned neck   Ribbed neck

FIGURE 3:16
Carriage bolts.

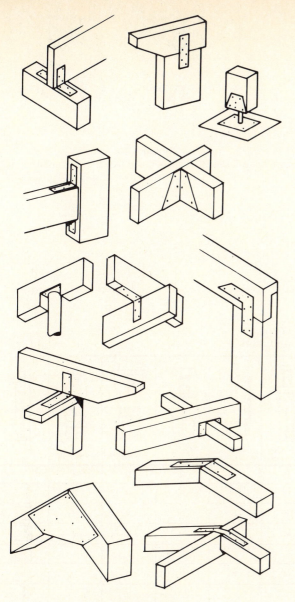

FIGURE 3:19
Metal fasteners and connectors for post-and-beam construction.
(Western Wood Products Association)

Staples

The use of staples is a very effective means of attaching 15-pound builder's paper to exterior sheathing or to a plywood roof prior to the application of the final siding or roofing material. Staples are also used to attach screening and to anchor acoustical ceiling tile.

Every staple has a *crown* that connects two *legs* each of which ends in one of three types of *points*. The leg and crown dimensions of a staple are stated first, followed by the type of point (Figure 3:20).

Staples may have *standard points* that penetrate directly into the material to be fastened, *convergent points* that criss-cross when driven so that they clinch the two

materials together, or *divergent points* that spread further apart when driven (Figure 3:21).

The primary reason for using a stapler is to reduce the time spent performing a specific operation. There are manual staplers and staplers operated by either electricity or compressed air (Figure 3:22).

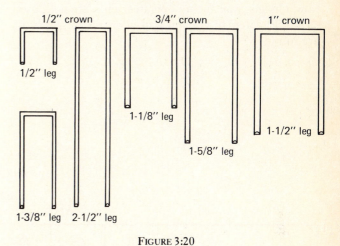

FIGURE 3:20

FIGURE 3:21
Three types of staple points and how they clinch after being driven.

FIGURE 3:22
(Courtesy of Bostitch Fastening Systems)

Wall Anchors

Table 3:4 presents the most common types of wall anchors and the building materials for which each one is best suited.

Figure 3.23 shows how the two types of toggle bolts open and tighten against the wallboard material and how a hollow wall anchor operates. They are both used to install an item to a hollow wall where no upright studs are found.

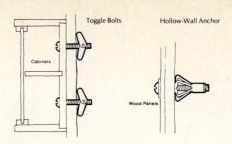

FIGURE 3:23
Toggle bolts, hollow wall anchor, and how they are installed in the wall cavity.

TABLE 3:4
Anchoring Selection Chart

	Plastic anchors	Nylon anchors	Hollow wall fasteners	Spring wing toggle bolts	Aluminium drive anchors	Machine screw anchors	Machine bolt anchors	Lead anchors	Zinc lag anchors	Hollow wall anchors
Brick	M	L		[M]	H	H	H	M	H	
Concrete	M	L			H	H	H	M	H	
Concrete block	M	L		[M]	H	H	H	M	H	[M]
Cinder block	M	L		[H]	M	M	M	M	M	[M]
Gypsum	M	L		[H]						[H]
Marble	M	L		[M]	H	H	H	M		[M]
Plaster	M	L	L							H
Dry wall	M	L	L							
Stone	M	L		[M]	H	H	H	M	M	[M]
Terrazzo	M	L		[M]	H	H	H	M		[M]
Tile	M	L		[M]				M		[M]
Accessories - bolt or screw required	Sheet metal or wood screw	None. Anchor complete	None. Anchor complete	None. Anchor complete	None. Anchor complete	Machine screw	Machine screw	Sheet metal or wood screw	Lag screw	None. Anchor complete
Tools	Wrench Screwdriver Drill	Drill Hammer	Drill Hammer	Drill Screwdriver	Drill Hammer	Drill Hammer Screwdriver Wrench Setting tools	Drill Wrench	Drill Hammer Screwdriver	Drill Wrench	Drill Screwdriver
Material	Special gold plastic	Nylon Steel Aluminium	Special nylon	Steel	Aluminium Steel	Zinc Lead	Zinc	Lead	Zinc	Steel

LEGEND			
	H - Best or heaviest duty anchor for this material	☐ - Anchor will work if material contains hollow	Hard materials: brick, concrete block, stone, etc.
	M - Good anchor for this material	BLANK SPACE - Not recommended	Soft materials:
	L - Light duty for this material		plaster, plasterboard, gypsum, fibreboard, etc.

QUESTIONS

1. Define the word *adhesive* and give three (3) examples of adhesives.

2. Select an adhesive and list its strengths, it weaknesses, its ability to resist water, and the ease with which it can be cleaned up.

3. Explain the basic difference in structure between a common nail and a finishing nail and tell where each is primarily used.

4. Explain the relationship between the penny designation and a nail's length to the number sold in one pound.

5. Explain why the hole drilled in the lower of two pieces fastened with wood screws should be smaller than the diameter of the screw being used.

6. List the two (2) basic categories used to classify wood screws.

7. What type of fastener would you use to attach a wood framing member to a concrete wall?

8. Explain the purpose for placing anchor bolts in a foundation wall.

9. Tell why galvanized nails must be used to attach exterior members together.

10. Explain the precautions that must be taken when masonry nails are being driven.

11. Explain how a lag screw and its companion lead shield work.

12. Compare three (3) different types of wall fasteners with respect to (a) use, (b) effectiveness, and (c) installation procedures.

4

BUILDING MATERIALS

Humankind derives many benefits from trees. They not only provide shade from the sun, protection from the wind, and a source of fuel for heat, but they also offer a variety of shapes and colors that beautify the landscape. However, their greatest contribution is when they are cut into uniform widths, lengths, and thicknesses and classified as lumber, the chief material of home construction.

Lumber is a designation that applies to: dimension lumber used primarily for framing purposes; items that hide the framing components (i.e., flooring, sheathing, paneling, and trim); and the timbers used as support posts and beams.

Every good carpenter knows which type of lumber to use in each particular phase of residential construction. Achieving this knowledge is no mean feat. The carpenter must be familiar with how a tree grows and how it is milled into lumber. He or she must know the differences between hard and softwoods and between heartwood and sapwood. A carpenter should be acquainted with how moisture content varies from wood to wood and how it may be altered when wood is seasoned, not to mention the defects commonly associated with the various species of trees. Finally, the carpenter must have a thorough working knowledge of the sizes and grades of lumber that are available.

TREE GROWTH

Trees produce their own food through the miracle of photosynthesis. Water in the soil is absorbed by the roots and is carried upward through the xylem tubes of the sapwood to the leaves. In the leaves, water combines with carbon dioxide which has entered the stomates from the surrounding air. Radiant energy in the form of sunlight is required to convert these two raw materials (water and carbon dioxide) into carbohydrates (plant food).

This food is carried to the various parts of the tree to be consumed in various biological activities or to be stored in the trunk.

The active growing area of a tree lies just beneath the bark and is called the *cambium*. The inner side of the cambium produces xylem cells, which become the new wood cells of the sapwood. On the bark side, the cambium produces phloem cells, which are involved in the transport of food throughout the tree (Figure 4:1).

The cambium area experiences two separate growing spurts, each producing a distinct layer; i.e., springwood and summerwood. Together, they comprise one annual ring. Because of ideal weather conditions which encourage rapid growth, the cells produced during the spring are large and thin-walled. Those produced during the summer, when growing conditions are not as favorable, are smaller, darker in color, and have thicker walls.

A tree's age can be determined by simply counting its annual rings (Figure 4:2). A tree's characteristic grain pattern, visible once it has been cut into lumber, is formed by its annual rings.

The trunk of a tree contains both sapwood and heartwood (Figure 4:3). The *sapwood*, located just inside the cambium layer, contains growing cells. Once these cells become inactive, they are transformed into *heartwood* and become darker in color. The main difference between the two is that the heartwood is better equipped to withstand adverse weather conditions.

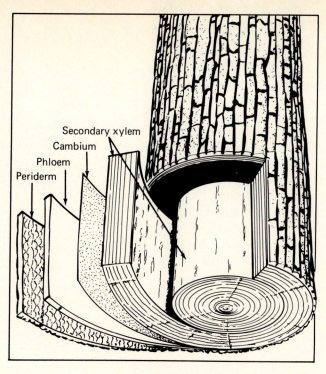

FIGURE 4:1

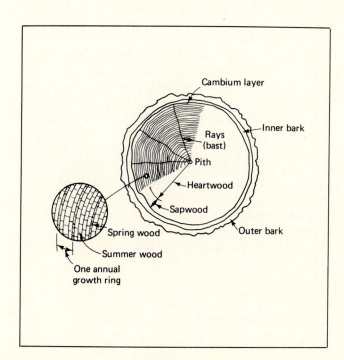

FIGURE 4:2

LOGS TO LUMBER

Before a log enters the sawmill, it must be thoroughly washed to remove any inert substances clinging to its bark that might dull the extremely sharp saw blade. As an added precaution, many mills debark the logs before

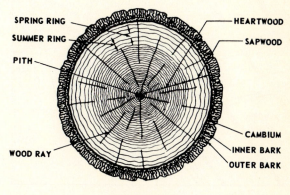

FIGURE 4:3

they are cut. Bark from pine trees is cut into small pieces, packaged, and sold as a plant mulch.

The log is positioned on a movable platform, called a *carriage,* which transports the log into the *headsaw.* The headsaw slices one board off each time the carriage transports the log past it until the log has been transformed into a pile of boards. Some lumbermills employ *gang saws* instead of the headsaw so that a log can be cut into boards in one single operation.

These cut boards, known as *green lumber,* are then conveyed to a series of *edger saws* which perform the double function of trimming off the rough edges and making the sides straight. After this, the boards are transported to the *trimmer saws.* They not only cut the boards to the desired length but also make each end square (Figure 4:4).

There are two methods employed to saw lumber into boards (Figure 4:5). The simpler method, *flat-grained* (softwoods) or *plain-sawed* (hardwoods), makes a cut where the angle formed between the annular rings and the board surface is less than 45 degrees.

The second method is not only more difficult to perform because the angle cut exceeds 45 degrees, but it is more expensive because a portion of the log is wasted. Boards cut by this method, known as *edge-grained* (softwoods) or *quarter-sawed* (hardwoods), have more attractive grain patterns, are more durable, and do not warp as easily as those cut by the first method.

Each cut board must be properly graded according to both size and quality with respect to its tree species.

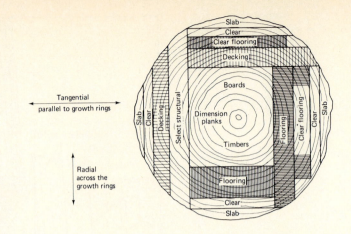

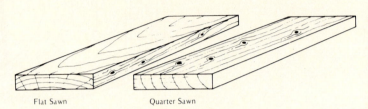

Flat Sawn   Quarter Sawn

FIGURE 4:5

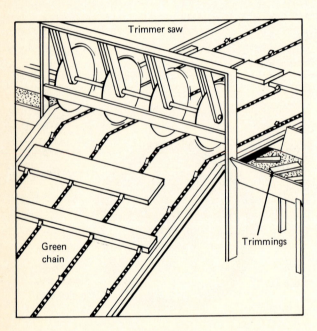

FIGURE 4:4
Sawmill. (Courtesy of American Forestry Institute)

LUMBER GRADING

The grading of hardwood lumber is based on the amount of usable lumber that exists in a specific cut piece. At least one side of the piece must be "clear" and free from rot, knots, or other defects. Most hardwoods are purchased from lumber yards that specialize in their sale. Table 4:1 shows the various hardwood grades.

Several different lumber-producing associations are responsible for the grading of softwoods. Because there are only slight differences among the criteria of the individual associations, the grading can be somewhat confusing. Figure 4:6 is an example of a grade stamp for machine stress-rated lumber.

In general, however, the two major grade categories for softwoods are *select* and *common*. Boards given the

TABLE 4:1
Hardwood Grades and Their Specifics

Grade	Description	Lumber size	% "Clear"
First and Seconds (FAS)	Highest grade, good both surfaces	6″ wide, 8′ long	$83\frac{1}{3}$%
Selects	Defects on one surface	4″ wide, 6′ long	75%
Number 1 common	Used for school shopwork	3″ wide, 6′ long	$66\frac{2}{3}$%
Number 2 common	Most pieces have their defects cut out and are sold as "shorts"	short lengths, narrow widths	50%
Number 3 common		3″ wide, 2′ long	$33\frac{1}{3}$%

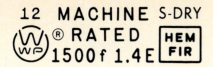

12 MACHINE S-DRY
WWP® RATED 1500f 1.4E HEM FIR

FIGURE 4:6
A typical machine stress-rated lumber mark. (Courtesy of Western Wood Products Association)

FIGURE 4:7
Air dried lumber.

select classification are used primarily for finished elements that will be visible, while those labeled common are for structural elements that will be hidden from view.

Table 4:2 gives the various grades of softwood lumber.

With both hardwoods and softwoods, a better grade has fewer defects and consequently costs more than a poorer grade of lumber.

DRYING (SEASONING) LUMBER

Before it can be used, all freshly cut "green" lumber must be dried by one of two methods so that the moisture content can be reduced to a low, stable level.

In the *air dried* method, green lumber is stacked in piles with spacers between the layers to allow adequate air circulation throughout the pile (Figure 4:7). The time required to reduce the lumber's moisture content to 12% to 18%, is related to the spacing between the individual boards as well as between the separate layers, the climatic conditions, and the species of lumber being dried; because of these variables, seasoning time may range from several weeks to six or seven months.

The *kiln dried* method is much quicker. Lumber is first stacked in piles and then placed in ovens where both the temperature and humidity levels are carefully controlled. Initially, the temperature is kept low and steam is used to maintain a high relative humidity level. The oven's temperature is gradually increased while the humidity level is reduced. Fans run constantly to ensure adequate air circulation within the oven during the entire drying time.

Kiln drying reduces the moisture content of most

TABLE 4:2
Softwood Grades and Their Specifics

Category/Grade	Description
Select	
B and Better (also called #1 or #2 clear)	Highest grade, clear both sides
C	Clear on one side
D	May have pin knots and other slight defects
Common	
1	These are the three grades
2	commonly available in most
3	lumber yards
4	These should *not* be used for
5	residential house construction
Factory and shop lumber Third clear (Factory Select) Number 1 shop Number 2 shop Number 3 shop	Used for millwork (window and door frames, trim)

lumber species to 7% to 10% in three to six days. An added advantage of this method is a marked reduction in the number and severity of the seasoning defects commonly associated with air dried lumber. This is the method used to dry all higher-grade lumbers.

LUMBER DEFECTS OR BLEMISHES

Any irregularity in a piece of wood that tends to impair its strength, durability, or appearance is called a defect. The quantity, type, severity, and the location of defects are all criteria used to establish the final grade of the lumber piece.

Table 4:3 presents the various lumber defects and a description of each.

Figure 4:8 shows two common lumber defects.

A board that shows any variation from a true, plain surface is called *warped*. This condition is usually directly related to a board's tendency to shrink during seasoning. There are four specific types of warping, which is not considered a true defect. They are

1. Bow—the board's width will not lie flat along its entire length

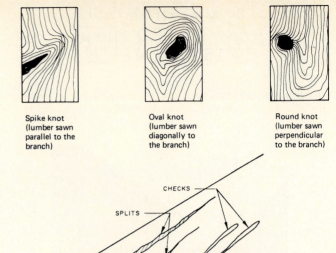

Spike knot (lumber sawn parallel to the branch)

Oval knot (lumber sawn diagonally to the branch)

Round knot (lumber sawn perpendicular to the branch)

FIGURE 4:8
Splits and checks are separations in the grain.

TABLE 4:3
Lumber Defects and Their Descriptions

Defect	Description
Check	A separation along the grain and *across* the annual growth rings. It occurs when the outer surface dries and shrinks faster than the inner portions.
Split	Similar to a check. It occurs at the end of the board.
Shake	A separation along the grain and *between* the annual growth rings. There are several types. 1. Ring shake—enlarges during seasoning 2. Wind shake—a separation of the fibers or annual rings during growth 3. Starshake—seen as cracks radiating outward from the tree's center
Knot	Occurs where side branches grew. There are several types. 1. Tight knot—formed when the branch's live fibers interlock with the tree's trunk 2. Loose knot—formed when the trunk's fibers grow around a dead branch 3. Spike knot—where the branch runs lengthwise through the lumber
Cross grain	Where wood fibers are twisted and do not run parallel with the board's edge. They are difficult to plane and chisel and are very prone to warping.
Wane	A defective edge that is not square, is missing wood along its edge, or has bark present
Honey combing	Occurs when the interior wood fibers separate along the wood rays
Pitch pocket	An internal cavity containing pitch, resin, and/or bark, making it difficult to properly finish the surface
Holes	These are caused by fungi, decay, or worms.

2. Crook—the board's edge will not lie flat along its entire length

3. Cup—one surface of the board's width is concave and the other is convex, over its entire length

4. Twist (wind)—a portion of the board's width appears to be twisted out of its natural parallel plane

These four types of warp are illustrated in Figure 4:9.

SOFTWOOD LUMBER CLASSIFICATIONS

Lumber for house construction falls into three major classifications and is graded according to its intended use. The categories are boards, dimension lumber, and heavy-duty structural members. The first two combined are sometimes referred to as yard lumber.

The carpenter must choose the grade that is most appropriate for a specific constructional use, since proper selection ensures not only sound construction but cost efficiency as well.

Table 4:4 presents the three major softwood lumber classifications and their subdivisions.

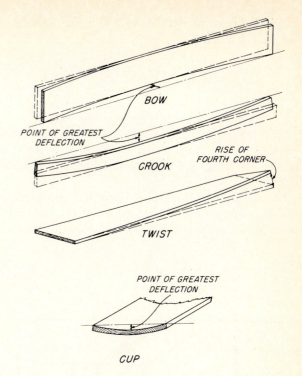

FIGURE 4:9

TABLE 4:4
Classifications of Softwood Lumber

Classification/Type	Thickness	Width	Uses	Nominal sizes
Boards				
Strips	Less than 2″	Less than 4″	Furring, shims, blocking bridging, stakes, battens	1 × 2, 1 × 3
Boards	Less than 2″	4″ or more	Exterior trim & sheathing shelving, valances, flooring paneling	1 × 4, 1 × 6 1 × 8 1 × 10, 1 × 12
Dimension lumber (framing, structural) Studs	2″ to 4″ Length 10′ or less	2″ to 4″	Wall studs, plates	2 × 2, 2 × 3, 2 × 4
Joists and planks	2″ to 4″	6″ or more	Joists, planks, headers, steps, stringers, rafters	2 × 6, 2 × 8, 2 × 10, 2 × 12
Heavy duty structural Posts	Thickness and width the same		Support columns, fencing, decking	4 × 4, 6 × 6
Beams Stringers	5″ or more	2″ more than thickness	Support columns	
Timbers	5″ or more	Not more than 2″ greater than thickness	Support columns	

Figure 4:10 displays the sizes and typical loading of structural lumber.

LUMBER MEASUREMENT

Board Feet

Most lumber is sold by the board foot. It is a unit of measure equal to the cubic contents of a piece of lumber 12″ long, 12″ wide, and 1″ thick (144 cubic inches).

To calculate the number of board feet in a piece of lumber, multiply the normal thickness (T) and width (W) in inches by the length (L) in feet, then divide by 12.

$$\frac{(T \times W \times L)}{12}$$

For example, a piece of lumber 8′ long, 1″ thick, and 12″ wide would contain 8 board feet. Lumber less than 1″ thick is figured as 1″. Consult Figure 4:11 for other examples.

To find the total board footage of several uniform pieces of lumber, use the following formula.

$$\frac{P \times T \times W \times L}{12} = \text{board feet}$$

P = Number of pieces
T = Thickness of wood in inches
W = Width of wood in inches
L = Length of wood in feet

You can estimate the cost of your project by multiplying the total number of board feet by the cost per board foot. For example, if you needed 60 pieces of 2 × 4 lumber, each 16′ long, and the cost per board foot was $37\frac{1}{2}$ cents, you would figure your total cost this way

$$\frac{60 \times 2 \times 4 \times 16}{12} \times \$0.375 = \$240^{00}$$

Table 4:5 presents the number of board feet in a variety of timber lengths.

Linear Feet

Some types of lumber, such as molding, dowels, and railings, are sold by the linear foot, where length is the only criterion of measurement. One linear or running

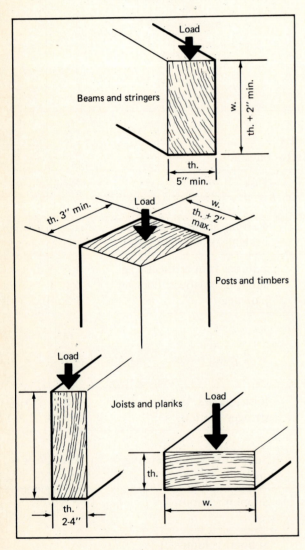

FIGURE 4:10
Sizes and typical loading of structural lumber.

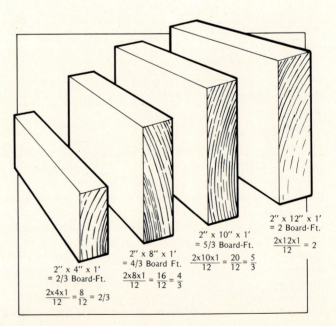

FIGURE 4:11

TABLE 4:5
Scale of Board Feet per Timber Length

Timber size	Length of timber (in feet)								
	8	10	12	14	16	18	20	22	24
1 × 4	$2\frac{2}{3}$	$3\frac{1}{3}$	4	$4\frac{2}{3}$	$5\frac{1}{3}$	—	—	—	—
1 × 6	4	5	6	7	8	—	—	—	—
1 × 8	$5\frac{1}{3}$	$6\frac{2}{3}$	8	$9\frac{1}{3}$	$10\frac{2}{3}$	—	—	—	—
2 × 4	$5\frac{1}{3}$	$6\frac{2}{3}$	8	$9\frac{1}{3}$	$10\frac{2}{3}$	12	$13\frac{1}{3}$	—	—
2 × 6	8	10	12	14	16	18	20	—	—
2 × 8	$10\frac{2}{3}$	$13\frac{1}{3}$	16	$18\frac{2}{3}$	$21\frac{1}{3}$	24	$26\frac{2}{3}$	—	—
2 × 10	$13\frac{1}{3}$	$16\frac{2}{3}$	20	$23\frac{1}{3}$	$26\frac{2}{3}$	30	$33\frac{1}{3}$	—	—
2 × 12	16	20	24	28	32	36	40	—	—
4 × 4	$10\frac{2}{3}$	$13\frac{1}{3}$	16	$18\frac{2}{3}$	$21\frac{1}{3}$	24	$26\frac{2}{3}$	—	—
4 × 6	16	20	24	28	32	36	40	—	—
6 × 6	24	30	36	42	48	54	60	66	72

foot is equivalent to one board foot only when the area (one square foot) is the same (for example, in a board 1 foot long, 2 inches thick, and 6 inches wide).

Normal Size

The dimensions given for a specific piece of lumber represents its size before it has been milled and dried (as it comes from the saw at the mill). Thus, a nominal 2 × 4 piece is actually $1\frac{1}{2}'' \times 3\frac{1}{2}''$. The thickness of 1″ stock becomes $\frac{3}{4}''$, while 2″ stock becomes $1\frac{1}{2}''$. The length of each piece of lumber, however, is not reduced by processing. Figure 4:12 shows both the nominal and actual sizes of the most commonly purchased lumber.

MANUFACTURED BOARDS

There are four common types of manufactured boards: plywood, hardboard, particle board, and wafer board. All four are made from wood components that have been treated so that the finished product no longer resembles wood in its natural state.

Manufactured boards not only provide the carpenter with a wide selection of usable products that lack many of the undesirable characteristics of solid wood, (e.g., warping, splitting, and twisting,) but also make it possible for the manufacturer to use wood scraps and pieces that would otherwise have no commercial value. These boards are stronger, more rigid, and are dimensionally more stable than natural wood. Their ability to cover a large area rapidly is applauded by all carpenters.

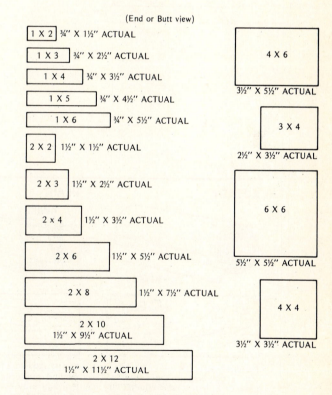

(End or Butt view)

1 X 2 — ¾″ X 1½″ ACTUAL
1 X 3 — ¾″ X 2½″ ACTUAL
1 X 4 — ¾″ X 3½″ ACTUAL
1 X 5 — ¾″ X 4½″ ACTUAL
1 X 6 — ¾″ X 5½″ ACTUAL
2 X 2 — 1½″ X 1½″ ACTUAL
2 X 3 — 1½″ X 2½″ ACTUAL
2 x 4 — 1½″ X 3½″ ACTUAL
2 X 6 — 1½″ X 5½″ ACTUAL
2 X 8 — 1½″ X 7½″ ACTUAL
2 X 10 — 1½″ X 9½″ ACTUAL
2 X 12 — 1½″ X 11½″ ACTUAL

4 X 6 — 3½″ X 5½″ ACTUAL
3 X 4 — 2½″ X 3½″ ACTUAL
6 X 6 — 5½″ X 5½″ ACTUAL
4 X 4 — 3½″ X 3½″ ACTUAL

FIGURE 4:12
Nominal and actual lumber sizes (end or butt view).

Plywood

A plywood panel consists of several thin sheets of softwood lumber that have been glued together under pressure. The layers (an odd-number of sheets) are placed together so that the grain of each sheet runs perpendicular to the grain of adjoining ones. This system provides extra strength without excessive bulk and reduces the possibility of warping. The two surface veneers (faces) are positioned so that the grain runs in the same direction (Figure 4:13).

The three methods used to manufacture plywood sheets are presented in Figure 4:14.

Interior and exterior plywoods differ in two basic ways: (1) exterior panels have waterproof glue so that they can be exposed to the weather and (2) there are

more possible grade combinations for interior plywood than for exterior panels. Exterior plywood is sold in grades A through D, with the grade of each surface clearly stamped on the panel (Figure 4:15). A notation of A/C on the back of a panel implies that the top exterior surface is grade A (smooth and sanded), while the back surface is grade C (suitable for sheathing and underlayment). Table 4:6 lists the characteristics of the various grades of plywood. Table 4:7 describes the types of plywood generally available and their most common uses.

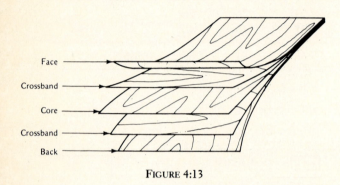

FIGURE 4:13

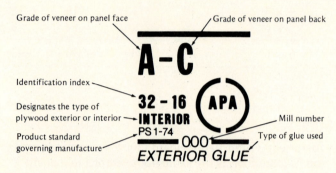

FIGURE 4:15
Information stamped on plywood panels.

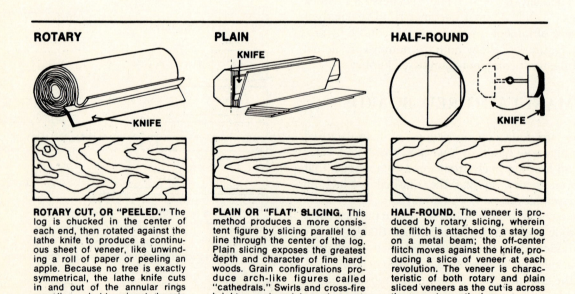

FIGURE 4:14
Three methods of making plywood sheets.

TABLE 4:6
Description of Plywood Grades

N	Smooth surface "natural finish" veneer. Select, all heartwood or all sapwood. Free of open defects. Allows not more than 6 repairs, wood only, per 4 × 8 panel, made parallel to grain and well matched for grain and color.
A	Smooth, paintable. Not more than 18 neatly made repairs, boat, sled, or router type, and parallel to grain, permitted. May be used for natural finish in less demanding applications.
B	Solid surface. Shims, circular repair plugs and tight knots to 1 inch permitted. Wood or synthetic patching material may be used. Some minor splits permitted.
C (Plugged)	Improved C veneer with splits limited to $\frac{1}{8}$ inch width and knotholes and borer holes limited to $\frac{1}{4} \times \frac{1}{2}$ inch. Admits some broken grain. Synthetic repairs permitted.
C	Tight knots to $1\frac{1}{2}$ inch. Knotholes to 1 inch and some to $1\frac{1}{2}$ inch if total width of knots and knotholes is within specified limits. Synthetic or wood repairs. Discoloration and sanding defects that do not impair strength permitted. Limited splits allowed.
D	Knots and knotholes to $2\frac{1}{2}$ inch width and $\frac{1}{2}$ inch larger within specified limits. Limited splits are permitted.

TABLE 4:7
Plywood and Its Applications
(Courtesy of American Plywood Association)

APA A–C

Typical
Trademark

APA
A-C   GROUP 1
EXTERIOR
___000___
PS 1-74

For use where appearance of only one side is important in exterior applications, such as soffits, fences, structural uses, boxcar and truck linings, farm buildings, tanks, trays, commercial refrigerators, etc. TYPE: Exterior. COMMON THICKNESSES, $\frac{1}{4}, \frac{3}{8}, \frac{1}{2}, \frac{5}{8}, \frac{3}{4}$.

APA B–C

Typical
Trademark

APA
B-C   GROUP 1
EXTERIOR
___000___
PS 1-74

Utility panel for farm service and work buildings, boxcar and truck linings, containers, tanks, agricultural equipment, as a base for exterior coatings and other exterior uses or applications subject to high or continuous moisture. TYPE: Exterior. COMMON THICKNESSES, $\frac{1}{4}, \frac{3}{8}, \frac{1}{2}, \frac{5}{8}, \frac{3}{4}$.

(Continued)

TABLE 4:7 (CONTINUED)

APA C–C Plugged

Typical
Trademark

```
APA
C-C PLUGGED
GROUP 2
EXTERIOR
000
PS 1-74
```

For use as an underlayment over structural subfloor refrigerated or controlled atmosphere storage rooms, pallet fruit bins, tanks, boxcar and truck floors and linings, open soffits, tile backing and other similar applications where continuous or severe moisture may be present. Provides smooth surface for application of resilient floor coverings and possesses high concentrated and impact load resistance. TYPE: Exterior. COMMON THICKNESSES, $\frac{3}{8}$, $\frac{1}{2}$, $\frac{19}{32}$, $\frac{5}{8}$, $\frac{23}{32}$, $\frac{3}{4}$.

APA 303 Siding

Typical
Trademark

```
APA
303 SIDING   6-S/W
24 OC   GROUP 2
EXTERIOR
000
PS 1-74  FHA-UM-64
```

Proprietary plywood products for exterior siding, fencing, etc. Special surface treatment such as V-groove, channel groove, striated, brushed, rough-sawn and texture-embossed (MDO). Stud spacing (Span Rating), and face grade classification indicated in trademark. TYPE: Exterior. COMMON THICKNESSES, $\frac{11}{32}$, $\frac{3}{8}$, $\frac{15}{32}$, $\frac{1}{2}$, $\frac{19}{32}$, $\frac{5}{8}$.

APA Texture 1–11

Typical
Trademark

```
APA
303 SIDING    6-S/W
16 OC   19/32 INCH
GROUP 2
EXTERIOR
T1-11   000
PS 1-74  FHA-UM-64
```

Special 303 siding panel with grooves $\frac{1}{4}''$ deep, $\frac{3}{8}''$ wide, spaced 4″ or 8″ o.c. Other spacings may be available on special order. Edges shiplapped. Available unsanded, textured and MDO. TYPE: Exterior. THICKNESSES, $\frac{19}{32}$ and $\frac{5}{8}$ only.

APA High Density Overlay (HDO)

Typical Trademark

```
HDO · A-A · G-1 · EXT-APA · PS1-74 · 000
```

Has a hard semi-opaque resin-fiber overlay both sides. Abrasion resistant. For concrete forms, cabinets, countertops, signs, tanks. Also available with skid-resistant screen-grid surface. TYPE: Exterior. COMMON THICKNESSES, $\frac{3}{8}$, $\frac{1}{2}$, $\frac{5}{8}$, $\frac{3}{4}$.

(Continued)

TABLE 4:7 (CONTINUED)

APA Medium Density Overlay (MDO)

Typical
Trademark

```
APA
M. D. OVERLAY
        GROUP 1
EXTERIOR
        000
    PS 1-74
```

Smooth, opaque, resin-fiber overlay one or both sides. Ideal base for paint, both indoors and outdoors. Also available as a 303 Siding. TYPE: Exterior. COMMON THICKNESSES, $\frac{11}{32}, \frac{3}{8}, \frac{1}{2}, \frac{5}{8}, \frac{3}{4}$.

APA Marine

Typical Trademark

MARINE · A-A · EXT-APA · PS1-74 · 000

Ideal for boat hulls. Made only with Douglas fir or western larch. Special solid jointed core construction. Subject to special limitations on core gaps and face repairs. Also available with HDO or MDO faces. TYPE: Exterior. COMMON THICKNESSES, $\frac{1}{4}, \frac{3}{8}, \frac{1}{2}, \frac{5}{8}, \frac{3}{4}$.

APA B–B Plyform Class I and Class II

Typical
Trademark

```
APA
PLYFORM
B-B  CLASS I
EXTERIOR
    000
  PS 1-74
```

Concrete form grades with high reuse factor. Sanded both sides and mill-oiled unless otherwise specified. Special restrictions on species. Class I panels are stiffest, strongest and most commonly available. Also available in HDO for very smooth concrete finish. In Structural I (all plies limited to Group 1 species), and with special overlays. TYPE: Exterior. COMMON THICKNESSES, $\frac{5}{8}, \frac{3}{4}$.

APA A–D

Typical
Trademark

```
APA
A-D   GROUP 1
INTERIOR
    000
PS 1-74 EXTERIOR GLUE
```

For use where appearance of only one side is important in interior applications, such as paneling, built-ins, shelving, partitions, flow racks, etc. TYPE: Interior. COMMON THICKNESSES, $\frac{1}{4}, \frac{3}{8}, \frac{1}{2}, \frac{5}{8}, \frac{3}{4}$.

APA B–D

Typical
Trademark

```
APA
B-D   GROUP 2
INTERIOR
    000
PS 1-74 EXTERIOR GLUE
```

Utility panel for backing, sides of built-ins, industry shelving, slip sheets, separator boards, bins and other interior or protected applications. TYPE: Interior. COMMON THICKNESSES, $\frac{1}{4}, \frac{3}{8}, \frac{1}{2}, \frac{5}{8}, \frac{3}{4}$.

(Continued)

TABLE 4:7 (CONTINUED)

APA Underlayment

Typical
Trademark

```
APA
UNDERLAYMENT
        GROUP 1
INTERIOR
        000
PS 1-74 EXTERIOR GLUE
```

For application over structural subfloor. Provides smooth surface for application of resilient floor coverings and possesses high concentrated and impact load resistance. TYPE: Interior. COMMON THICKNESSES, $\frac{3}{8}, \frac{1}{2}, \frac{19}{32}, \frac{5}{8}, \frac{23}{32}, \frac{3}{4}$.

APA C–D Plugged

Typical
Trademark

```
APA
C-D PLUGGED
        GROUP 2
INTERIOR
        000
PS 1-74 EXTERIOR GLUE
```

For built-ins, wall and ceiling tile backing, cable reels, walkways, separator boards and other interior or protected applications. Not a substitute for Underlayment or APA Rated Sturd-I-Floor as it lacks their puncture resistance. TYPE: Interior. COMMON THICKNESSES, $\frac{3}{8}, \frac{1}{2}, \frac{19}{32}, \frac{5}{8}, \frac{23}{32}, \frac{3}{4}$.

APA A–A

Typical Trademark

A-A · G-1 · INT-APA · PS1-74 · 000

Use where appearance of both sides is important for interior applications such as built-ins, cabinets, furniture partitions; and exterior applications such as fences, signs, boats, shipping containers, tanks, ducts, etc. Smooth surfaces suitable for painting. TYPES: Interior, Exterior. COMMON THICKNESSES, $\frac{1}{4}, \frac{3}{8}, \frac{1}{2}, \frac{5}{8}, \frac{3}{4}$.

APA A–B

Typical Trademark

A-B · G-1 · INT-APA · PS1-74 · 000

For use where appearance of one side is less important but where two solid surfaces are necessary. TYPES: Interior, Exterior. COMMON THICKNESSES: $\frac{1}{4}, \frac{3}{8}, \frac{1}{2}, \frac{5}{8}, \frac{3}{4}$.

APA B–B

Typical Trademark

B-B · G-2 · INT-APA · PS1-74 · 000

Utility panels with two solid sides. TYPES: Interior, Exterior. COMMON THICKNESSES, $\frac{1}{4}, \frac{3}{8}, \frac{1}{2}, \frac{5}{8}, \frac{3}{4}$.

(Continued)

TABLE 4:7 (CONTINUED)

APA Decorative

Typical
Trademark

APA DECORATIVE GROUP 2 INTERIOR 000 PS 1-74	Rough-sawn, brushed, grooved, or striated faces. For paneling, interior accent walls, built-ins, counter facing, exhibit displays. Can also be made by some manufacturers in Exterior for exterior siding, gable ends, fences and other exterior applications. Use recommendations for Exterior panels vary with the particular product. Check with the manufacturer. TYPES: Interior, Exterior. COMMON THICKNESSES, $\frac{5}{16}, \frac{3}{8}, \frac{1}{2}, \frac{5}{8}$.

APA Plyron

Typical Trademark

PLYRON -INT-APA · 000

Hardboard face on both sides. Faces tempered, untempered, smooth or screened. For countertops, shelving, cabinet doors, flooring. TYPES: Interior, Exterior. COMMON THICKNESSES, $\frac{1}{2}, \frac{5}{8}, \frac{3}{4}$.

Hardboard

When wood chips are exposed to high pressure steam, they explode into fibers. These fibers are then refined, felted, and finally compressed in heated hydraulic presses to produce a smooth, "man-made" panel. These untempered panels do not require any fillers or adhesives, since the natural lignin in the wood is sufficient to hold the pressed wood fibers together.

To form tempered hardboards, the fibers are impregnated with a special adhesive and then baked. This procedure makes the panels harder, darker in color, and more moisture-resistant than untempered panels.

Hardboard is 4′ wide; it may be purchased in lengths of 8′, 10′, and 12′, and in thicknesses ranging from $\frac{1}{8}″$ to $\frac{5}{16}″$. Pegboard is a popular example of hardboard. Any cutting of a hardboard panel should be done with a fine-toothed saw blade with the good, or finished, side of the panel facing upward.

Untempered hardboard is used on the interior for underlayment, drawer bottoms, cabinet backs and decorative wall panels. Tempered hardboard is used as exterior siding in a variety of patterns and textures.

Particleboard

Particleboard, another "man-made" product, is composed primarily of wood chips, shavings, and veneer clippings that are processed and bonded together to produce a smooth, uniform panel with endless uses, e.g. shelving, cabinets, countertops, and floor underlayment.

Urea formaldehyde is the bonding material used for interior panels, while phenolic resin glues are used to bond exterior panels. The use intended and the thickness required (up to $1\frac{7}{16}$ inches) will help determine the type of panel to select (Figure 4:16).

FIGURE 4:16

Waferboard

Waferboard (Figure 4:17) is a structural 4′ × 8′ panel that is composed of wood wafers bonded together with a phenolic resin. It is manufactured in two thicknesses, $\frac{7}{16}$″ and $\frac{1}{2}$″, and can be used as roof, wall, or floor sheathing and as soffit material.

Except for floor sheathing where only the $\frac{1}{2}$″ size placed 16″ o.c. is recommended, either thickness can be used when the underlying structural members are positioned 24″ o.c.

It is imperative that there is proper spacing between both the end and edge joints of adjacent panels of waferboard to allow for expansion.

FIGURE 4:17

QUESTIONS

1. Explain the basic difference between springwood and summerwood with respect to their formation and appearance.

2. Compare how plain-sawed boards differ from those that have been quarter-sawed.

3. When graded, softwood lumber is labeled as either select or common. Explain how these grades differ in appearance and useability.

4. Compare air dried to kiln dried lumber with respect to speed of drying and frequency of seasoning defects.

5. Select one common lumber defect and describe its appearance.

6. When nailing a cupped board to a flat surface, e.g., a deck, should its cupped surface be facing up or down?

7. How many board feet are there in a 12′ long 2″ × 4″?

8. What are the actual measurements of a nominal piece of 1″ × 8″ pine?

9. List three (3) advantages that manufactured boards have over natural solid wood pieces.

10. Explain the basic difference in sawing technique employed between a power saw cut made on a plywood panel and one made on a hardboard panel.

11. List three (3) hardwoods and three (3) softwood tree species.

12. List three (3) defects of wood that carpenters must always check for.

13. Explain the differences between plywood, particleboard, waferboard, and hardboard.

14. Explain the difference between a board foot and a linear foot measurement. Are they ever equal?

5

SITE SELECTION AND PLANNING

Every potential house builder should realize, before embarking upon the critical task of selecting a building site, that there are inherent problems associated with any site. The builder's job is to blend the house with its surroundings, with a minimum of problems and expense. This feat is normally accomplished by either searching for the ideal site to complement a specific house style or designing the house to fit the existing landscape.

The builder must evaluate the existing environmental features of the site (e.g., drainage, topography, compass orientation, available plant material), plus be aware of the local building codes, before deciding whether or not the actual construction is economically feasible (Figure 5:1).

The local climatic conditions could also have a dramatic effect on the energy efficiency of a home built on a specific site, and, if unfavorable, may discourage its construction. These conditions might be markedly different from those influencing the building site directly across the street. Climatic conditions include the sun, and how its rays strike the house, the wind and its direction, the amount of precipitation in the area, the proximity of the house to bodies of water, and both the altitude and geographic latitude of the location.

TOPOGRAPHY

Building a house on a level site is easier and less expensive than building where there are sloping contours. Because of its flatness, terraces and retaining walls are not required, but water drainage can present a serious problem. This should be checked out thoroughly before

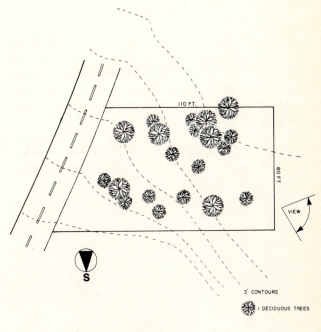

FIGURE 5:1
Sample building site. (Source: Northeast Solar Energy Center)

the site is purchased. Excavated soil can be used to raise the existing grade near the foundation, so that surface water runs away from, rather than towards, the house and possibly into the basement (Figure 5:2).

A gentle sloping site makes an excellent house site. Entrance walkways and driveways are easily traversed and drainage problems are slight. Soil excavated for the foundation is used to ensure the necessary grade away from the house (Figure 5:3).

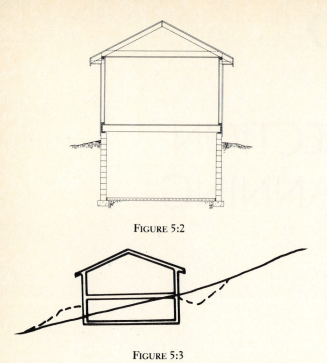

FIGURE 5:2

FIGURE 5:3

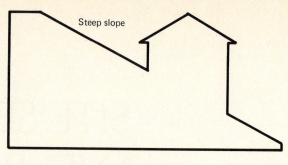

FIGURE 5:4
Steep slope.

this nature require the construction of an elaborate series of wooden decks or terraces to enable the owners and their guests to traverse the grounds. The access to the house as well as to the garage location may present additional problems for the builder to solve.

DRAINAGE

Obvious wet plots should be avoided whenever possible. Be suspicious of level plots that are surrounded by higher ground or are lower than adjoining properties. They will be bathed repeatedly by natural surface water.

Subsurface drainage, soil texture, and the existence of bed rock are best determined by taking test borings throughout the entire plot area.

A building plot higher than the street will not be affected by water runoff from the street, while a plot located below street level will probably require an extensive drainage system to keep the house dry during periods of precipitation (Figure 5:5). The placement of a swale, or diversion ditch, is necessary to ensure a dry basement.

It is possible to construct a house on a plot that is either permanently or occasionally "too wet" by eliminating the basement and by appropriate grading.

An ideal house site is one where the excavated soil can be repositioned to provide the necessary grading around the foundation. Those less than ideal would require either additional fill or total removal of excess excavated material from the site, and thus would involve additional expenses.

As a builder, you should pay special attention to any building site that is already well-defined by permanent, existing features such as an established street grade, storm drains, rock crops, and steep banks. Their presence may seriously affect your ability to grade the site properly.

Sites having steep slopes should be avoided. They might seem appealing to the nonprofessional. However, the problems they present, both to the builder during construction and later to the occupants, can only be solved by very expensive measures (Figure 5:4). Sites of

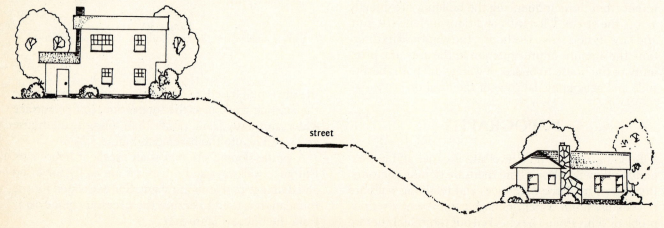

street

FIGURE 5:5

BUILDING CODES

Long before purchasing a plot, it is imperative to become thoroughly familiar with the local building code requirements with respect to house setback, distance from boundary lines, utility hookups, and sewage disposal. It is extremely unpleasant, after construction has begun, for a builder to have to apply for a variance from the town board due to an oversight or incorrect assumption. The desired constructional outcome will definitely be affected if the variance is denied (Figure 5:6).

Road construction may be necessary when developing virgin land. If you learn beforehand who is responsible for building roads, as well as maintaining them (i.e., builder or local government), this knowledge will help you to determine the feasibility and expense of developing land.

HOUSE DESIGN
AND ORIENTATION

For a modern home to be energy efficient year-round, it must be constructed with certain design features and be properly positioned on the lot (Figure 5:7). The builder who realizes the importance and effectiveness of proper house orientation and design for energy con-

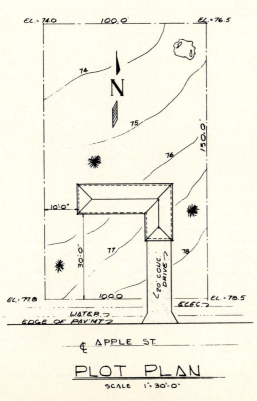

FIGURE 5:6

FIGURE 5:7
Some basics of house orientation.

servation will be better able to adequately assess the suitability of the proposed site.

The roof line of the house should lie in an east-west orientation (Figure 5:8). If it is expected that solar panels will eventually be mounted on the roof, site selection is then restricted to those that would allow the rear roof to face dead south or within 20° of south (Figure 5:9). The panels can be placed on the front side of a south-facing roof, but, for purely cosmetic reasons, the preference is to place them on the rear roof (Figure 5:10).

Note that the tilt angle of the solar collector should be equal to or be within ± 15° of the local latitude. Table 5:1 can be used to convert the slope or pitch of a roof to its approximate roof angle. A roof with a $\frac{3}{8}$ pitch has a 37° roof angle.

A site having a southeast- or south-facing slope is best suited to capitalize on the beneficial effects of the sun's rays. Sites having north-facing slopes should not be considered (Figure 5:11).

The majority of the windows should face south. By allowing solar heat into the house, south-facing windows help reduce the heating load in northern homes during the winter months. Very few windows should be positioned on the north side of the home.

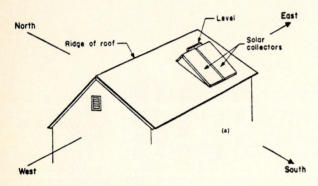

FIGURE 5:8

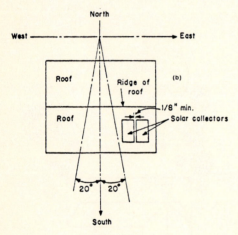

FIGURE 5:9

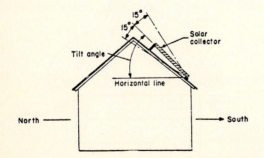

FIGURE 5:10

Collector tilt angle: equal to the local latitude or within local latitude ± 15°.

TABLE 5:1
Roof Angle

Slope	Pitch	Angle (approx.)
2/12	1/12	9°
3/12	1/8	14°
4/12	1/6	18°
5/12	5/24	22°
6/12	1/4	26°
7/12	7/24	30°
8/12	1/3	33°
9/12	3/8	37°
10/12	5/12	39°
11/12	11/24	42°
12/12	1/2	45°
13/12	13/24	48°
14/12	7/12	50°
15/12	5/8	52°
16/12	2/3	54°
17/12	17/24	55°
18/12	3/4	57°
19/12	19/24	59°
20/12	5/6	60°

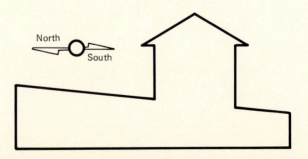

FIGURE 5:11

Roof overhangs, 30″ to 48″, located on the south side of the house prevent the summer sun from entering the windows (a cooling effect), but allow the winter sun to enter and help heat the house (Figure 5:12).

Positioning an attached garage or carport on the southwest or west side of the house helps reduce the ill effects caused by sustained exposure to the sun (Figure 5:13). It is not advisable to install a concrete patio in this same area because it will absorb solar heat and radiate it back into the house. A wooden deck would be a much better choice.

A detached garage or carport might be a necessity on steep sloping plots. Either of these should be positioned as close to the house as is architecturally feasible so that the connecting walkway is safe and easy to traverse.

LANDSCAPING CONCERNS

For aesthetic and energy conservation reasons, a comprehensive landscape plan must be devised for the entire building site (Figure 5:14). The first step in the plan is to make every possible attempt to save all existing plant material. The second step would be to divide the site into three basic outdoor living areas (Figure 5:15).

The *public* area is that portion of the overall landscape between the house and the street. The *service* area may be beside the house or in an area of the side or rear lawn set aside for that purpose. The vegetable garden and compost pile are the two most common features of this area. The *private* area, usually located behind the house, is for the owner's use and pleasure with features designed to meet the family's desired activities. These features must be strategically positioned to ensure convenient and logical traffice flow throughout the area as well as to provide lasting beauty.

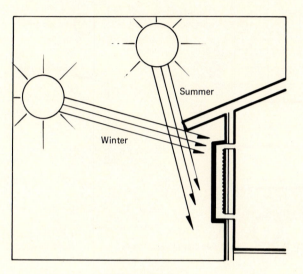

FIGURE 5:12

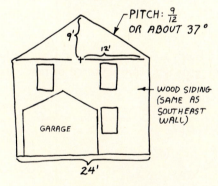

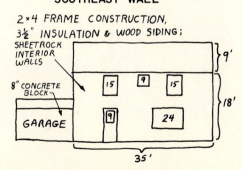

FIGURE 5:13

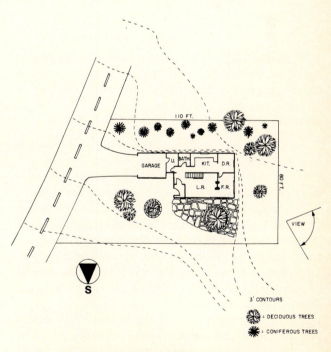

FIGURE 5:14
Completed building site plan. (Source: Northeast Solar Energy Center)

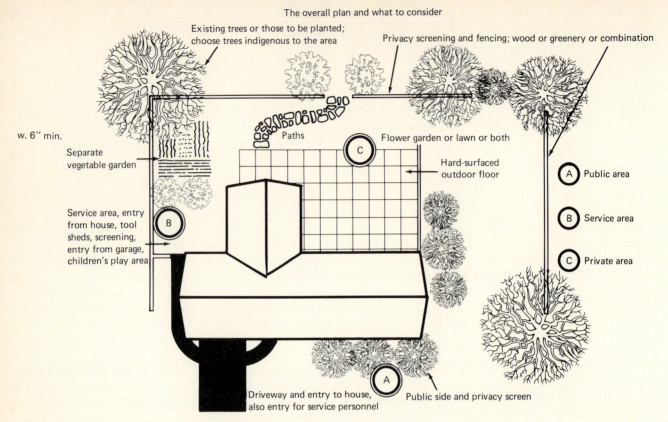

FIGURE 5:15
Overall plan and what to consider.

In each of the three areas, sound landscape principles must be employed in an effort to achieve a perfect mix of beauty, privacy, and energy efficiency.

Deciduous Trees

These trees are excellent providers of shade from the sun and shelter from the wind during the summer months; and because they lose their leaves, they also allow penetration of the sun's rays during the winter.

For maximum effect, they should be planted on the south side of the house (SE to SW is the ideal range), and far enough away from the house to ensure that autumn leaves fall on the ground, and not in gutters or downspouts. Some of the lower branches may have to be removed to enable the sun's rays to reach the house during the winter months.

In addition to providing shade, the leaves of deciduous trees help cool temperatures in another way. The continuous evaporation of water produces air currents in the vicinity of the tree that tend to reduce the temperature of the surrounding air by as much as 15° to 20°F (Figure 5:16).

Select a hardy tree that thrives in your area and provide it the growing room required to reach its mature height.

FIGURE 5:16

Evergreen Trees

Because they keep their foliage the year round, evergreens are an excellent shelter from the wind, which they force upwards over their tops, rather than letting it pass through their close-knit branches (Figure 5:17). They are extremely effective as wind breaks when given a triangular spacing (Figure 5:18), and they are able to deflect the wind two to five times their height on the windward side and 10 to 20 times their height on the leeward side, thus providing a house with effective insulation from the wind.

Their height on the north side is of no concern, but on the south side, they must never be allowed to reach a height that would block out any of the all important winter sun.

Evergreen hedges are excellent for screening undesirable views and for reducing street noises. When planted singly or in clumps at the corners of a house, they help reduce the velocity of the wind by redirecting its path.

Lawns

A thick green lawn, because of its evaporative cooling properties, is able to negate roughly 50 percent of the solar radiation it receives and reflect an additional 20 percent back into the air above it, while only absorbing approximately 5 percent of the total radiation received. This is why grass always feels cooler under foot than blacktop or concrete, and why it is so effective in preventing solar-radiated heat from entering the house.

Ground Covers

Ground covers and low shrubs require less maintenance than grass but are equally as effective at dissipating solar radiation. Snow cover during the winter months can be beneficial because it reflects solar light and heat toward the home.

Water Features

Areas close to large bodies of water are much cooler during the summer and warmer in winter than areas that are distant from water. This is due to a *sea breeze*

FIGURE 5:17

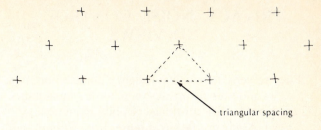

FIGURE 5:18

during the day flowing from the body of water to replace air that rises away from the hot land areas, plus a *land breeze* at night which flows from the land towards the body of water.

Swimming pools, fountains, and small ponds will all exert a cooling effect on the house and its surroundings. However, to obtain the maximum cooling potential from any water source, be sure to position them upwind from the house.

FENCES

They are frequently installed to increase the privacy of the home, but can also provide effective protection from the wind. The fence type selected should allow for some penetration of air; the louvered styles are the most appropriate.

When placed on the south side of the house, they can be positioned at the proper distance to allow the winter sun to reach the house, since their height never changes like an evergreen hedge or windbreak does. It is important to take into account the factors of maintenance and longevity when selecting a fence since most do not last as long as evergreen hedges do.

UTILITIES

As a builder, you must be fully aware of the utilities that are available for a specific building site. Questions, such as will the site be served by town water and sewer lines or will a well and septic system be needed, must be answered long before any decision concerning the site can be reached. In those locations where the builder is offered a choice between the two systems, the expenses associated with each type of installation must be examined.

Homes served by town water and sewage treatment should be positioned on the site to limit the length of the underground lines. This is accomplished by using the minimum setback. A home's interior can be designed so that the plumbing facilities are concentrated within a relatively small area to reduce the need and expense of piping.

PRIVACY, NOISE CONTROL, AND VIEWS

Plant screens, fences, and hedges eliminate undesirable views and provide privacy for a home's inhabitants. They also offer the advantage of noise reduction.

They not only muffle street noises from without, but also help contain the noises the residents make within the confines of the property that might be offensive to their neighbors.

Strategic placement of plant materials, as well as choosing the proper location for the service area, can provide the home's inhabitants with an acceptable view of their property from any vantage point, within or outside the house.

QUESTIONS

1. List three (3) environmental features that should be checked before a building site is considered acceptable.

2. List three (3) local building code regulations that must be known in advance of the actual construction.

3. In which compass direction should potential building sites slope to capitalize on the sun's rays?

4. List three (3) design features that will make a house more energy efficient.

5. List three (3) benefits derived from planting trees on a building site.

6. List two (2) landscape features that are very effective at providing protection from the wind.

6

CONSTRUCTION DOCUMENTS

A builder who contracts with a client to construct a specific house must be able to fully interpret the architect's building plans, understand and agree with the written specifications related to the structure, and be familiar with the building codes for that area.

BUILDING PLANS

The builder must recognize the various symbols and architectural abbreviations on building plans, and have a working knowledge of the other trades so that he can interpret the plans and adequately prepare for the other tradesmen.

An architect uses a three-sided ruler, known as an architect's scale (Figure 6:1), to scale down building plans.

Each side of the ruler is divided into two different scales, with one running from left to right, the other running right to left.

Each scale has divisions of equal units, each representing a foot in length. The $\frac{1}{2}''$ scale implies that $\frac{1}{2}$ inch on the plan represents one foot in length (Figure 6:1).

Views and Lines

A building plan is a scaled-down version of the structure to be built. Of a total of six possible views—front, top, right side, left side, rear, and bottom—a plan represents only the views necessary to describe the object's shape (Figure 6:2). Each view consists of the lines that are visible when the object is viewed from that direction. These lines are called *visible* or *working lines*.

Surfaces that are hidden from a particular view are represented on the plan by dotted lines known as *hidden object lines*. Measurements between two distinct points are expressed by a line that is capped at each end by an arrowhead, with the distance noted midway between the arrowheads. This type of line is called a *dimension line*. An *extension line* is used to complement the dimension line; it projects $\frac{1}{8}$ inch beyond the dimension line but does not actually touch the drawing (Figure 6:3).

Closely drawn parallel lines that form an angle with the working lines are designed to show the viewer what could be observed if a slice were made through the part being shown and a section removed. These lines are called *section lines*.

Section views show pertinent information that is better presented by itself than included in either an elevation or a plan view (Figure 6:4). They are usually drawn

75

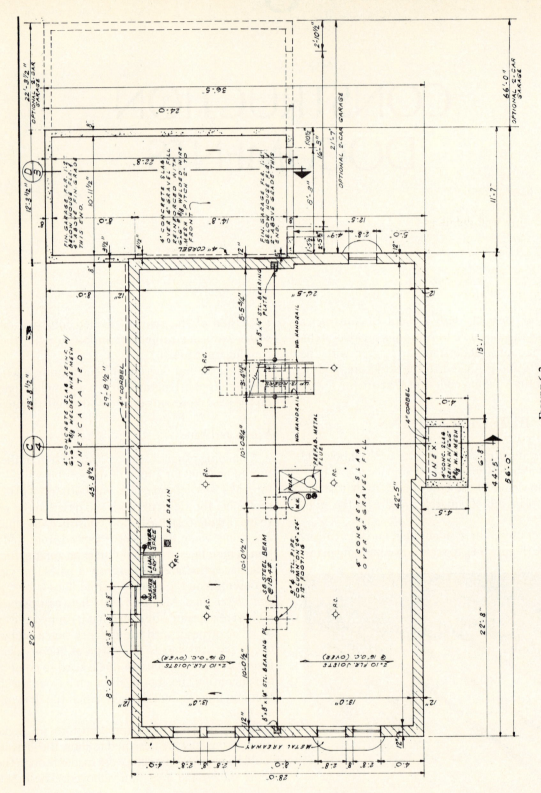

FIGURE 6:2
Foundation plan (top view).

76

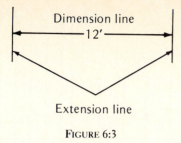

FIGURE 6:3

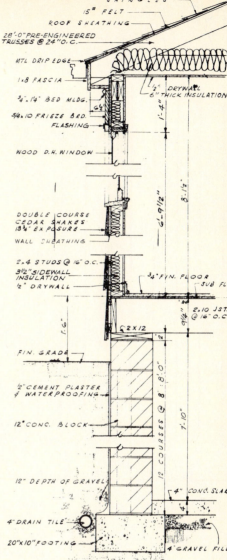

to a much larger scale than plan views so that they can include structural details.

Wall section views (Figure 6:5) are cross-sectional drawings showing wall detail and usually extend from the footing or foundation base to the roof frame area.

Scale

All architectural plans must be drawn to scale so that the length of a line on the plan is directly proportional to the actual length of the object it depicts. A $\frac{1}{4}$" scale ($\frac{1}{4}$ inch equals one foot) implies that $\frac{1}{4}$" on the plan or drawing equals 1' of the object being constructed. House framing plans are normally drawn to a $\frac{1}{4}$" scale. (Figure 6:6).

Although a plan is drawn to scale, the dimensions stated on it are those of the actual object. Measurements less than 1' are expressed in inches; measurements greater than 12" are expressed in feet and inches except for 16" o.c. or 24" o.c.

FIGURE 6:4

FIGURE 6:5

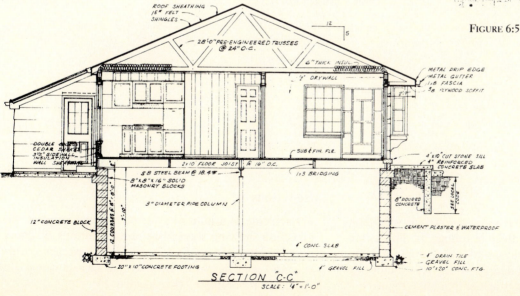

SECTION "C-C"
SCALE: ¼" = 1'-0"

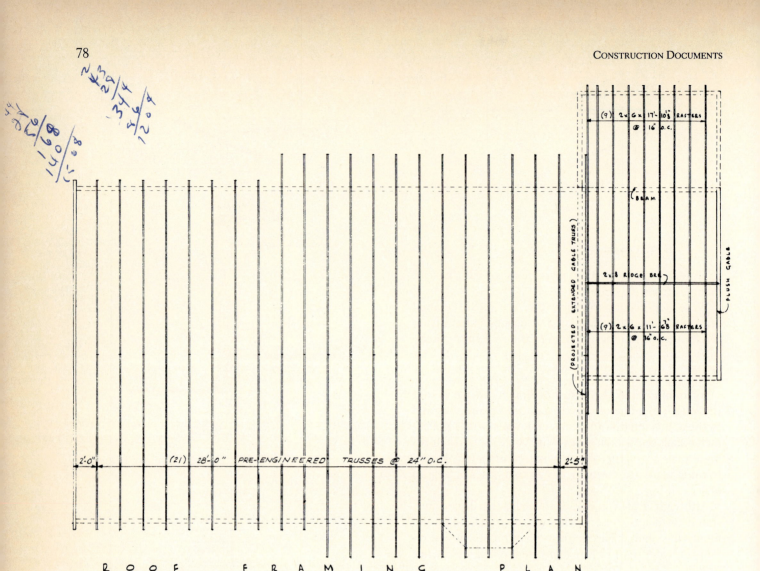

FIGURE 6:6
Roof framing plan.
(Courtesy of Home Planners Inc.
16310 Grand River Avenue
Detroit, Michigan 48227)

Floor Plans

A building's floor plan is a scaled-down top view showing the layout of all individual rooms, its dimensions, closet locations, stairways, and other important features, such as windows, doors, and the location of electrical, plumbing, and heating components (Figure 6:7).

Elevations

Each side of a house has its own elevation plan. An elevation drawing shows the exact locations of doors and windows, the roof pitch, suggested exterior covering, and wall height as they would appear when the structure is completed (Figures 6:8–13).

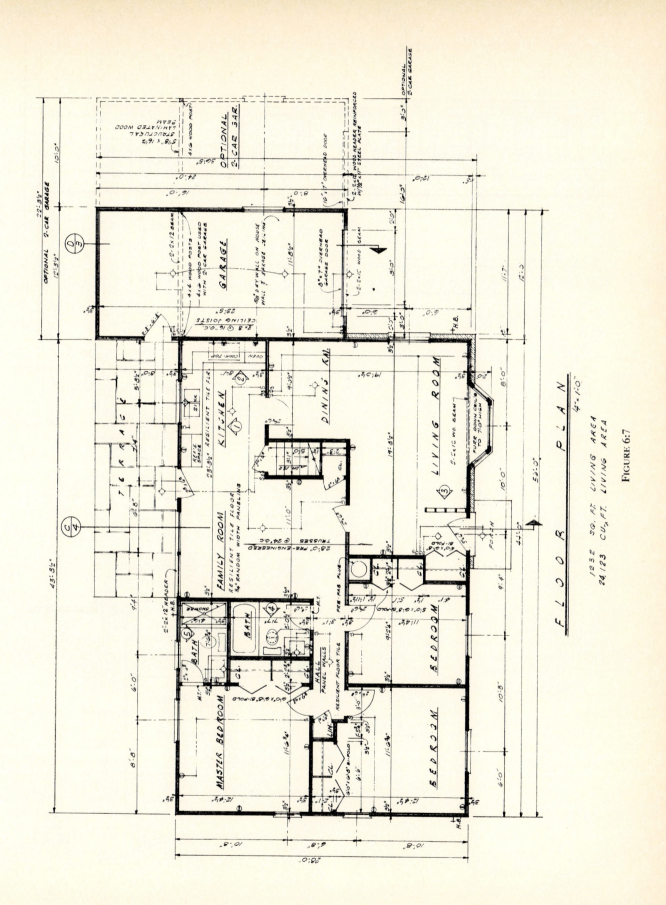

FLOOR PLAN
¼"=1'-0"

1232 SQ. FT. LIVING AREA
24,123 CU. FT. LIVING AREA

FIGURE 6:7

79

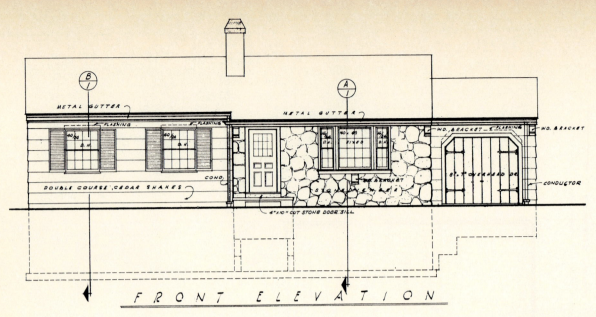

FRONT ELEVATION

FIGURE 6:8

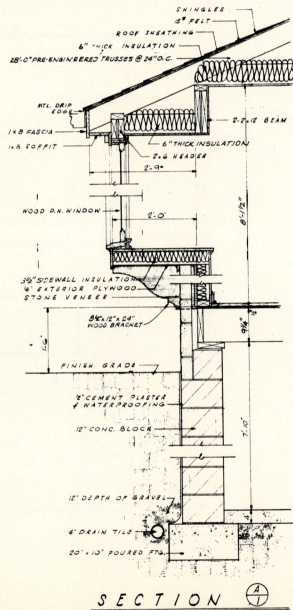

SECTION Ⓐ/1

FIGURE 6:9

A section view $\frac{A}{1}$ through the living room window.

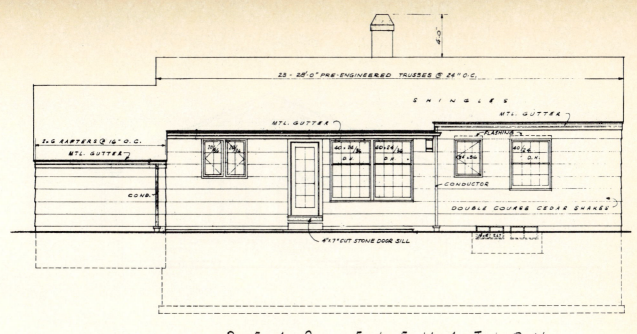

REAR ELEVATION

FIGURE 6:10

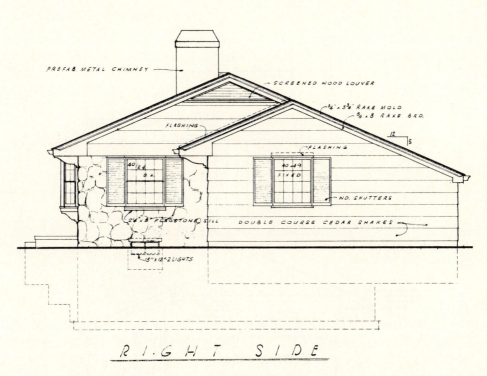

RIGHT SIDE

FIGURE 6:11

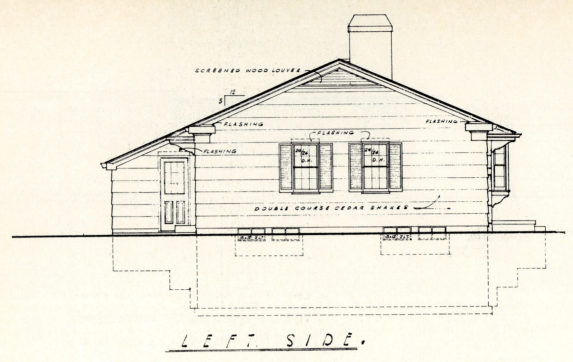

SCREENED WOOD LOUVER

12
5

FLASHING

FLASHING

FLASHING

FLASHING

20/24
D.H.

24/24
D.H.

DOUBLE COURSE CEDAR SHAKES

L E F T   S I D E.

FIGURE 6:12

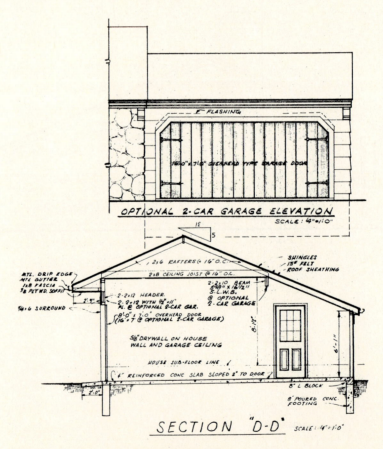

FLASHING

16'-0" x 7'-0" OVERHEAD TYPE GARAGE DOOR

OPTIONAL 2-CAR GARAGE ELEVATION
SCALE: ¼" = 1'-0"

18
5

SHINGLES
15# FELT
ROOF SHEATHING

2x6 RAFTERS @ 16" O.C.

ATL. DRIP EDGE
MTL. GUTTER
1x8 FASCIA
⅜ PLYWD. SOFFIT

2x8 CEILING JOIST @ 16" O.C.

2-2x10 BEAM
5⅛" x 16½" G.L.W.B.
@ OPTIONAL
2-CAR GARAGE

2-2x12 HEADER.
2-2x12 WITH ½"x11"
PL. @ OPTIONAL 2-CAR GAR.

2'-1"

¾x6 SURROUND

8'-0" x 7'-0" OVERHEAD DOOR
(16'x7' @ OPTIONAL 2-CAR GARAGE)

⅝" DRYWALL ON HOUSE
WALL AND GARAGE CEILING

8'-1½"

6'-1½"

HOUSE SUB-FLOOR LINE

4" REINFORCED CONC. SLAB SLOPED 2" TO DOOR

8" L BLOCK

2'-0"

8" POURED CONC.
FOOTING

SECTION "D-D" SCALE: ¼" = 1'-0"

FIGURE 6:13

Framing Diagrams

Another segment of the building plans consists of the framing diagrams which deal with all the segments of the structure's frame. They include

1. Floor joist framing plan
2. Elevation plans for each of the four sides showing how the two gable ends are framed, as well as the rough openings for doors and windows
3. Ceiling joist plan showing where they meet over a bearing partition
4. The roof sheathing plan (Figure 6:14).

Plan Symbols

Because of the limited amount of space available on building plans, architects use symbols to represent a variety of structural items. There are symbols for building materials, door openings, plumbing fixtures and electrical components (Figures 6:15–18). To comprehend a building plan, a carpenter must be thoroughly familiar with all the symbols used.

The architect must also use abbreviations on building plans because of the limited space available. Figure 6:19 lists the most common architectural abbreviations.

Through the use of symbols and abbreviations, the architect can provide the carpenter with a wealth of information on a scaled-down drawing or building plan.

WRITTEN SPECIFICATIONS

The specification outline (Figure 6:20), which accompanies the building plans, is designed to give specifics with respect to the size and quality of the materials to be used and a thorough description of the work to be performed. Designed primarily as a means of protecting the client, the specifications leave little question in the minds of the builder and the subcontractors as to what is expected of them. Figure 6:21 is an example of the specifications related to the wood framing of a house.

Optional items that may be included with the written specifications are a comprehensive lumber and millwork list, anticipated dates for completion, payment schedules, and identification of the types of insurance the builder must carry.

The specifications should state whether a bid or performance bond is required by the builder. They should also state whether the client or builder is responsible for obtaining the necessary building permits and alerting the building inspector when a certain phase of the construction has been completed.

The specifications, when adhered to completely by the builder, will enable the client or owner to obtain the best possible mortgage terms. They are very effective in providing the money lenders with a thorough insight into the quality of materials used as well as the construction methods employed.

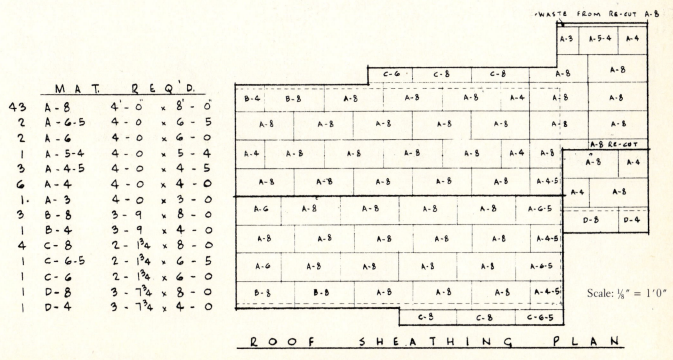

	M A T.	R E Q'D.					
43	A - 8	4' - 0"	x	8' - 0"
2	A - 6-5	4 - 0	x	6 - 5
2	A - 6	4 - 0	x	6 - 0
1	A - 5-4	4 - 0	x	5 - 4
3	A - 4-5	4 - 0	x	4 - 5
6	A - 4	4 - 0	x	4 - 0
1	A - 3	4 - 0	x	3 - 0
3	B - 8	3 - 9	x	8 - 0
1	B - 4	3 - 9	x	4 - 0
4	C - 8	2 - 1³⁄₄	x	8 - 0
1	C - 6-5	2 - 1³⁄₄	x	6 - 5
1	C - 6	2 - 1³⁄₄	x	6 - 0
1	D - 8	3 - 7³⁄₄	x	8 - 0
1	D - 4	3 - 7³⁄₄	x	4 - 0

Scale: ⅛" = 1'0"

ROOF SHEATHING PLAN

FIGURE 6:14

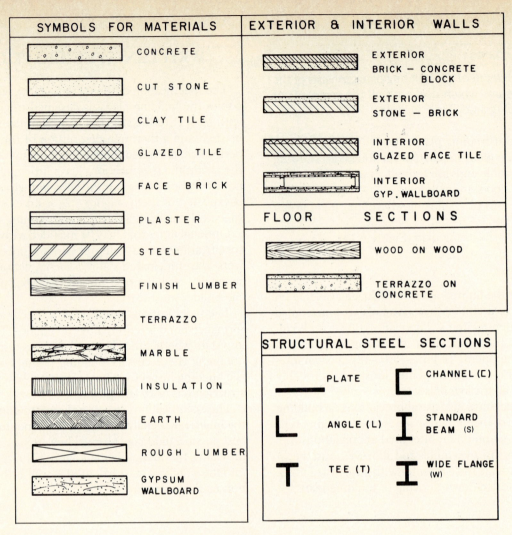

SYMBOLS FOR MATERIALS	EXTERIOR & INTERIOR WALLS
CONCRETE	EXTERIOR BRICK – CONCRETE BLOCK
CUT STONE	EXTERIOR STONE – BRICK
CLAY TILE	INTERIOR GLAZED FACE TILE
GLAZED TILE	INTERIOR GYP. WALLBOARD
FACE BRICK	**FLOOR SECTIONS**
PLASTER	WOOD ON WOOD
STEEL	TERRAZZO ON CONCRETE
FINISH LUMBER	**STRUCTURAL STEEL SECTIONS**
TERRAZZO	PLATE CHANNEL (C)
MARBLE	ANGLE (L) STANDARD BEAM (S)
INSULATION	TEE (T) WIDE FLANGE (W)
EARTH	
ROUGH LUMBER	
GYPSUM WALLBOARD	

FIGURE 6:15
Symbols used for building materials.

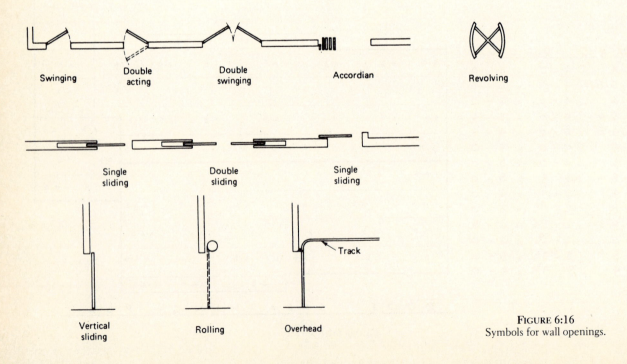

Swinging   Double acting   Double swinging   Accordian   Revolving

Single sliding   Double sliding   Single sliding

Vertical sliding   Rolling   Overhead   Track

FIGURE 6:16
Symbols for wall openings.

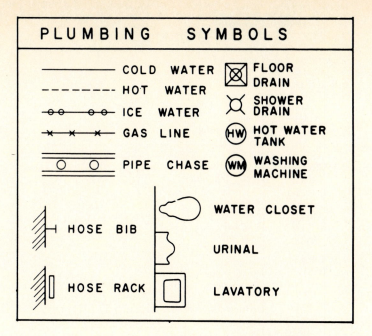

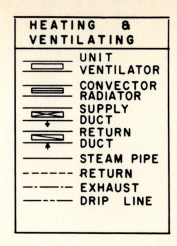

FIGURE 6:17
Symbols for plumbing fixtures, appliances, and other mechanical equipment.

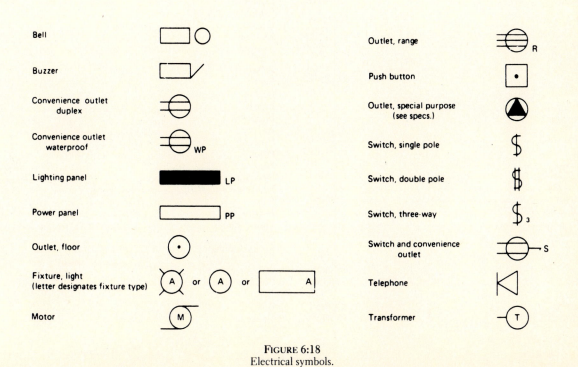

Bell		Outlet, range	R
Buzzer		Push button	
Convenience outlet duplex		Outlet, special purpose (see specs.)	
Convenience outlet waterproof	WP	Switch, single pole	
Lighting panel	LP	Switch, double pole	
Power panel	PP	Switch, three-way	
Outlet, floor		Switch and convenience outlet	S
Fixture, light (letter designates fixture type)	A or A or A	Telephone	
Motor	M	Transformer	T

FIGURE 6:18
Electrical symbols.

Access panel . . . AP
Acoustic . . . ACST
Addition . . . ADD.
Adhesive . . . ADH
Aggregate . . . AGGR
Air condition . . . AIR COND
Air dried . . . AD
Alternating current . . . AC
Aluminum . . . AL
Ampere . . . AMP
Anchor bolt . . . AB
Approved . . . APPD
Approximate . . . APPROX
Architectural . . . ARCH
Area . . . A
Asbestos . . . ASB
Asphalt . . . ASPH
Automatic . . . AUTO
Average . . . AVG

Balcony . . . BALC
Basement . . . BSMT
Bathroom . . . B
Bathtub . . . BT
Beam . . . BM
Bearing . . . BRG
Bedroom . . . BR
Bench mark . . . BM
Between . . . BET.
Bevel . . . BEV
Blocking . . . BLKG
Board . . . BD
Board foot . . . BF
Boiler . . . BLR
Both sides . . . BS
Brick . . . BRK
British thermal units . . . BTU
Bronze . . . BRZ
Broom closet . . . BC
Building . . . BLDG
Building line . . . BL
Bundle . . . BDL

Cabinet . . . CAB.
Casing . . . CSG
Caulking . . . CLKG
Cast concrete . . . C CONC
Cast iron . . . CI
Catalog . . . CAT.
Ceiling . . . CLG
Cement . . . CEM
Center . . . CTR
Centerline . . . CL
Center to center . . . C to C
Ceramic . . . CER
Circle . . . CIR
Circuit . . . CKT
Circuit breaker . . . CIR BKR
Circumference . . . CIRC
Cleanout . . . CO
Clear . . . CLR
Closet . . . CL
Coated . . . CTD
Column . . . COL
Common . . . COM
Concrete . . . CONC
Conduit . . . CND
Construction . . . CONST
Contractor . . . CONTR

Corrugate . . . CORR
Courses . . . C
Cross section . . . X-SECT
Cubic foot . . . CU FT
Cubic inch . . . CU IN.
Cubic meter . . . m³
Cubic yard . . . CU YD

Damper . . . DMPR
Dampproofing . . . DP
Dead load . . . DL
Decking . . . DKG
Degree celsius . . . °C
Degree fahrenheit . . . °F
or DEG. F
Design . . . DSGN
Detail . . . DET
Diagonal . . . DIAG
Diagram . . . DIAG
Diameter . . . DIA
Dimension . . . DIM
Dining room . . . DR
Dishwasher . . . DW
Ditto . . . DO.
Division . . . DIV
Door . . . DR
Double . . . DBL
Double hung . . . DH
Down . . . DN
Downspout . . . DS
Drain . . . DR
Drawing . . . DWG
Drop siding . . . DS
Dryer . . . D

East . . . E
Electric . . . ELEC
Elevation . . . EL
Enamel . . . ENAM
Entrance . . . ENT
Equal . . . EQ
Equipment . . . EQUIP.
Estimate . . . EST
Excavate . . . EXC
Existing . . . EXIST.
Exterior . . . EXT

Fabricate . . . FAB
Feet . . . (') or FT
Feet board measure . . . FBM
Finish . . . FIN.
Fireproof . . . FPRF
Fixture . . . FIX.
Flashing . . . FL
Floor . . . FL
Floor drain . . . FD
Flooring . . . FLG
Fluorescent . . . FLUOR
Foot . . . (') or FT
Footing . . . FTG
Foundation . . . FDN
Full size . . . FS
Furred ceiling . . . FC

Galvanize . . . GALV
Galvanized iron . . . GI
Garage . . . GAR
Gas . . . G
Gauge . . . GA

FIGURE 6:19
Architectural abbreviations.

Girder	G	Obscure	OB
Glass	GL	On center	OC
Grade	GR	Opening	OPNG
Grade line	GL	Opposite	OPP
Gypsum	GYP	Overall	OA
		Overhead	OVHD
Hall	H		
Hardware	HDW	Panel	PNL
Head	HD	Parallel	PAR.
Heater	HTR	Part	PT
Height	HT	Partition	PTN
Horizontal	HOR	Penny (nail size)	d
Hose bib	HB	Permanent	PERM
Hot water	HW	Perpendicular	PERP
House	HSE	Piece	PC
Hundred	C	Plaster	PL
		Plate	PL
I beam	I	Plumbing	PLMB
Inch	(") or IN.	Pound	LB
Insulate	INS	Precast	PRCST
Interior	INT	Prefabricated	PREFAB
Iron	I	Preferred	PFD
Joint	JT	Quality	QUAL
Joist	JST	Quantity	QTY
Kiln dried	KD	Radiator	RAD
Kilogram	kg	Radius	R
Kip (1000 lb.)	K	Random	RDM
Kitchen	KIT	Random lengths	RL
		Range	R
Laminate	LAM	Receptacle	RECP
Laundry	LAU	Reference	REF
Lavatory	LAV	Refrigerate	REF
Left	L	Refrigerator	REF
Length	LG	Register	REG
Length overall	LOA	Reinforce	REINF
Light	LT	Required	REQD
Linear	LIN	Return	RET
Linen closet	L CL	Riser	R
Live load	LL	Roof	RF
Living room	LR	Room	RM
Long	LG	Rough	RGH
Louver	LV	Round	RD
Lumber	LBR		
		Safety	SAF.
Machine stress rated	MSR	Sanitary	SAN
Main	MN	Scale	SC
Manhole	MH	Schedule	SCH
Manual	MAN.	Second	(") or SEC
Material	MATL	Section	SECT
Maximum	MAX	Select	SEL
Medicine cabinet	MC	Service	SERV
Membrane	MEMB	Sewer	SEW.
Metal	MET.	Sheathing	SHTHG
Meter (the instrument)	M	Sheet	SH
Meter (metric length)	m	Shower	SH
Minimum	MIN	Side	S
Minute	(') or MIN	Siding	SDG
Miscellaneous	MISC	Similar	SIM
Mixture	MIX.	Sink	S
Model	MOD	Soil pipe	SP
Modular	MOD	South	S
Moisture content	MC	Specification	SPEC
Molding	MLDG	Square	SQ
Motor	MOT	Square meter	m²
		Stairs	ST
Natural	NAT	Standard	STD
Newton	N	Steam	ST
Nominal	NOM	Steel	STL
North	N	Stock	STK
Not to scale	NTS	Storage	STG
Number	NO.		

FIGURE 6:19 (CONTINUED)

Street . . . . ST	Unfinished . . . . UNFIN
Structural . . . . STR	Urinal . . . . UR
Supply . . . . SUP	
Surface . . . . SUR	Valve V
Switch . . . . SW	Vapor proof . . . . VAP PRF
Symmetrical . . . . SYM	Vent pipe . . . . VP
System . . . . SYS	Ventilate . . . . VENT.
	Vertical . . . . VERT
	Vitreous . . . . VIT
Tangent . . . . TAN.	Volt V
Tar and gravel . . . . T & G	Volume . . . . VOL
Tarpaulin . . . . TARP	
Tee . . . . T	Washing machine . . . . WM
Telephone . . . . TEL	Water closet . . . . WC
Television . . . . TV	Water heater . . . . WH
Temperature . . . . TEMP	Waterproofing . . . . WP
Terra-cotta . . . . TC	Watt . . . . W
Terrazzo . . . . TER	Weather stripping . . . . WS
Thermostat . . . . THERMO	Weatherproof . . . . WP
Thick . . . . THK	Weep hole . . . . WH
Thousand . . . . M	Weight . . . . WT
Through . . . . THRU	West . . . . W
Timber . . . . TBR	Width . . . . W
Toilet . . . . T	Window . . . . WDW
Tongue and groove . . . . T & G	With . . . . W
Total . . . . TOT.	Without . . . . WO
Tread . . . . TR	Wood . . . . WD
Tubing . . . . TUB.	Wrought iron . . . . WI
Typical . . . . TYP	Yard . . . . YD

FIGURE 6:19 (CONTINUED)

General Instructions, Suggestions and Information . . . . Section I
Excavating and Grading . . . . Section II
Masonry and Concrete Work . . . . Section III
Sheet Metal Work . . . . Section IV
Carpentry, Millwork, Roofing, and Miscellaneous Items . . . . Section V
Lath and Plaster or Drywall Wallboard . . . . Section VI
Schedule for Room Finishes . . . . Section VII
Painting and Finishing . . . . Section VIII
Tile Work . . . . Section IX
Electrical Work . . . . Section X
Plumbing . . . . Section XI
Heating and Air Conditioning . . . . Section XII

FIGURE 6:20
Sample contents of a specification outline.

SECTION V — CARPENTRY, MILLWORK, ROOFING AND
MISCELLANEOUS ITEMS

1. **General:** The contractor shall furnish and install all work under this heading as specified herein and shown on the plans. All work shall be done by mechanics skilled in the trade and workmanship shall be of the best. No finish woodwork or flooring is to be stored in the building until the plastering is finished and thoroughly dry, when plaster is used.

   All framing lumber to be: 1500f or better.
       Specie:_____

2. **Wood Framing:** The carpenter shall do all cutting and fitting of his woodwork required by the Plumbing, Heating, and Electrical contractors or any other mechanics to install their work.

   Cutting of the floor joists to a depth of 1/6 of the joist depth shall be permitted except in the middle third of the span. No stud shall be cut more than ½ its depth to receive piping and duct work.

   All joists to be sized to width and framed on the ends with material of the same size as the joists.

   Trimmers and headers of double joists shall be put around all chimneys, stairways, fireplace openings, etc. Allow 2″ clearance around chimney masonry. Place double joists or ladder joist construction under all partitions running parallel to same. Bridging to be 1 x 3 white pine or manufactured metal bridging. All rows of joists 12 feet and under to have one row of bridging. Use 2 nails at each end of bridging. Spans of joists of over 12 feet to have two rows of bridging.

   All exterior wall openings and interior load-bearing partition openings to be framed with headers designed to carry the imposed loading conditions at each opening. Double studs shall be used at sides of all load bearing wall openings.

   Single studs may be used at sides of all non-load bearing partition openings.

   All outside and inside corners to be formed with 3 pieces of 2 x 4 spiked together.

   Furnish and install all grounds and/or nailers of sufficient size for plastering or drywall work for nailing wood base, trim, etc., and as required by other trades. Provide all furring and blocking as required for architectural features. Plates to be 2 x 4 doubled at top on all load-bearing partitions and so installed to tie into intersecting partitions. Where roof trusses are used, partitions may have single top plates.

   Sizes of framing members for roof construction to be as shown and erected in accordance with the detailed drawings. Where roof trusses are specified, the trusses are to be of an approved design and fabrication required for various geographical areas. Roof trusses shown on the drawings are diagrammatic. Contractor to provide specific structural truss details for roof construction.

FIGURE 6:21

BUILDING REGULATIONS

Building Permit

   Most localities require that the owner file for and be granted a building permit before any construction is started. This procedure must be followed for new house construction as well as for additions to existing structures.

   To obtain a building permit, an application must be filed with the local building department. Drawings pertinent to the proposed structure and the written specifications must be submitted with the application, usually in duplicate. Plans for a new house may be rejected if not accompanied by an architect's seal.

   Most building departments use a sliding scale, related to the total square-foot area of the structure, as the means of assessing the fee for the building permit.

Some construction projects cannot be accepted by the building department as requested because they conflict with the zoning regulations (e.g., set back, property boundary lines). To avoid altering the building plans, a request for a variance must be presented to the zoning board. If it is granted, the building permit application can then be filed.

Building Codes

Each locality uses some standard building code as the basis for specifications governing any new construction. They are designed to protect the eventual owner(s) by preventing use of improper construction methods and faulty materials.

The building code regulates how a house is constructed and is designed to protect the health, safety, and welfare of the community.

A building code used by a specific locality will contain the following:

1. The structural requirements for the building
2. The material specifications
3. The expected standards of performance by the subcontractors

4. Information concerning and governing
   a) ventilation
   b) room sizes
   c) fire resistance
   d) electrical code
   e) plumbing and sanitary code

Building Inspection

With new home construction, the city or town building inspector must visit the site periodically to inspect and either reject or approve the following phases of the construction: footings, foundation, framing, rough wiring and plumbing, and finished wiring and plumbing.

Certificate of Occupancy

Before a new home can be occupied, it must have the following items thoroughly inspected and approved by the building department: the entire house, the electrical system, the plumbing system, the heating system, the sanitary system, and the water supply. If all are approved, the owner is issued a certificate of occupancy by the building department.

QUESTIONS

1. Explain what the term "1/8 inch scale" implies on a building plan.

2. What are the two (2) basic characteristics of every dimension line?

3. List the four (4) most important framing diagrams required to construct any residential home.

4. If a carpenter were building your home, what are the three (3) most important items that you would incorporate in the written specifications?

5. Assume that you want to make an addition to your present dwelling. Contact your local building inspector to ascertain what will be needed to obtain a building permit for the new construction.

6. Explain what is required by most building inspectors before a C.O. can be issued.

7

EXCAVATION OF THE SITE

There are many factors that affect the excavation for the foundation of a residential home. Each one must be considered as a separate entity and as an integral component of the final overall plan.

INITIAL LAYOUT SURVEY

For most existing building lots, this important procedure would have already been performed by a licensed surveyor who has driven stakes into the ground signifying where the exact boundaries are. Consulting the survey should help to locate these stakes or, when absent, to determine the boundaries from other reference marks noted on the survey (Figure 7:1). The survey of the plot must be examined to make certain that the foundation will conform to the local building code with respect to the distance from the street and the adjacent property boundary lines (Figure 7:2).

UNDERLYING SUBSTRATA

Test borings or core samples should be taken to learn the composition of the material to be excavated as well as the bearing strength of the soil. It is possible to gain insight into what lies beneath the surface by learning what your neighbors experienced during their excavations. This information, however, is never as conclusive as taking your own test borings.

The core samples will provide the excavator with a clear picture of the underlying materials, which could vary from loose sand, to hard clay mixed with rocks, to solid rock requiring blasting. Knowing this information in advance also enables you to properly design the foundation, and to anticipate the ease or difficulty associated with both the excavation and any subsequent need to shore up the foundation. For example, excavating a sandy soil area should be relatively easy, but shoring the same area would present some problems.

Learning that the water table in the proposed excavation area is high or that the soil moisture content is excessive may either force you to relocate the foundation or to alter the design of the home and thus its foundation.

EXISTING TERRAIN

After the underlying substrata have been deemed suitable for excavation, tentative stakes outlining the foundation's approximate location can be placed so that the immediate area can be cleared of trees. Too often, plot developers, intent on building houses as rapidly as possible, ignore the landscape entirely and remove all the trees. It is very important, however, to retain as much as possible of the existing trees and plantings.

Sites with existing slopes and rugged terrain will require some form of initial rough grading as a preliminary step to the excavation.

It is advisable to define the driveway area at this time in preparation for the delivery of the various building supplies (Figure 7:3).

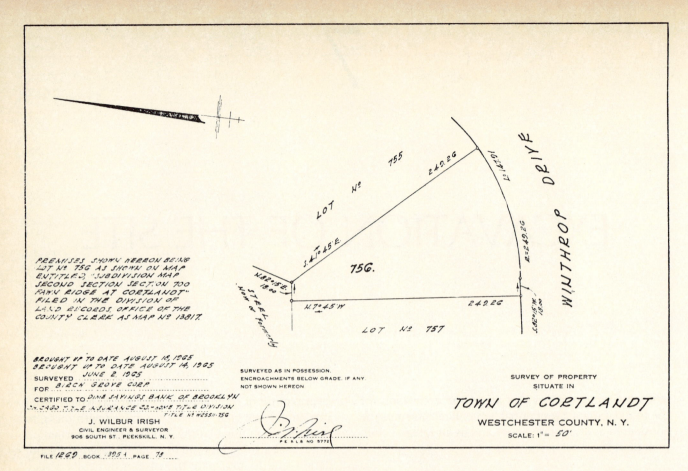

SURVEYED AS IN POSSESSION.
ENCROACHMENTS BELOW GRADE, IF ANY,
NOT SHOWN HEREON

SURVEY OF PROPERTY
SITUATE IN

TOWN OF CORTLANDT
WESTCHESTER COUNTY, N. Y.
SCALE: 1" = 50'

FIGURE 7:1

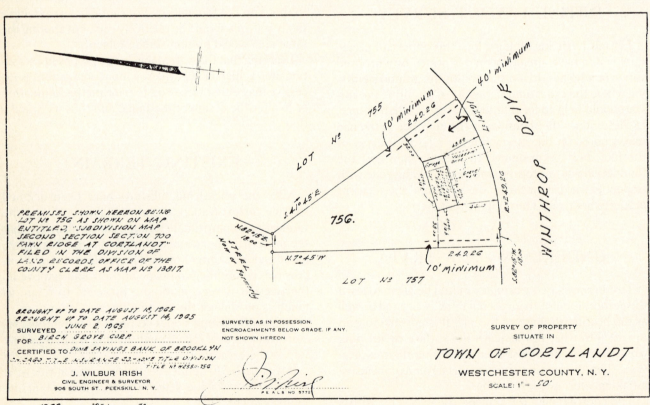

SURVEYED AS IN POSSESSION.
ENCROACHMENTS BELOW GRADE, IF ANY,
NOT SHOWN HEREON

SURVEY OF PROPERTY
SITUATE IN

TOWN OF CORTLANDT
WESTCHESTER COUNTY, N. Y.
SCALE: 1" = 50'

FIGURE 7:2

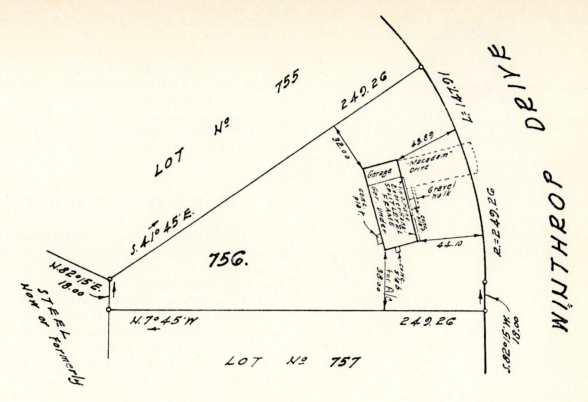

FIGURE 7:3

HOUSE PLACEMENT

The building lines can be established once the site has been cleared. Their exact lengths are obtained from the building plans. Using the surveyor's stakes as a guide and paying close adherence to the local building codes with respect to setback and distances from adjacent property lines, wooden stakes are driven at the corners of the anticipated foundation. Each stake is topped with a nail or tack to anchor the nylon strings that will be stretched between the stakes. These strings form the building lines (Figure 7:4).

The procedure for staking out a rectangular foundation on a level site is relatively simple. The first corner stake, or reference stake, is driven into the ground at the proper location. A steel tape measure (Figure 7:5) is used to determine the approximate location of two more corner stakes. To make certain that these two stakes form a 90 degree angle with the first stake, use the 3-4-5 triangle measurement method or some multiple of it (Figure 7:6). This procedure is repeated again to locate the fourth corner stake and to square the other two sides.

Checking the two diagonal distances with the tape measure is the final means of confirming whether or not the stake layout is square. If they are equal, the plot is square and batter boards can be erected (Figure 7:7).

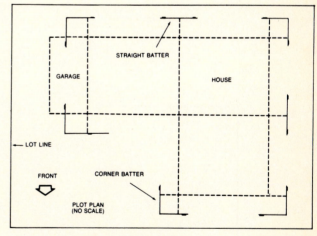

FIGURE 7:4

Staking foundation corners on sloping land is best accomplished with a builder's level or a transit and a leveling rod, a steel tape, and a plumb bob (Figure 7:8). The level is positioned directly over the first stake by aligning the plumb bob to the nail driven into the top of the stake.

After the desired distance has been measured with the tape, the builder's level is used to sight on the leveling rod placed at this point. Its elevation is noted before the second corner stake is driven into the soil and topped with a nail.

Steel tape

FIGURE 7:5

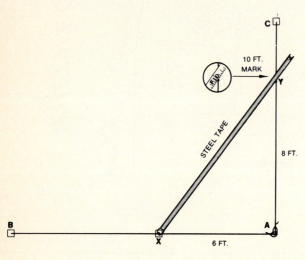

FIGURE 7:6

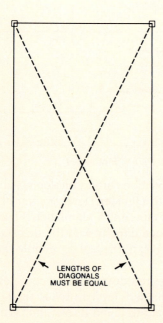

FIGURE 7:7
If AC = BD, the plot is square.

The level is rotated 90° and the procedure is repeated to locate the third stake (Figure 7:9).

The level is then moved and positioned directly over the third stake. Once it is plumb and level, it is first used to sight back on the initial corner stake before it is rotated 90° to locate the fourth stake (Figure 7:10).

To check for square, you can either move the level to the fourth stake and repeat the above operation or you can check for equal diagonals with the steel tape.

DESIGN OF THE HOUSE

The style of the house, its geographical location (e.g., a cold or warm climate), the slope of the site, and its existing substrata all have a determining effect on the extent of the excavation.

The excavation for a full basement will be substantial compared to that for a slab floor or crawl space. The excavation for either of the latter two is usually limited to a trench for their footings and walls.

The footing to support the walls in any of the three foundation types must always extend below the frost line, which varies from one geographical location to another (1′ in warm climates to 4′ or 5′ in cold climates).

The excavation for a crawl space foundation needs to be at least 3′ deep, as measured from the base of the floor joists to the ground level. The entire ground area inside a crawl space must be covered with 4- or 6-mil polyethylene, which serves as a vapor barrier.

A concrete slab foundation is chosen whenever the drainage of water is a problem. It must be at least 4″ thick and be poured over a 5″ to 6″ bed of gravel. The slab should be positioned 8″ to 12″ above the final exterior grade. For proper drainage, the earth must slope away from the foundation. Whenever the core samples show extensive bedrock very close to the surface, a concrete slab foundation is the only solution.

DETERMINING THE DEPTH OF THE EXCAVATION

The architectural plans will provide you with information on the required depth of the foundation and its relationship to the other floor levels. All foundations should extend at least 8″ above the final finished grade to prevent the wood frame members resting on the foundation from absorbing excessive soil moisture. (Figure 7:11).

To determine the depth of excavation on land with sloping contours, select the highest point of the excavation area as the reference point. From this point, you

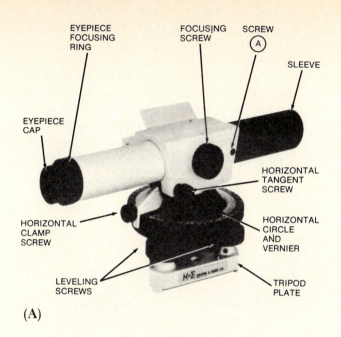

(A)

EYEPIECE FOCUSING RING

FOCUSING SCREW

SCREW (A)

SLEEVE

EYEPIECE CAP

HORIZONTAL TANGENT SCREW

HORIZONTAL CLAMP SCREW

HORIZONTAL CIRCLE AND VERNIER

LEVELING SCREWS

TRIPOD PLATE

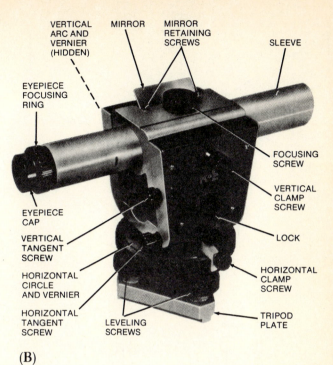

(B)

VERTICAL ARC AND VERNIER (HIDDEN)

MIRROR

MIRROR RETAINING SCREWS

SLEEVE

EYEPIECE FOCUSING RING

FOCUSING SCREW

VERTICAL CLAMP SCREW

LOCK

HORIZONTAL CLAMP SCREW

TRIPOD PLATE

EYEPIECE CAP

VERTICAL TANGENT SCREW

HORIZONTAL CIRCLE AND VERNIER

HORIZONTAL TANGENT SCREW

LEVELING SCREWS

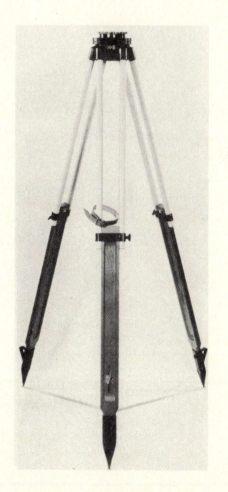

(C)

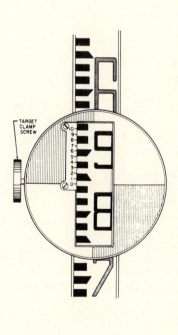

TARGET CLAMP SCREW

FIGURE 7:8
(A) Builder's Level. (B) Transit Level. (C) Tripod. (D) Leveling Rod
(close up).

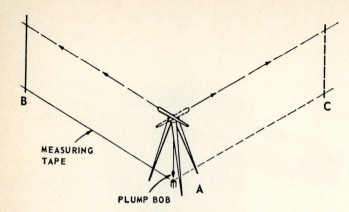

FIGURE 7:9

can determine how much soil must be removed, how deep the excavation must be, and what areas should be filled in with soil (Figure 7:12). This information can be obtained by using the transit and leveling rod or by stretching level string lines between the corner posts (Figure 7:13). Once obtained, the information should be transferred to graph paper showing in cross section the existing terrain, the proposed footing, the proposed foundation, and the final desired grade (Figure 7:14).

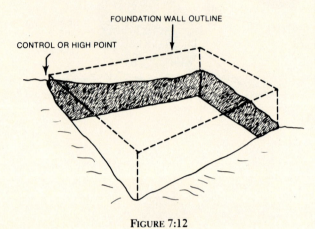

FIGURE 7:12

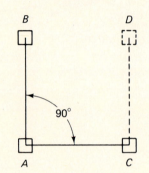

FIGURE 7:10

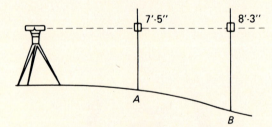

FIGURE 7:13

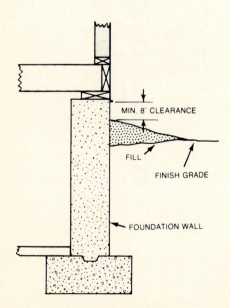

FIGURE 7:11

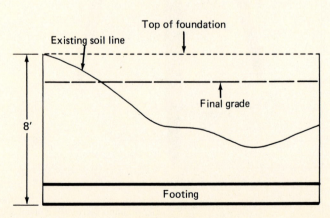

FIGURE 7:14

For a full basement, the entire floor area must be excavated to the base of the footings before it can be covered with gravel. This procedure ensures adequate drainage and also provides the base for the eventual concrete floor (Figure 7:15).

EXCAVATION SITE PREPARATION

Batter boards are needed to retain the foundation lines while the excavating is being done. At each corner, drive three 2 × 4 stakes at least 4′ outside the building line so they form a square with the corner stake. As shown in Figure 7:16, 1 × 6 ledger boards are then fixed to the stakes, horizontal and level, at a height approximating the top of the foundation. The building lines are raised to the top of the ledger boards and aligned with a plumb bob to the layout stake beneath, so that the ledger board can be marked with a saw kerf showing the

exact spot where the two lines cross. They will remain intact during the actual excavation.

Sufficient space should be left outside the eventual foundation wall (usually 2 feet) to allow for the convenient installation of forms for both footings and foundation, plus the drain pipes on the outside of the foundation wall (Figure 7:17). The area immediately surrounding the forms must be made accessible for the delivery of gravel and concrete.

Where to pile the surplus soil removed during the excavation should be decided before the excavation begins. Putting it in areas where fill will be needed to satisfy grade requirements will help reduce the time required when the final grade is established.

Top soil should be skimmed off the soil surface and stockpiled in an isolated location so that it can be reused when the final grade and lawn area are established.

From careful examination of the architectural plans, you will learn where to excavate for the sewer or septic lines (Figure 7:18), the septic fields (Figure 7:19), dry wells, and incoming water lines. Excavating for each of these, directly after the foundation has been completed, will be more efficient and should cost less than if done at a later time. Also, if difficulties with these excavations should occur, they are more easily handled early in the construction sequence than after the house has been completed.

Either a transit or a builder's level can be used to adjust the grade stakes for the footings after the excavation has been completed. The instrument is placed in a central location and is used to check each grade stake, with the leveling rod, for the desired elevation (Figure 7:20).

It is possible for an excavation to appear "to be level," yet when checked with the transit and leveling rod, it shows marked fluctuations in its surface. When this happens, the grade stakes, driven at each location, must be labeled with a notation indicating whether soil must

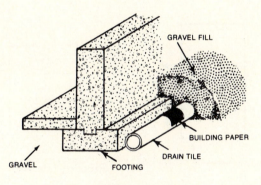

FIGURE 7:15
Drain tile.

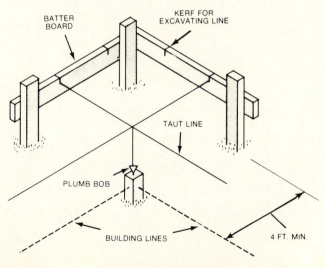

FIGURE 7:16

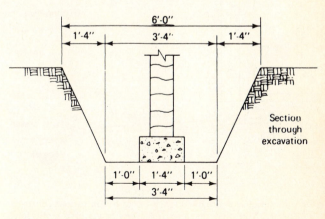

FIGURE 7:17

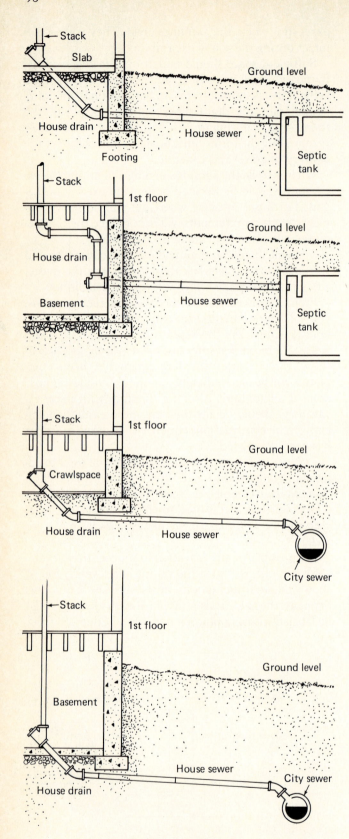

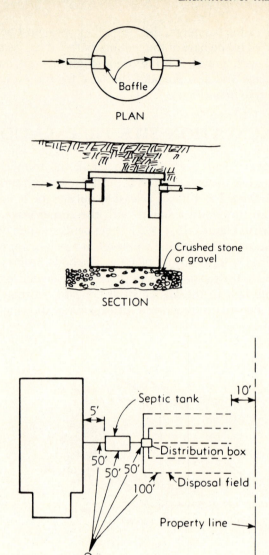

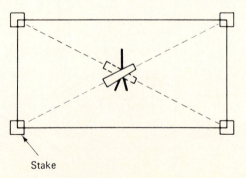

Figure 7:19
(A) A typical septic-tank installation. (B) Plan and installation of a disposal field.

Figure 7:18
Plans for a house drain and a house sewer to suit various construction needs. Note that the plans for a city sewer or a septic tank are interchangeable.

Figure 7:20

be added or removed at that point (Figure 7:21). To correct an uneven surface, some hard work with a pick and shovel will be necessary. Areas filled in with soil must be thoroughly tamped with a gasoline powered tamper or compressed by two or three hard rainfalls before the concrete is added to the footings.

EXCAVATING

Most excavations are done with a power shovel called a back-hoe (Figure 7:22), or a bulldozer, except when conditions are not favorable for their use. Excavation by hand tools is the only alternative. Even when power tools are used, every excavation requires a certain amount of hand work.

Developing a plan of attack prior to the excavation should make the work run smoothly, be beneficial to some of the future operations such as final grading, and prevent mistakes. The following are the items for you to consider in your plan of attack:

1. Does the site area consist of uneven terrain? If so, rough grading the area before the excavation might be beneficial.

2. Where will the excess soil be placed? It is advisable to skim off the topsoil from the excavation site and stockpile it for future use. The extra subsoil is then deposited in any low lying areas that require build-up.

3. How deep and extensive will the excavation be? As mentioned earlier, this is determined by house style, and water table. On uneven terrain, however, filling in the surrounding low areas with a final resulting slope away from the foundation may lessen the excavation depth.

4. How will the foundation be constructed? For a full basement, the excavation must extend at least 2′ beyond all building lines to provide sufficient space to erect the foundation. Crawl space areas, as well as poured concrete slabs, only require the removal of sufficient soil to install the footing and walls. Some localities even allow concrete footings to be cast into an open trench of firm soil, 3′ deep and 18″ wide (Figure 7:23).

5. In addition to the foundation, what other services will be required of the back-hoe?

Planning should also include knowing in advance where the septic tank and related fields (Figure 7:19) will be located, or where to excavate for both sewer and water lines. Careful perusal of the building plans, examination of the building site, and close adherence to the local building codes will provide this information and enable you to avoid the costly mistake of improper placement.

In many localities, a building inspector will reject plans for the proposed foundation until a percolation test has been performed on the soil and the bulk of the septic or sewer system has been installed.

A percolation test determines how well surface and subsurface water drains, or percolates, through the soil. Sandy soils drain rapidly while heavy clay soils do not. To be effective, the soil must be able to absorb the effluent from the septic system.

FIGURE 7:22
An excavation with a 14′ backhoe. (J. I. Case Co.)

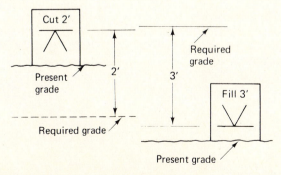

FIGURE 7:21

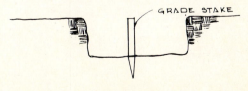

FIGURE 7:23

QUESTIONS

1. Explain why it is important to take test borings of the potential building site before any excavation takes place.

2. What precautions must be exercised when placing the house on the building lot?

3. Assuming a rectangular foundation, what are the two methods you should use to guarantee that the area staked out is square?

4. Contact your local building inspector to learn the depth of the frost line in your area.

5. Explain why a concrete slab floor must be positioned higher than the final exterior grade of the dwelling.

6. Explain the procedures required to prepare a sloping plot for a level foundation.

7. Explain why it is important that low areas of the proposed foundation that have been filled in with soil must be thoroughly compressed before concrete can be poured in the footing areas.

8. How many degrees are there in (a) a circle, (b) a right angle, (c) in a rectangle, (d) in a triangle?

9. Explain the procedures used to set up and level a transit.

8
FOUNDATIONS AND OTHER MASONRY FEATURES

The foundation requirements for any house are related to its style, its overall size, the layout of its rooms, and its geographical location. For example, a house built on a concrete slab will not require a foundation as extensive as one having a full basement situated below ground level. In areas subject to freezing temperatures, the base of the footing of the foundation wall must extend below the frost line to protect it from damage due to freezing (Figure 8:1).

Most local building departments require that the footings for a residential foundation extend a specific depth below grade. This measurement is usually determined by acknowledging the national government's requirements (Table 8:1) for certain cities throughout the country and then blending pertinent local information, such as soil conditions, depth of frost penetration and climatic conditions throughout the winter months, to arrive at the specified depth.

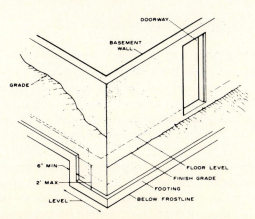

FIGURE 8:1

TABLE 8:1
Government Requirements for Depth of Footings in Selected Locations

Area	Depth of footings (inches)
Albany, New York	42
Albuquerque, New Mexico	18
Anchorage, Alaska	42
Bangor, Maine	48
Camden, New Jersey	30
Charleston, West Virginia	24
Columbus, Ohio	32
Fort Worth, Texas	6
Greensboro, North Carolina	12
Helena, Montana	36
Knoxville, Tennessee	18
Los Angeles, California	12
Minneapolis, Minnesota	42
Omaha, Nebraska	42
Seattle, Washington	16
Shreveport, Louisiana	18
Tampa, Florida	6

Local regulations may vary; check building code

FOOTINGS

The standard design of most residential wall footings calls for a thickness equal to and a width twice the thickness of the wall they support (Figure 8:2). The base of the footing must be firm and free of loose debris and standing water.

The footing area must be tamped by a pneumatic gasoline-powered tamper or be compacted by two or

101

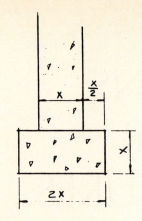

FIGURE 8:2
Note: If w = 8 inches, 2w = 16 inches.

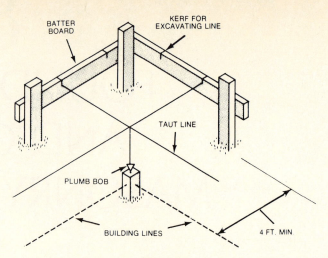

FIGURE 8:3

more heavy rainfalls before the footings recieve the concrete. This should eliminate the possibility that a portion of the footings will settle sometime in the future and cause damage to the foundation wall.

Footing Forms

After excavation, the building lines are again stretched between the batter boards so that the corner stakes can be relocated with a plumb bob. The outside dimensions and the diagonal lengths are then checked to make certain that they are equal before the footing forms can be set and nailed in place to the support stakes.

The outside form is the first to be positioned. It must be located outside the building lines at a distance equal to half the wall's thickness. Grade stakes, driven along the entire footing line, are used to level the top of the outside form. The top surface of each stake is placed at a specified distance below the leveled building line so that it can be used to dictate the top of the footing and also to ensure the desired thickness (Figure 8:3).

After the outside form has been properly braced with stakes positioned at 3′ intervals and leveled, spacers, equal in length to the footing's expected width, are used to locate the position of the inside form (Figure 8:4). They are left in place while the inner form is braced and leveled with outer form counterparts, and are removed only when the concrete is being added to the forms (Figure 8:5).

Installing Footings

Footings for each of the piers used to support the central girder, as well as the one required to adequately support a chimney, should be constructed and poured at the same time as the wall footings (Figure 8:6). Footings can have either a *flat* or a *keyed* top (Figure 8:7). A keyed top is formed by embedding an oiled 2 × 4

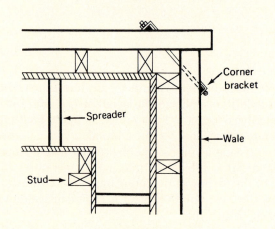

FIGURE 8:4
Top view of footing form.

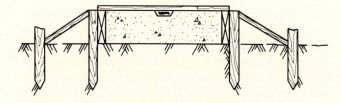

FIGURE 8:5
Form boards are staked and leveled.

in the top surface of the newly poured concrete and removing it once the concrete has become firm. Its design provides a better mortar bond with the wall above then a flat top footing. It also helps control water seepage into the basement through the construction joint and restricts the lateral movement of basement walls under pressure.

Before the concrete is added to the forms, they must be thoroughly checked for correct location, proper depth and width, sturdiness, and accurate measurement. To prevent the heavy concrete from causing the top of each form to spread more than its bottom

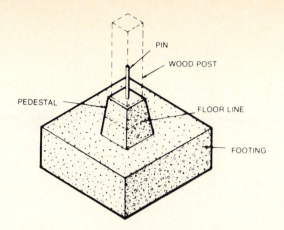

FIGURE 8:6

FIGURE 8:9
Ready-mix truck preparing to pour into foundation forms.

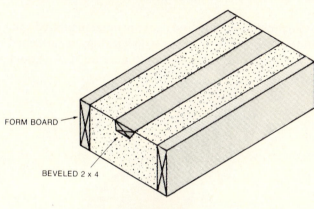

FIGURE 8:7
Keyed top.

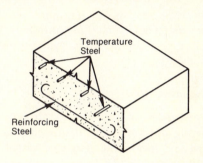

FIGURE 8:10
Steel reinforcing rods are used to strengthen footings.

(resulting in waste of concrete) short pieces of 2 × 4, called *ties,* are used to connect the outside and inside forms together (Figure 8:8).

The concrete for the footings can either be mixed at the site or delivered in a ready-mix truck. The advantages of the ready-mix concrete are the uniformity of its mixture and the speed with which it fills the forms (Figure 8:9).

Footings are strengthened by the addition of steel reinforcing rods (Figure 8:10).

Steel rods (rebar) are also set in foundation walls (Figure 8:11) and over door and window openings (Figure 8:12) to add strength to the concrete.

FOUNDATION WALLS

Foundation walls may be constructed of concrete, concrete block, or treated wood. The overall cost of the installation for each of these wall types must be com-

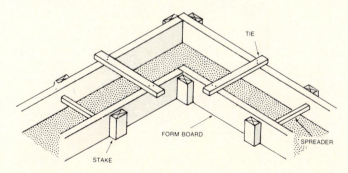

FIGURE 8:8
The stakes in the footing form are placed below the form boards.

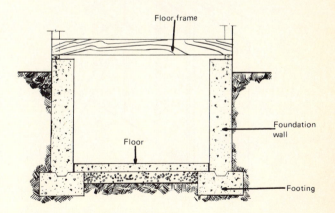

FIGURE 8:11
Cross section of a typical foundation (end view).

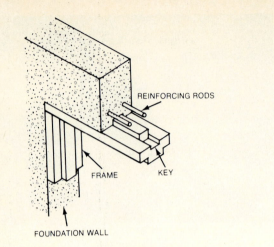

FIGURE 8:12
Reinforcing rods are used over door and window openings.

pared, especially when the foundation will be built by a subcontractor, to ascertain which type should be used in the installation.

Poured Concrete

Walls constructed of poured concrete require forms that are strong and well built, adequately braced, properly aligned and level, and that have smooth, defect-free-surfaces (Figure 8:13). Tall walls, in excess of 4′, will require 2 × 4 wales and some form of patented wall tie to help prevent the sides of the form from spreading outward (Figure 8:14). Most wall ties are very efficient at maintaining the desired wall thickness, since they resist both expansion and collapse of the upright forms.

Prefabricated form panels can be used to form the foundation walls. The individual units are designed to

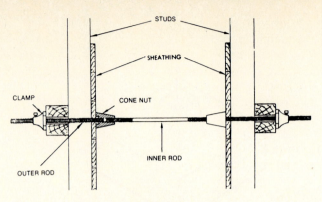

FIGURE 8:14

be quickly and easily joined together, properly braced, and adequately strengthened with wall ties. (Figure 8:15).

To provide window and door openings in poured concrete walls, special forms or wooden stops are attached to the panels or regular forms (Figure 8:16). Basement

FIGURE 8:15

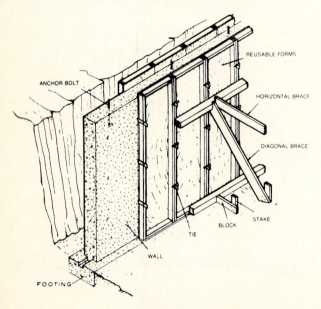

FIGURE 8:13

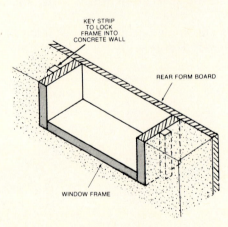

FIGURE 8:16
The basement frame is permanently cast in the wall. The key holds frame in place.

windows are conveniently positioned at the top of the foundation so that the wooden sill plate not only encloses the top of the window but also distributes the weight of the structural components located above the window (Figure 8:17).

Anchor bolts should be placed, at intervals of 4′ or less, in the top surface of the concrete. They are used to anchor the nominal 2″ thick sill plate to the foundation wall; the threaded end must extend at least 2″ above the leveled concrete top (Figure 8:18).

Concrete Block Foundations

The total cost of materials for a concrete block wall is less than that for a poured concrete wall. The time needed to construct the block wall, however, is usually much greater, making the overall cost of each very similar in most localities.

The standard stretcher block measures 8″ × 8″ × 16″ when assembled with a $\frac{3}{8}$″ mortar joint. Some walls may be 10″ or 12″ wide instead of 8″.

FIGURE 8:17

To aid you in the construction of a concrete block wall, the following is the suggested sequence to follow:

1. Make certain that the blocks are thoroughly dry before you start work. Also, any wall that is started but not completed by the end of the day must be covered to protect it from the rain.

2. Place the first course of blocks on the footing for proper fit and position. Mark the position of each block with respect to where it should be located on the footing.

3. Mix your mortar (one part masonry cement to three parts mason sand) and deposit a quantity with your trowel at one of the corners. Its consistency should be thick enough so that it can be easily applied to both the block edges and to the ends of those blocks that will soon become part of the wall (Figure 8:19).

4. Set the corner block in the mortar and check it for level, (both ways—along its length, across its width), and for straightness (Figure 8:20). Next, set the two blocks adjacent to the corner block, making certain that they are also level and square with the corner block. Repeat this same procedure at each corner.

5. After each corner and its adjacent blocks have been set and leveled, the balance of the first course (row of blocks) can be set. Continue to check that you are consistently using the same $\frac{3}{8}$″ mortar joint between each block.

6. Return to the corners and build each one up several courses, making certain to stagger the direction of the corner block with each course. The handle of your trowel can be used to make the necessary position and leveling adjustments.

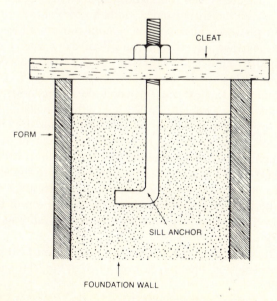

FIGURE 8:18

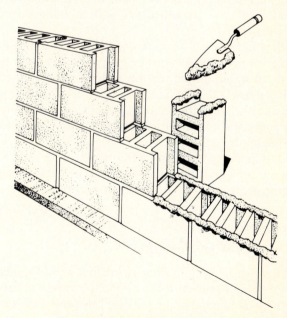

FIGURE 8:19

FIGURE 8:20

The components of the wall can be formed into panels (panelized) in a fabricating plant, transported to the site, and then assembled, or they can be framed at the site. Using either method, the time required to build the entire foundation is much less than that needed to construct a masonry foundation. This type foundation is easily constructed by carpenters and thus eliminates the need for masons.

Because the wooden components of this foundation are not affected by adverse weather conditions, they can be installed at any time throughout the year. The same is not true for masonry foundations, which require dry weather and temperatures above freezing.

Complaints concerning the absence of warmth in a concrete wall basement are usually eliminated when a wood foundation is installed, because the space between the upright studs can be completely filled with insulation. Also, the portion of the exterior plywood wall, below the final grade, is covered with 6-mil polyethylene plastic. Its primary function is to serve as a vapor barrier, but it has the added advantage of helping to eliminate the infiltration of air, cold or warm, that is present in the surrounding soil (Figure 8:24).

After the subsoil in the excavated basement area has been leveled, it is covered with 4″ to 6″ of gravel. A 2 × 6 or 2 × 8 pressure-treated base, or footing plate, is placed directly on the leveled gravel and is then topped with a treated bottom plate of the same size. The upright studs, also pressure-treated and either 2 × 4s or 2 × 6s, are attached to the bottom plate with silicon bronze, copper, or galvanized nails.

The exterior pressure-treated plywood sheathing, having a minimum thickness of $\frac{5}{8}″$, is nailed to the upright studs. It is covered with a 6-mil polyethylene film just before the foundation is backfilled. The film serves as a moisture barrier and helps to seal the joints in the plywood. To prevent accidental movement by, or damage to the polyethylene, an adhesive or a wood cleat is used to attach the film's top edge to the plywood (Figure 8:24).

Joints in the exterior plywood sheathing must be completely sealed with caulk prior to the installation of the polyethylene vapor barrier. Provisions for the proper drainage of the foundation (1) must be well thought out, (2) must adhere to the local building codes, and (3) must be completely assembled before the area surrounding the foundation is backfilled with soil.

The floor frame is positioned directly on top of the wall's double top plate and a load-bearing wall (Figure 8:25).

The wood footing for a crawl space area must extend below the frost line and rest on a bed of gravel (Figure 8:26).

7. Once the corner blocks have been squared and leveled, the center blocks (stretchers) can be placed (Figure 8:21). Plastic "block clips" are positioned on each corner block with a string drawn taut between them. Their purpose is to help you keep the wall both straight and level (Figure 8:22).

8. Repeat all the above procedures until the wall is completed. Remember to make the necessary adjustments as you build the wall and before the mortar becomes too hard.

9. The mortar joints, once hardened somewhat, are compressed and smoothed off with a round-edged grooving tool.

10. After the wall has been completed, the top cores in the stretcher blocks must be plugged with mortar to keep out water; or, the top row of blocks are replaced with cap blocks, having a solid top (Figure 8:21). Placing screening between the last two courses is an easy means of filling these cores, conserving mortar, and installing the anchor bolts.

11. Crawl spaces require vents in each of the walls to remove any moisture that develops in these areas (Figure 8:23).

All-weather Wood Foundation

The installation of a pressure-treated wood foundation, instead of the customary poured concrete or concrete block foundation, is preferred by some builders.

Dimensions shown are actual unit sizes. A 7⅝" x 7⅝" x 15⅝" unit is commonly --
known as an 8" x 8" x 16" concrete block.
Half length units are usually available for most of the units shown below. See
concrete products manufacturer for shapes and sizes of units locally available.

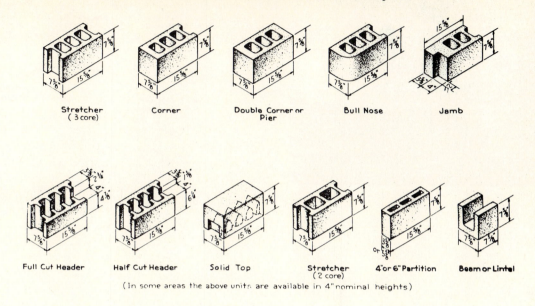

Stretcher (3 core)   Corner   Double Corner or Pier   Bull Nose   Jamb

Full Cut Header   Half Cut Header   Solid Top   Stretcher (2 core)   4"or 6" Partition   Beam or Lintel

(In some areas the above units are available in 4" nominal heights)

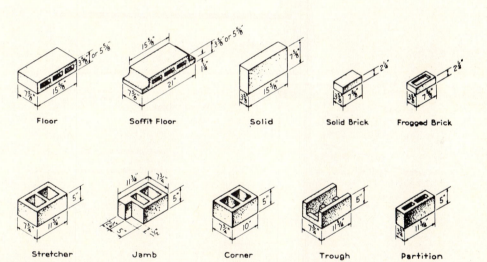

Floor   Soffit Floor   Solid   Solid Brick   Frogged Brick

Stretcher   Jamb   Corner   Trough   Partition

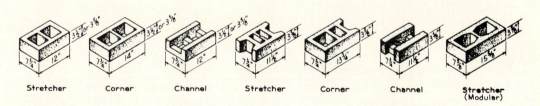

Stretcher   Corner   Channel   Stretcher   Corner   Channel   Stretcher (Modular)

FIGURE 8:21
Concrete blocks of various shapes and sizes.

107

FIGURE 8:22

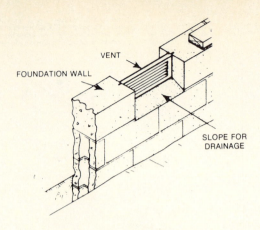

FIGURE 8:23
A foundation vent in a crawl space masonry wall.

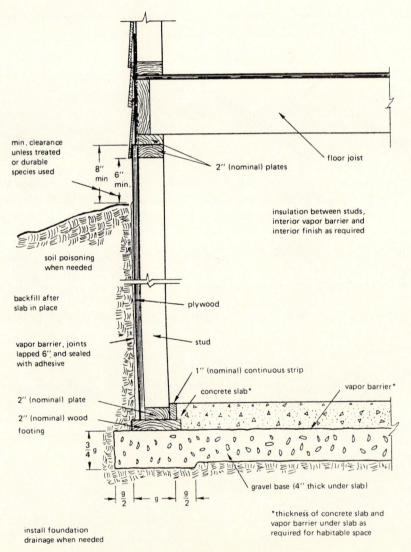

FIGURE 8:24
Basement construction details for wood foundation.

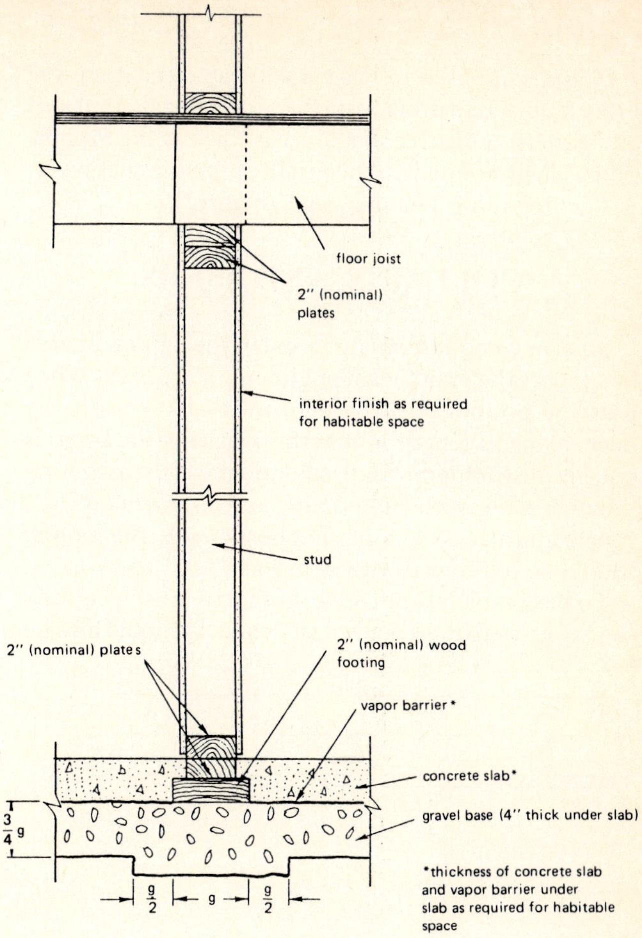

floor joist

2″ (nominal) plates

interior finish as required for habitable space

stud

2″ (nominal) plates

2″ (nominal) wood footing

vapor barrier *

concrete slab*

gravel base (4″ thick under slab)

$\frac{3}{4}$ g

*thickness of concrete slab and vapor barrier under slab as required for habitable space

$\frac{g}{2}$ — g — $\frac{g}{2}$

FIGURE 8:25
Details for load-bearing wall.

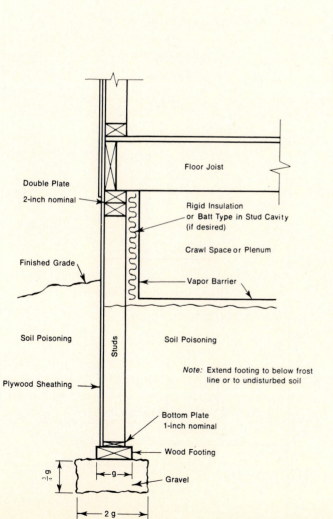

Double Plate
2-inch nominal

Floor Joist

Rigid Insulation
or Batt Type in Stud Cavity
(if desired)

Crawl Space or Plenum

Finished Grade

Vapor Barrier

Soil Poisoning

Soil Poisoning

Note: Extend footing to below frost
line or to undisturbed soil

Plywood Sheathing

Studs

Bottom Plate
1-inch nominal

Wood Footing

$\frac{3}{4}$ g

g

Gravel

2 g

FIGURE 8:26
A section through a pressure-treated crawl-space wall. Note: Extend
footing to below frost line or to undisturbed soil.

Pilasters

Pilasters are used to make a vertical foundation wall more rigid and to provide support for the ends of beams and girders. A pilaster should be as thick as the foundation wall, positioned on the inside of the foundation on 25′ centers, and at least 2′ wide (Figure 8:27).

COLUMNS AND PIERS

Steel or wood *columns* are used in the basement area to support the main beam or girder (Figure 8:28). They must be positioned at definite intervals beneath the beam along its length so that the weight of the beam is equally distributed. Each column requires a footing (Figure 8:29) with minimum measurements of 2′ square and at least 8″ thick. For heavy loads, the footing should be reinforced with steel rods.

To be completely effective, a column must be plumb so that its weight is evenly distributed. It should also be anchored to the floor to prevent it from moving (Figure 8:30). Some builders prefer to bury one end of a steel-concrete-filled Lally column directly into the concrete footing.

Crawl space areas usually rely on *piers* constructed of solid concrete or mortar-filled concrete blocks instead of columns to support the girder (Figure 8:31). Every pier must also be placed on an adequate footing, which is reinforced when necessary, and project at least 18″ above grade.

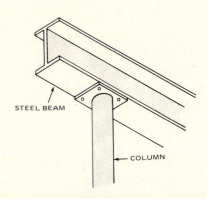

FIGURE 8:28
An I-beam supported by a pipe or Lally column.

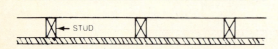

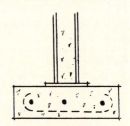

FIGURE 8:29
Reinforced column footing.

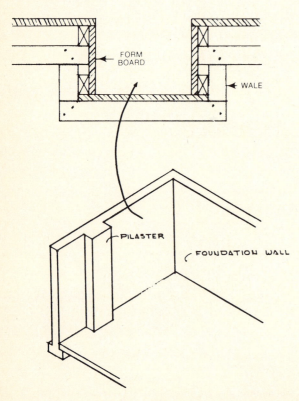

FIGURE 8:27

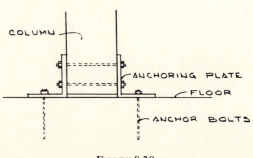

FIGURE 8:30
Column anchorage.

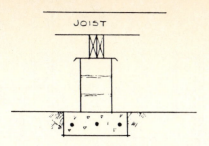

FIGURE 8:31

TERMITE PROTECTION

There are several measures that must be taken to ensure that the foundation area and the house structure above it are protected from termites.

A sheet metal shield should be installed between the foundation wall, column, or pier and the sill plate or girder. It should cover the entire top of the foundation and protrude from the edge of the wall (Figure 8:32).

Another means of protection is to use solid concrete blocks in the top course of every hollow concrete block foundation.

Using pressure-treated lumber for the structural members close to the ground, where moisture may present a problem, should also deter termites.

WATERPROOFING THE EXTERIOR SURFACE

Most local building codes require that the exterior surfaces of foundation walls be waterproofed, especially those areas that are to be located below the finished grade.

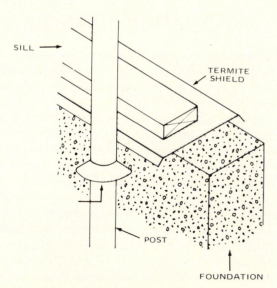

FIGURE 8:32
Installation of termite shield on exterior wall.

Poured concrete walls are usually covered with either a bituminous waterproofing material or painted with a foundation sealer.

Concrete block walls are "parged" with one or two coats of $\frac{1}{4}$" thick Portland cement plaster (Figure 8:33), plus the bituminous waterproofing material when the surrounding soil mass is extremely wet (Figure 8:34).

Water-repellent vermiculite is often added to the cores of concrete blocks as a means of insulating the walls against the cold temperatures experienced in northern areas (Figure 8:35).

To avoid the possibility of water entering the basement in the area where the exterior wall joins the footing, a cove of plaster should be applied. (Figure 8:34).

Styrofoam insulation, when applied to the exterior surface of a poured concrete, concrete block, or wood frame wall, provides superior resistance to moisture penetration. Its insulating effectiveness lasts for many

FIGURE 8:33
Scratching the foundation-wall parging.

FIGURE 8:34
Bituminous water proofing applied over the parged coat.

FIGURE 8:35

years, even when installed below ground level. It can also be applied directly to the interior surface of concrete walls (Figure 8:36).

CURTAIN DRAINS

Once the foundation walls have been waterproofed, curtain drains can be installed. Plastic perforated PVC pipe, 4″ in diameter, is positioned around the entire foundation at the base of the footing. These drains should have a minimum slope of 1″ in 20′ and eventually lead to either a permanently open drain or to connection with the septic system. The plastic pipe is covered with 6″ to 8″ of ¾″ gravel after the entire installation has been completed (Figure 8:37).

CONCRETE FLOORS

There are several preparatory steps that must be completed before a residential home can have a concrete slab floor installed. They are the following:

1. All utilities lines must be positioned below the ground at a depth that will resist freezing temperatures. Their connections, however, must be above the anticipated slab height.
2. Stumps, roots, and large stones must be removed before the subsoil is leveled.
3. Gravel or coarse stones must be placed on the leveled subsoil to ensure adequate drainage.
4. A vapor barrier must be placed on top of the gravel. Its purpose is to stop any water beneath it from penetrating the slab.
5. To reduce the possibility of heat loss, rigid insulation will be required along the inside surface of the walls.
6. Before the actual pour, welded wire fabric must be set in the form—positioned in the middle of the anticipated depth of the slab (usually 4″) and lapped at least 6″. This practice both seals the poured area and strengthens it (Figure 8:38).

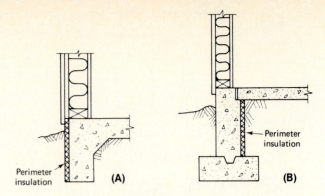

FIGURE 8:36
(A) Outside foundation walls. (B) Inside masonry walls.

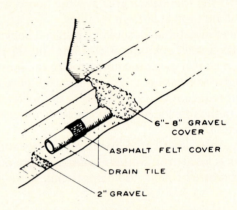

6″- 8″ GRAVEL COVER
ASPHALT FELT COVER
DRAIN TILE
2″ GRAVEL

FIGURE 8:37
Curtain drain placement at footing for 8″ basement wall.

7. Whenever the area receiving the concrete is extra large, it may be necessary to section the area into smaller units to ensure a level floor.

FOUNDATION ADJUNCTS

Foundations for entrance platforms and steps, as well as for fireplaces bordering an exterior wall, should be considered as an integral component of the main foundation.

Lengths of reinforcing bar (rebar) are used to connect these adjuncts to the main foundation. They usually have one end firmly anchored to the main foundation. The other end is allowed to jut out in the area where the adjunct foundation will be located (Figure 8:39).

This procedure helps prevent the problems of separating or sinking away from the main foundation, which are common with steps, platforms, and fireplace footings that are not attached properly to the main foundation.

The forms for all these structures should be constructed of 2″ nominal lumber and should be extremely well braced to avoid the possibility of bulges in the final

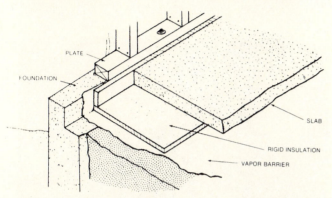

FIGURE 8:38

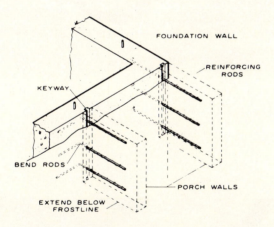

FIGURE 8:39

structure. Step risers are usually constructed of 2 × 8's (to give a $7\frac{1}{2}$" riser height), which are set at a 15° angle to provide a slight overhang and beveled at their base to allow for ease of troweling the stair tread (Figure 8:40).

WALKWAYS AND PATIOS

Neither of these structures should be considered until the house is completed. By then, the foundation will have been backfilled and the final grade will have been established. Then, you can stake out positions, ex-

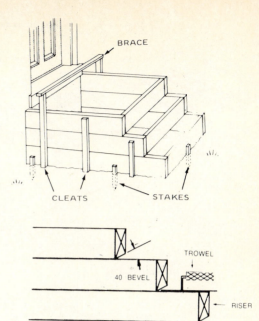

FIGURE 8:40

cavate, and construct the forms. Make certain that one side of the form is lower than the other side for proper water runoff ($\frac{1}{4}$" per foot). Before the forms receive the concrete, however, it is advisable to compress the soil with a power tamper or wait for two or three heavy rains to thoroughly settle it.

A walkway 4' wide is large enough to accommodate traffic leading to the front door, while one having a 3' width is sufficient for side or rear entrances.

Once the forms have been constructed, positioned, and adequately braced, 2" to 3" of gravel should be spread on the bottom of the excavation area for drainage and protection against heaving in those areas subject to freezing conditions.

Welded wire fabric should be cut and positioned within the confines of the form before the concrete is poured. Placing the wire on concrete blocks helps to ensure that it will remain in the middle of the concrete slab's depth (Figure 8:41).

Temporary dividers are used to separate extra-wide flat surfaces into workable segments. They are removed once the area has been leveled and the resulting space is filled with concrete (Figure 8:42).

As soon as the form is filled with concrete, a strikeoff board is used to level the concrete within the form, using a back-and-forth motion (Figure 8:43).

A wood float is used to smooth the surface, and to "float" any apparent aggregates below the surface to produce a nonskid surface. Either a bull float or a rented motorized circular float is used on large surface areas (Figure 8:44). A street broom is used to produce a rough finished surface (Figure 8:45) ("broom finish"), while a

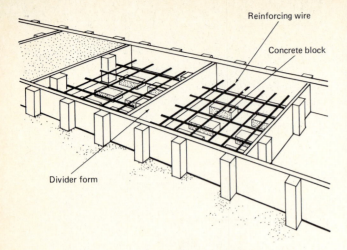

FIGURE 8:41

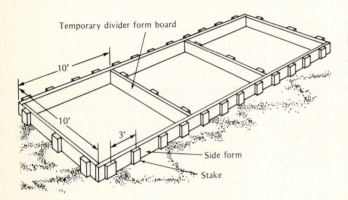

FIGURE 8:42

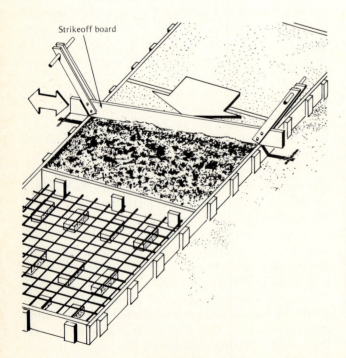

FIGURE 8:43

(A)

(B)

FIGURE 8:44
(A) Bull float. (B) Motorized circular float.

FIGURE 8:45
Use a stiff brush to give it a rough, skid-resistant surface; pull the brush across the slab toward you. For a smoother surface, make the final finish with a wood float.

steel trowel produces a very smooth finish (Figure 8:46). An edger should be used just inside the perimeter of the form to produce a rounded edge (Figure 8:47). Joints used to control expansion and contraction within an expanse of concrete are formed with a groover.

The surface of the finished concrete area must be kept moist for five to seven days after the concrete has been poured to allow ample time for it to cure thoroughly. The methods commonly employed to accomplish this important task are to (1) sprinkle the concrete surface occasionally during the daylight hours with a hose or (2) cover the entire surface with plastic, burlap, sand, straw, or soil once it has become completely hard.

FIGURE 8:46
Steel finishing trowel.

FIGURE 8:47
Edging the concrete.

QUESTIONS

1. Explain the two (2) procedures used to check the tentative house building lines for square.

2. List three (3) items that must be considered prior to the excavation of the foundation.

3. What should the dimensions (thickness and width) be for a footing designed to support a 10″ concrete block wall?

4. How many cubic yards of concrete will be needed to fill a footing whose total length is 148 feet (24′ × 50′) having a width of 16″ and a thickness of 8″?

5. Explain the reasons why concrete blocks are staggered with each course, both at the foundation's corners and with each successive row.

6. What is the recommended procedure used to waterproof the exterior surface of a concrete block wall?

7. Explain why curtain drains should be placed at the base of the foundation where it meets the top of the footing rather than either at a higher or lower depth.

8. Why is rigid insulation placed against the inside of the foundation wall prior to the pouring of a concrete slab floor?

9. Explain how the size of most footings is determined.

10. What is a lintel and where is it used?

11. Explain why concrete forms are oiled before they receive the concrete.

12. Give the dimensions of a standard stretcher concrete block.

9

FLOOR FRAMING

Once the foundation for the home has been completed, the floor can be framed. Some builders prefer to have the foundation backfilled with soil just before or directly after the floor is framed. This provides the workers easy access to the building site and allows delivery trucks to deposit materials close to where they will be used.

FLOOR JOISTS

The structural design of all floor frames is basically the same. *Joists* are the main ingredient. They run parallel across the width of the structure. Their butt ends rest on a *sill plate* that has been anchored to the foundation. In platform construction, a *header* is used to connect the butt ends and the sill plate together in the form of a box (Figure 9:1).

In balloon construction, both the butt ends of the joists and the upright studs rest on a double sill plate and are connected together with a solid wood firestop (Figure 9:2).

No sill plate should rest directly on the foundation wall. It is advisable to position a termite shield, and then a sill sealer, on top of the foundation wall before the sill plate is installed (Figure 9:3). The shield protects the house against termite invasions while the sealer is designed to seal out drafts.

FLOOR TRUSSES

Some builders prefer to use floor trusses for the floor frame in residential structures instead of the conventional joist-framing system (Figure 9:4).

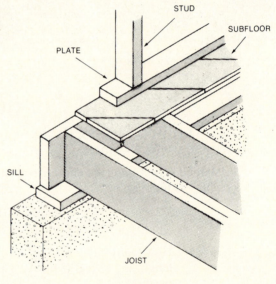

FIGURE 9:1

The advantages of floor trusses over the joist system are

1. They are able to span longer lengths (Table 9:1).
2. Their use eliminates support beams and bearing walls (Figure 9:5).
3. They provide adequate openings to run ductwork, plumbing pipes, and electrical wires (Figure 9:6).
4. They can be spaced farther apart (24″, rather than 16″, o.c.).
5. The cost of both the materials and the labor to construct and install them is much less.
6. They can be easily installed to fit a variety of different bearing and cantilever conditions (Figure 9:7).

TABLE 9:1
Floor Truss Spans

12″ Depth	14″ Depth	16″ Depth	18″ Depth	20″ Depth	22″ Depth	24″ Depth	Spacing
20′0″	21′7″	23′4″	25′5″	27′0″	28′4″	29′8″	24″ o.c.
20′0″	23′4″	26′8″	28′5″	30′1″	30′11″	32′4″	19.2″ o.c.
20′0″	23′4″	26′8″	30′0″	32′10″	34′6″	36′1″	16″ o.c.

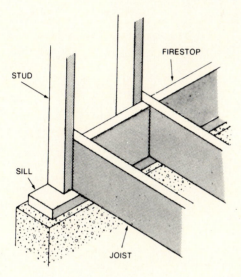

FIGURE 9:2

FIGURE 9:4
(Courtesy of Hydro-Air Engineering, Inc.)

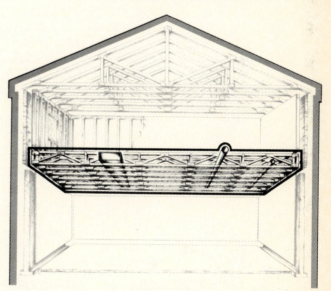

FIGURE 9:5

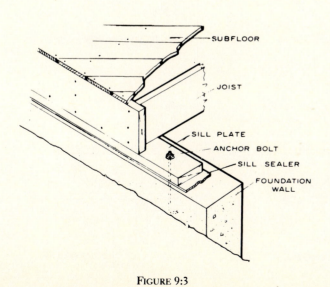

FIGURE 9:3

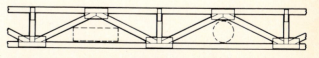

FIGURE 9:6

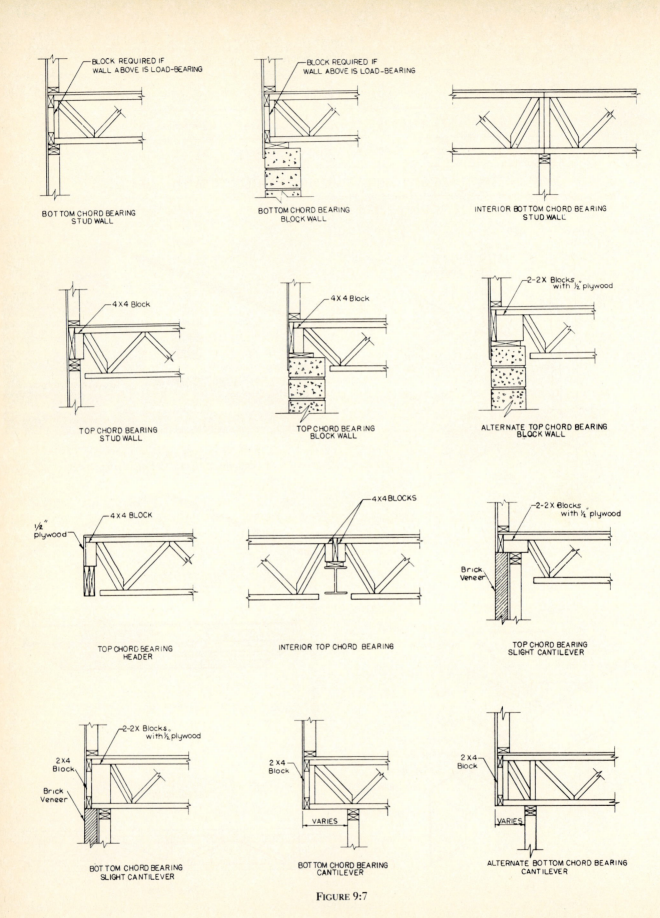

FIGURE 9:7

118

CONSTRUCTING THE JOIST FRAME

Sill Plate

The technique used to align the sill plate to its anchor bolts becomes relatively simple, provided all the installation considerations are recognized. They are the following:

1. The sill plate must be set back from the foundation edge a distance equal to the thickness of the sheathing that will be used. A chalk line is used to snap a line marking this setback.

2. The sill plate is then placed either flat or on edge on the foundation wall so that the location of each bolt can be correctly marked (Figure 9:8).

3. The diameter of the drilled hole must be *larger* than that of the $\frac{1}{2}''$ bolt to allow for some lateral adjustments. Its center can be determined by taking an accurate measurement from the setback line to the center of the bolt.

4. Position the longest sill plate on the longest foundation wall when only one plate is required. When a double plate is needed, the two plates must overlap each other at the corners (Figure 9:9). Also, check the corners for square.

Girders

Girders or beams are needed whenever the span between the foundation walls is greater than the length of the joists. They can be solid wood, built-up lumber (two or more pieces of 2″ thick nominal lumber nailed together), or steel I beam (Figure 9:10).

Whenever a girder is built up, the joints of the individual pieces must be staggered and be joined over a *support post*. The posts can be either steel or wood, but must always rest on a firm concrete foundation (Figure 9:11).

Always make certain that at least 4″ of the beam butt end is resting on the foundation wall. Its top surface should be level with the top of the sill plate (Figure 9:12). An exception would be when notched joists are positioned on ledger strips to provide more headroom in the basement (Figure 9:13).

It is advisable to consult the local building code with respect to both girder and joist size long before construction actually begins. Girder size is dependent not only upon the weight to be supported, but also on the length of its span.

In general, as the length of the span increases, the girder size increases. It is possible to minimize the size required by increasing the number of support posts, and making certain that they are spaced at uniform intervals beneath the girder.

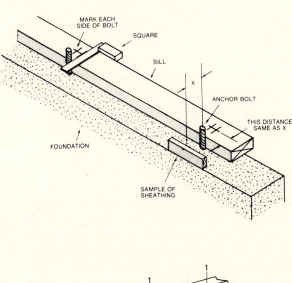

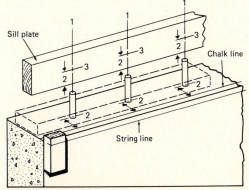

1. Sill bolt location
2. Distance from line to center bolt
3. Drill hole for sill bolt

FIGURE 9:8

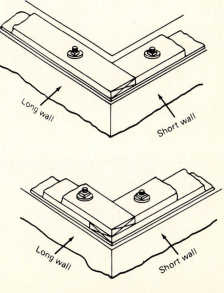

FIGURE 9:9

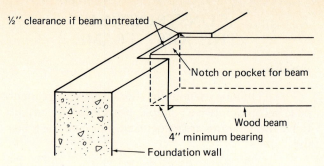

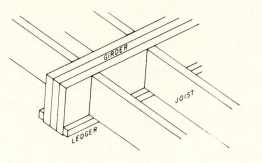

FIGURE 9:12

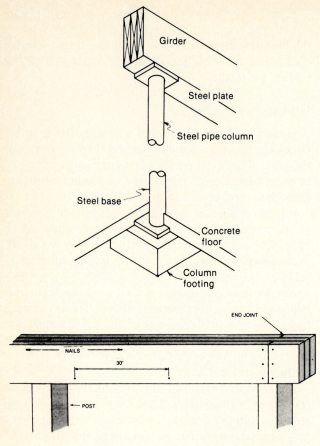

FIGURE 9:10

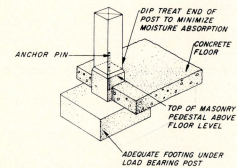

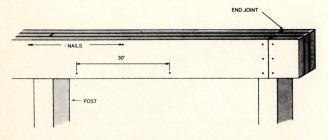

FIGURE 9:11

FIGURE 9:13

Local building code officials use tables to determine the proper girder size for a specific residential dwelling. They make a thorough examination of your structural diagrams to calculate the total floor load for the structure. On that basis, they may specify larger joists or a different joist spacing than you have listed on your diagrams.

This calculation, based on the spacing between the support posts and the species of wood used, will determine the most suitable solid wood girder. For example, a solid 4 × 8 girder (actual size $3\frac{1}{2}'' \times 7\frac{1}{2}''$), supported at 7' intervals, will support 3,570 pounds, while the same girder, supported at 9' intervals, will support only 2,950 pounds. Also, a built-up girder consisting of two 2 × 8s nailed together (actual size $3'' \times 7\frac{1}{2}''$) should not be expected to support the same amount of weight as the solid girder because it is $\frac{1}{2}''$ thinner than the solid girder (Figure 9:14).

Joist Size

The correct joist size is dependent upon the expected load it will carry (for example, a live load of 40 lb./sq. ft.), the length of its span, the spacing between adjacent members (12″, 16″ or 24″ o.c.), and the grade or species of lumber used.

Table 9:2 presents the safe spans of four different sized joists at three different spacings for a live load of 40 lb./sq. ft.

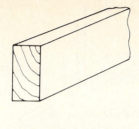
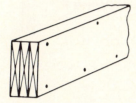

FIGURE 9:14
(A) Solid and (B) built-up girders, butt end view.

TABLE 9:2
Span Limits of Floor Joists Carrying 40
lb./sq. ft. Live Load

Joist size	Joist spacing (o.c.)	Span limit
2 × 6	12″	10′2″
	16″	9′3″
	24″	8′0″
2 × 8	12″	13′6″
	16″	12′4″
	24″	10′10″
2 × 10	12″	17′0″
	16″	15′6″
	24″	13′8″
2 × 12	12″	20′5″
	16″	18′9″
	24″	16′6″

The Basic Frame

The joists can be laid out once the sill plates have been anchored and squared to the foundation, and the girder has been properly positioned and supported. In platform construction, the header is marked with a framing square to indicate the position of each joist (Figure 9:15). In balloon construction, the double sill plate is marked (Figure 9:16).

You must sight along the length of each joist before placing it in its nailing position to make certain the crown points upwards (Figure 9:17).

It is also imperative that the edge of each joist that rests on the foundation is square before it is nailed into

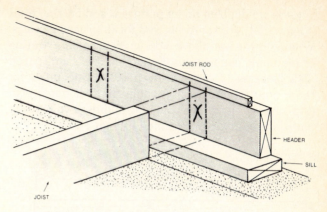

FIGURE 9:15

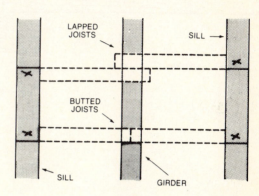

FIGURE 9:16
Note that the sill is doubled in balloon construction.

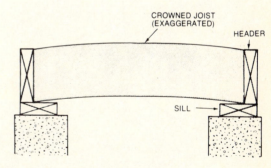

FIGURE 9:17

place. The edge that rests on and overlaps the girder (spliced joists) does not need to be square. The squaring must be done to avoid problems as construction progresses (Figure 9:18).

Joists should overlap the girder by at least 4″ (Figure 9:19). Because of this overlap, each header will have its marks vary by the width of the joist (Figure 9:20). One *stringer*, or terminal joist, will be needed at each end of the floor frame. When nailed to the outside of the indented joist, it makes the frame into a neat, rectangular-shaped box that will provide the proper support, along its entire perimeter, for the subfloor (Figure 9:20).

Joists are doubled under partitions and bearing walls (Figure 9:21), for stairwells, and for fireplace openings. Whenever the wall above is designed to contain heating or plumbing connections, the double joists are spaced with blocking to provide the necessary openings for these units (Figure 9:22). A complete floor frame plan (top view), with stairwell opening and the two types of partition supports, is presented in Figure 9:23.

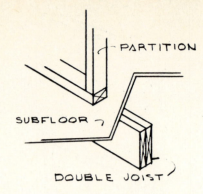

FIGURE 9:21

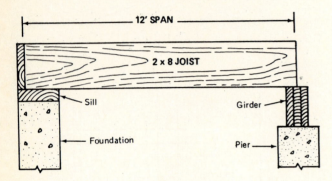

FIGURE 9:18

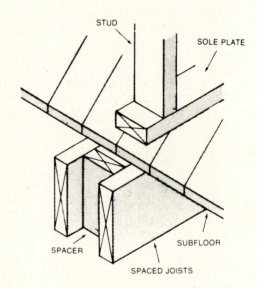

FIGURE 9:22

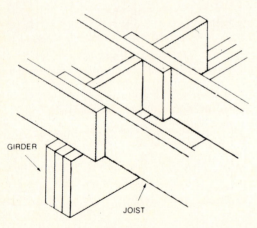

FIGURE 9:19

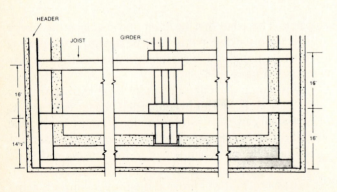

FIGURE 9:20

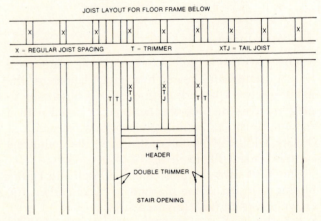

FIGURE 9:23

OPENINGS

A specific nailing sequence must be followed when framing a stairwell or fireplace opening. Begin by nailing the first trimmers (A) to the header. They define the width of the rough opening. Occasionally, a regular joist can be used as one of the trimmers.

The first headers (B) are then nailed into place. Make certain they are square with the trimmers. Also, allow room for the second headers so that the length of the rough opening is correct.

The tail joists (C) are then positioned between the headers and are nailed into place. They are usually located in the same spot as that of a full-length joist.

The second headers (D) are installed next, and, finally, the second trimmers (E) are nailed into place (Figure 9:24).

It is important to realize that all the members of the frame opening must be cut and installed square, so that each one fits tightly in its location. This will minimize the loss of strength due to the opening.

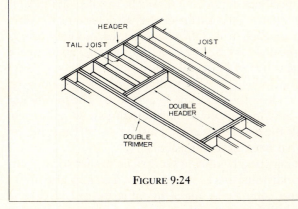

FIGURE 9:24

FLOOR EXTENSIONS

The design of some homes specifies an extension of the floor frame beyond the foundation walls. To frame this extension, the joists must overhang, or cantilever, the foundation. The direction the joists are running in the vicinity of the extension will determine which one of two framing patterns to use.

One framing pattern simply uses longer joists to overhang the foundation (Figure 9:25). The basic rule to follow for cantilevered joists is that they can extend beyond the foundation only one third ($\frac{1}{3}$) of their overall length. For example, a 6' extension requires joists that are at least 18' long.

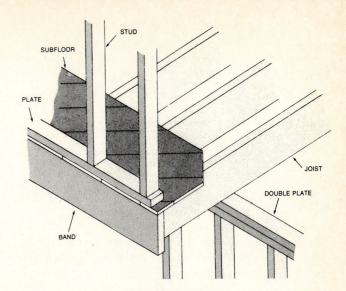

FIGURE 9:25

When the joists run parallel to the foundation wall, the framing pattern is more complex. The length of the desired overhang must be established first. This length will determine the length of the cantilevered joists for example, a 3' overhang must employ joists that are 9' long (Figure 9:26).

The cantilevered joists can be attached to a full-length joist with framing anchors, or by employing a ledger strip (Figure 9:27). Note that the ledger must be positioned along the top edge of the joist because the force at this point is upward.

It is possible to nail this type of floor frame extension without using anchors or a ledger strip, if the proper nailing sequence is followed. Begin by nailing the cantilevered joists to the full-length joist. Make certain to double the joists on both sides of the frame for additional strength. Then, nail a second full-length joist to

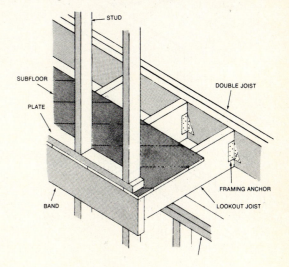

FIGURE 9:26

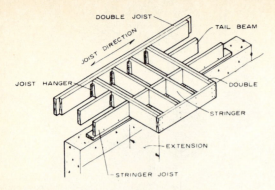

FIGURE 9:27

the first one. Complete the frame by nailing the outside header in place, toe-nailing the cantilevered joists to the sill plate, and, finally, inserting solid blocking between the joists to help strengthen the area.

BRIDGING

Bridging is placed between joists to help distribute loads, to restrict the tendency of the joists to twist or warp, to stiffen the floor frame and to help maintain a level floor.

Regular, or cross, bridging employs two short pieces of 1 × 2 or 1 × 3 lumber placed in a diagonal pattern between each joist, forming an X (Figure 9:28). Manufactured metal cross bridging can also be used. The bridges are installed in pairs (Figure 9:29).

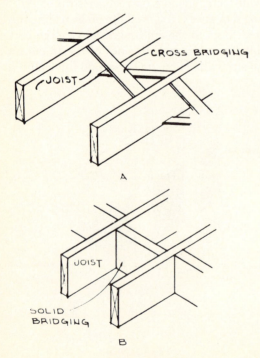

FIGURE 9:28

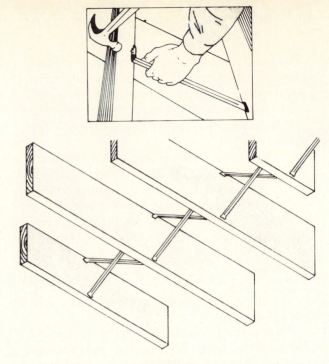

FIGURE 9:29

Since lines of bridging should be installed at intervals more than 8′, it is first necessary to determine the position of each run. Once determined, a chalk line is snapped across the tops of the joists to guarantee a straight line of bridging.

The individual pieces alternate their positions on each side of the chalk line (Figure 9:30). The top ends of the bridging pieces are nailed with two 8d nails each, while the lower ends are not fastened until the subflooring has been nailed in place.

It may be advisable to set up a jig so that you can cut the individual pieces more rapidly. You may also prefer to use commercial metal bridging instead of cutting wooden ones.

Solid bridging, or blocking, can also be used and

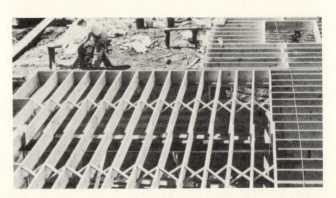

FIGURE 9:30

should be staggered along the chalk line (Figure 9:31) to make it easy to nail them into place.

JOIST CUTOUTS

In most residential dwellings, the utility lines are designed to run parallel to the joists for the majority of their runs. On occasion, however, the joists must be cut or notched to minimize the length of the runs and keep the material costs down.

If possible, a cut in a joist should be made near its neutral axis (Figure 9:32). Provided the cut does not exceed one fourth of the total depth of the joist, the joist will not suffer loss of strength.

A cut made in a joist near the center of its span is the most detrimental. To compensate for the reduction in strength, it will be necessary to install additional trimmers, headers, or joists.

It is wise to make the cut from the top edge of the joist. When performed, the reduction in joist strength is equivalent to the depth removed. For example, a 2″ deep cut in a 2 × 6 joist reduces the joist's strength to that of a 2 × 4 (Figure 9:33).

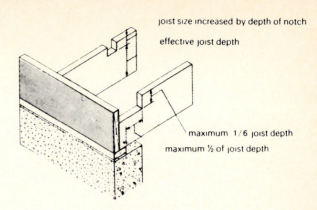

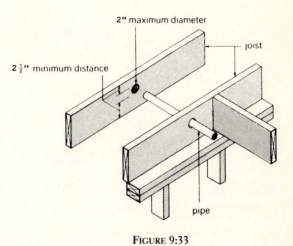

FIGURE 9:33

FIGURE 9:31

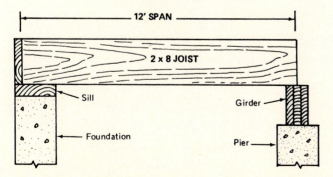

FIGURE 9:32
The neutral axis is an imaginary line running lengthwise, and equidistant between the top and bottom surfaces.

SUBFLOORING

As soon as the floor frame has been completed, the subflooring can be nailed into place. The subflooring connects all the floor frame components together, making the structure more rigid, and serves as a base for both the finished floor and for all the framing components assembled above it.

The floor frame can be covered with plywood, tongue-and-groove flooring, shiplap or end-matched boards. Using 4 × 8 sheets of plywood is preferred by many builders because they can be installed rapidly. In addition, an individual sheet dimensions 48″ × 96″ are compatible with any of three common joists settings (12″, 16″, or 24″ o.c.). Each sheet should be placed so that its long dimension (also its surface grain direction) is at a right angle to the joists. Also make certain that the end joints of the pieces are staggered in each succeeding row (Figure 9:34).

It is important to remember that when plywood subfloors run perpendicular to the joists, the finished floor must be laid at a right angle to the subfloor.

The American Playwood Association has found that gluing $\frac{5}{8}$″ plywood panels to 2 × 8″ joists in conjunction

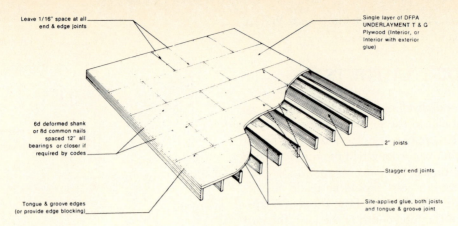

Leave 1/16" space at all
end & edge joints

Single layer of DFPA
UNDERLAYMENT T & G
Plywood (Interior, or
interior with exterior
glue)

6d deformed shank
or 8d common nails
spaced 12" all
bearings or closer if
required by codes

2" joists

Stagger end joints

Tongue & groove edges
(or provide edge blocking)

Site-applied glue, both joists
and tongue & groove joint

FIGURE 9:34

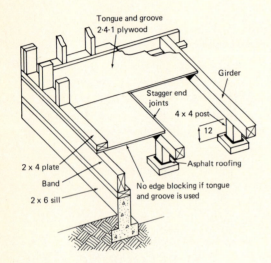

Tongue and groove
2·4·1 plywood

Girder

Stagger end
joints

4 x 4 post

12

2 x 4 plate

Asphalt roofing

Band

No edge blocking if tongue
and groove is used

2 x 6 sill

FIGURE 9:35

with some nailing, when compared to the normal nailing procedure, has increased the stiffness of the joists by 25% and has eliminated squeaks, the popping of nails, and the bounce or deflection associated with heavy foot traffic.

NONJOIST FLOORING

There is an alternative to the joist flooring system. It employs wooden girders spaced 48″ o.c. and supported by piers for the basic floor frame. Over this are laid sheets of 2-4-1 plywood, a special $\frac{11}{8}$″ thick tongue-and-groove material, as the floor surface (Figure 9–35). The face grain of each panel must be placed at right angles to the underlying girders before it can be nailed into place with 8d threaded nails.

QUESTIONS

1. Explain the basic difference in framing design between platform and balloon construction with respect to how the floor joists are nailed to the sill plate.

2. List three (3) requirements that every support beam must satisfy.

3. List three (3) factors that affect the correct joist size for a residential home.

4. List three (3) places where floor joists must be doubled.

5. Explain how you would properly frame out the opening for a cellar stairway.

6. How many feet can a joist 12′ long extend or overhang the foundation?

7. List three (3) reasons for placing bridging between adjacent floor joists.

8. Under what conditions are cuts made in floor joists detrimental and what are the techniques used to restore this loss in strength?

9. Explain how floor joists can be supported.

10. How is the sill plate fastened to the home's foundation?

11. Explain the relationship between the size of the beam and the distance between its support posts.

10

WALL FRAMING

There are two methods used to frame exterior residential walls: platform framing and balloon framing. In platform framing, the walls are built on top of a subfloor that has been nailed to the underlying floor joists (Figure 10:1). In balloon framing, the exterior walls are constructed of studs that run continuously from the

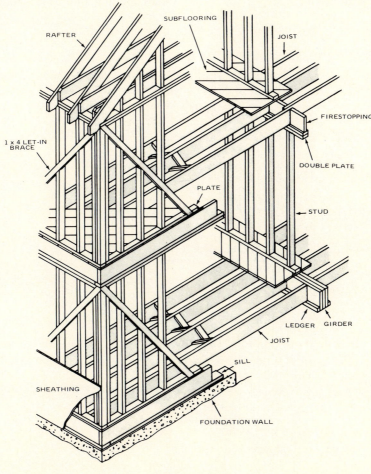

FIGURE 10:1
Platform framing.

foundation sill plate to the rafter plate. These studs support the floor joists, rather than rest on them (Figure 10:2).

The terminology associated with platform wall framing and its various components is presented in Figure 10:3. It is important for every carpenter to know all the framing terms and the structural purpose for each component.

The *sole plate*, usually a 2 × 4, is nailed to the subfloor and serves as the supportive base for the wall frame. The *top plate*, also a 2 × 4, is the top structural member in the framed wall. It is doubled after the walls have been assembled, raised into position, and leveled. Its purpose is to tie the two adjacent walls together, as well as to provide additional support for ceiling joists and roof rafters (Figure 10:4).

The *full upright studs* are usually 2 × 4s, but may be 2 × 6s in colder climates where 6″ of insulation is desired, or in bathroom walls that will house the plumbing lines (Thus the sole plate and top plate must also be 2 × 6s.). They are usually spaced at intervals of either 16″ or 24″ to accommodate any of the various 4′ × 8′ wall covering materials (Figure 10:5).

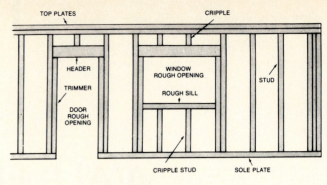

FIGURE 10:3

Headers must be installed above all door and window rough openings. They are designed to support and distribute the weight of the ceiling and roof components.

Headers must always rest on two *trimmer studs*, which are shorter than full studs. They not only define the width, but also strengthen the sides of the rough opening. A header's length is equal to the width of the rough opening, plus 3″ (1½″ for each trimmer stud). It is

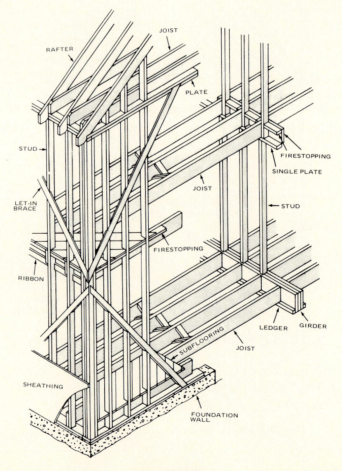

FIGURE 10:2
Balloon framing.

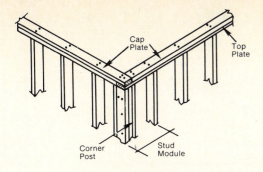

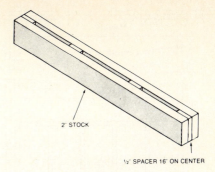

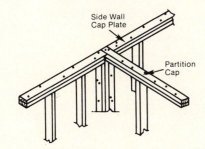

FIGURE 10:4
(A) Wall corner. (B) Partition intersecting outside wall.

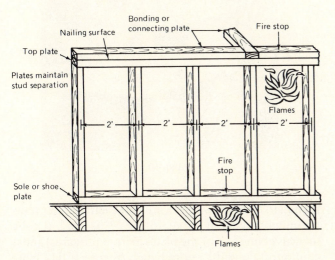

FIGURE 10:5

constructed by nailing together two pieces of 2″ (nominal) lumber, with short pieces of $\frac{1}{2}$″ plywood used as spacers between them (Figure 10:6). The resulting thickness of the header equals that of the upright 2 × 4 wall.

The width of the nominal lumber used for the header is dictated by the size of the opening. For example, a 2 × 6 is sufficient for narrow openings, while a 2 × 12 is used for any opening 8′ or wider. Some contractors prefer to use only 2 × 12 headers for all exterior wall rough openings. The 2 × 12 is positioned just beneath the top plate to ensure a uniform height for all windows and doors, as well as to reduce the time spent in measur-

ing, cutting, and nailing the *cripple studs* into place (Figure 10:7).

The rough opening height for each window must be known in advance so that the supporting *rough sill* can be properly positioned, and the underlying *cripple studs* can be cut to size and installed.

Most door frames are constructed to accommodate a finished door height of 6′ 8″ (80″). The height of the frame rough opening should be $2\frac{1}{2}$″ more than this, and the width should be $2\frac{1}{2}$″ more than whatever the width of the finished door is. A header, supported by trimmer studs and possibly filled in with upper cripple studs, is sufficient to frame a doorway. The sole plate must be cut out after the wall construction has been completed and before the door jamb is installed.

Since most exterior sheathing materials as well as interior wall coverings are purchased as 4′ × 8′ sheets, the carpenter must maintain the same stud spacing, 16″ or 24″, along the entire length of the wall. This minimizes the amount of cutting that is required for the covering material (Figure 10:8). It is quite possible that both the door and window openings will not be in complete agreement with the stud layout. This condition, however, should not affect or alter the prescribed spacing, since it can be maintained by the proper positioning of the cripple studs (Figure 10:9).

MASTER LAYOUT

The architectural plans must be consulted prior to the construction of any wall section to learn the exact location of all windows and doors, as well as their rough opening dimensions. Once these are known, the carpenter selects a straight 2 × 4 (master stud pattern) or story pole, and transfers all pertinent heights to it, as shown in Figure 10:10. This provides a standardized reference for the various stud lengths and reduces the possibility of incorrect measurements.

Then, the carpenter must lay out the plates for each wall. The sole plate and top plate are placed side by side (or the top plate is placed on top of the sole plate), and

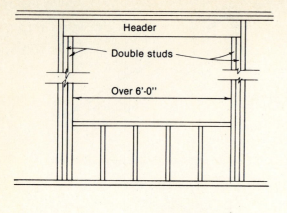

Header
Double studs
Over 6'-0"

Top plate
Cripples
Header

FIGURE 10:7

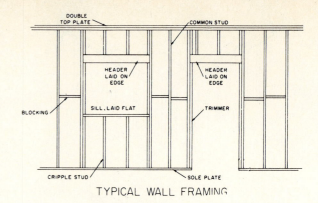

DOUBLE TOP PLATE   COMMON STUD
HEADER LAID ON EDGE   HEADER LAID ON EDGE
BLOCKING   SILL, LAID FLAT   TRIMMER
CRIPPLE STUD   SOLE PLATE

TYPICAL WALL FRAMING

SYMBOLS:
X – COMMON STUD
T – TRIMMER STUD
C – CRIPPLE STUD

SOLE PLATE LAYOUT

FIGURE 10:9

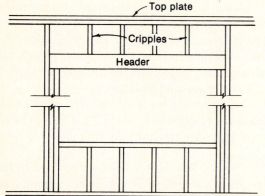

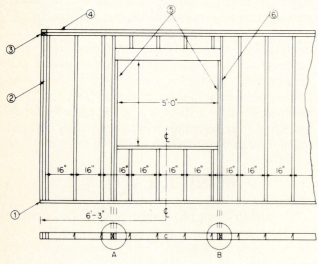

5'-0"
₵
16" 16" 16" 16" 16" 16" 16" 16" 16"
6'-3"   ₵
A   B

FIGURE 10:8

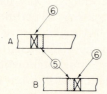

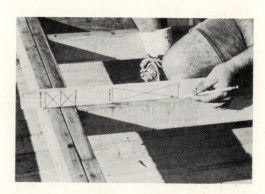

FIGURE 10:10

then marked to show the location of each stud. A predetermined spacing of either 16″ or 24″ is used. It commences at one end of the plates (corner post area) and is continued along the entire length (Figure 10:11). Note that the first stud position is marked 15$\frac{1}{4}$″ from the corner.

Next, the center lines for all window and door openings must be marked on the plates so that the location of the trimmer studs, which outline the rough openings, can be marked. These, in turn, dictate the location of the adjacent full studs.

Stud marks that fall between two trimmer studs indicate placement of cripple studs beneath a window's rough sill, or deletion of studs if located below a door header.

Carpenters mark the location of full studs with an X, cripple studs with a C, and trimmer studs with a T to make sure the different studs are put in the right locations (Figure 10:12).

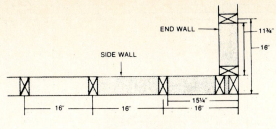

FIGURE 10:11
Regular stud layout.

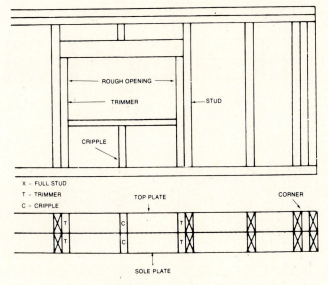

FIGURE 10:12

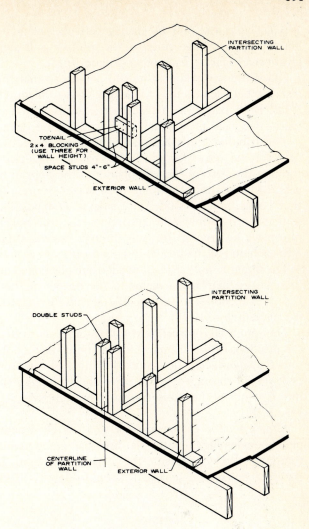

FIGURE 10:13

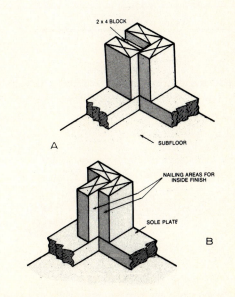

FIGURE 10:14

The plans will also help the carpenter to pinpoint where the interior partitions will intersect the exterior walls, and to determine whether additional full-length studs or blocking between adjacent studs will be necessary (Figure 10:13).

Exterior corners can be assembled in several ways. They may be constructed as separate components; positioned before the rest of the wall frame is installed, or built after a side wall and corresponding end wall have been raised and properly braced. A minimum of three full-length studs are required to construct an exterior corner. Figure 10:14 shows two different exterior corner designs.

ASSEMBLING WALL SECTIONS

To avoid the expenditure of time required to measure and cut each full-length stud, it is advisable to purchase either full 8'-long precision end cut studs or "precuts" that measure 7' 8" in length.

Headers should be cut and assembled to the length specified on the plans, as well as the rough sill piece for each window.

Begin your assembly by placing both the sole plate and top plate on edge on top of the subfloor with their layout marks facing each other (Figure 10:15).

Space the two plates apart and parallel so that the distance between them is the length of the full-length studs. Both plates are nailed to the studs with 16d nails—two nails in each end of the stud (Figure 10:16).

The trimmer studs are placed on their respective marks on the sole plate and attached by two nails driven through the base of the sole plate into the bottom of each trimmer stud. These are then nailed securely to their companion full stud. The header is set in position so that it rests firmly on each trimmer stud. It is nailed into place by driving 16d nails through the full studs into each end of the header (Figure 10:17).

The upper cripple studs, if needed, are positioned on their layout marks, toe nailed to the header at their base, and anchored at their top by 16d nails through the top plate (Figure 10:18).

The rough sill is placed next to the sole plate so that the positions of the cripple studs can be marked on its underlying surface. These studs are attached with 16d nails driven through the sole plate and the rough sill into the ends of the studs (two nails for each end). See Figure 10:19.

After a section of the exterior wall has been completely assembled, with its various components thoroughly nailed, measure the diagonal distances. If they are equal, the section is square (Figure 10:20).

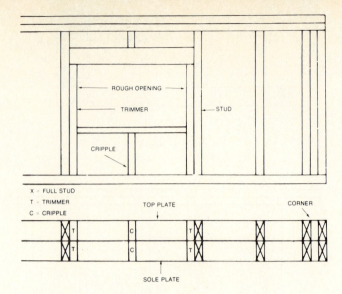

FIGURE 10:17

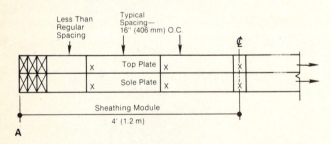

FIGURE 10:15
Procedure for laying out the sole and top plates for the wall to contain the corner post framing.

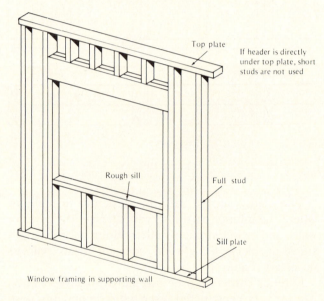

Window framing in supporting wall

FIGURE 10:18

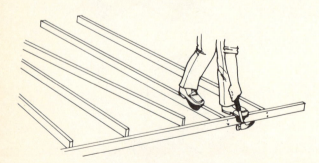

FIGURE 10:16
Place studs and plates on edge during assembly—a flat and clean surface is essential.

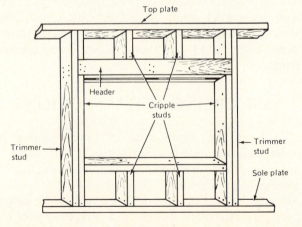

FIGURE 10:19

FIGURE 10:20
Wall frame is checked. If diagonals are equal, the section is square.

Then, the completed section can be raised to a vertical position (Figure 10:21). The size of the section and the carpenter's discretion determine whether or not it will be covered with sheathing before it is raised into position. It is important to know that all exterior sheathing must be attached to the frame before any roof framing is started.

Once a section is raised, its sole plate must be properly aligned with the chalk line snapped on the subfloor. Only then is it nailed to the underlying floor frame and joists with 20d nails. The section can be braced with extra 2 × 4s. Each brace, placed on a 45° angle, has one end attached to a stud near the top plate and the other end nailed to a 2 × 4 cleat, which has been nailed to the subfloor (Figure 10:22).

FIGURE 10:21
Erecting a partition wall.

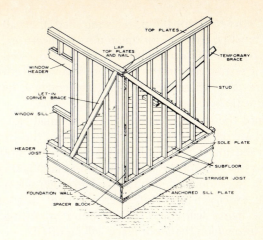

FIGURE 10:22

After the adjacent wall sections have been raised into position and sufficiently braced, the corners can be plumbed. To accomplish this, some braces may have to be loosened, moved slightly, and then renailed (Figure 10:23). Doubling the top plate at this time will help prevent the various sections from shifting out of their plumbed positions. Many carpenters leave these braces intact until the house is completely enclosed.

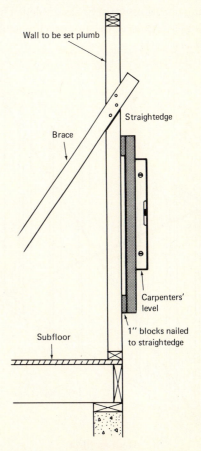

FIGURE 10:23
A carpenter's level and straightedge are used to plumb a wall.

Whenever diagonal bracing is required, the wall section must be checked for squareness by measuring the length of its diagonals. This is done while the wall section is resting flat on the subfloor. The wall section is square only when the diagonals are equal in length (Figure 10:24).

Once the section has been squared, small blocks of wood are nailed to the subfloor to prevent the section from shifting. The 1 × 4 let-in brace is then positioned across the studs of the wall section for marking (Figure 10:25). Each stud should be marked on either side of the brace exactly where it crosses the stud diagonally. These marks will guide the carpenter in cutting the slots into which the let-in brace will be fitted.

Adjust the depth of cut on your circular saw ($\frac{3}{4}''$ deep) and then not only saw on the pencil marks but also make several additional cuts between the two diagonal marks. Remove the cut material with a wood chisel (Figure 10:26).

PARTITIONS

After the complete exterior wall frame has been raised, plumbed, and thoroughly braced, the interior load-bearing partitions must be assembled. They are the ones that are required to support the ceiling joists (Figure 10:27).

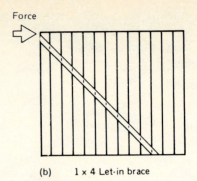

(b)   1 × 4 Let-in brace

FIGURE 10:25
Bracing the wall to resist forces.

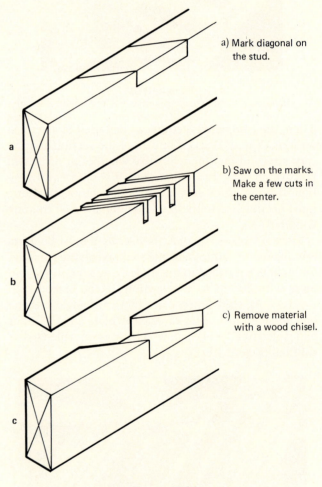

a) Mark diagonal on the stud.

b) Saw on the marks. Make a few cuts in the center.

c) Remove material with a wood chisel.

FIGURE 10:26
Steps to cut slots in studs and plates for diagonal bracing.

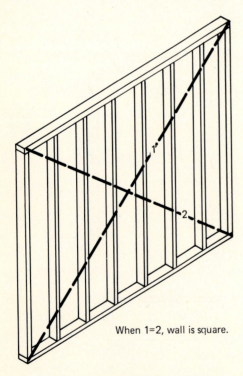

When 1=2, wall is square.

FIGURE 10:24
A wall can be checked for squareness by measuring its diagonals. Wall is square if distance 1 = 2.

The nonbearing partitions, whose primary purpose is to define the outlines of the various rooms and closet areas, can be omitted until the home's exterior has been thoroughly enclosed and sealed from the weather (Figure 10:28).

The framework of all interior partitions and intersect-

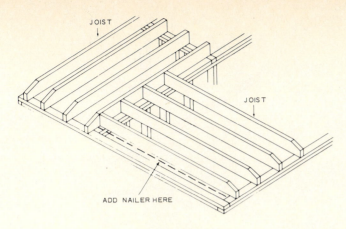

JOIST

JOIST

ADD NAILER HERE

FIGURE 10:27

ing walls must be constructed so that there are adequate nailing surfaces at all corners, both inside and outside. The purpose is to ensure that the installation of interior wall covering material will run smoothly (Figure 10:29).

Headers for doorways located in nonbearing walls can be constructed with 2 × 4s, while those for bearing walls require at least a 2 × 6 header (Figure 10:28).

Any partition that intersects with the exterior wall frame must be first checked for square in its alignment with the exterior wall, before it is securely nailed to the frame (Figure 10:30).

Bearing partitions that are located between and run parallel to the underlying joists should be supported by adding either a doubled joist directly beneath the partition or installing solid bridging, at 12″ intervals, be-

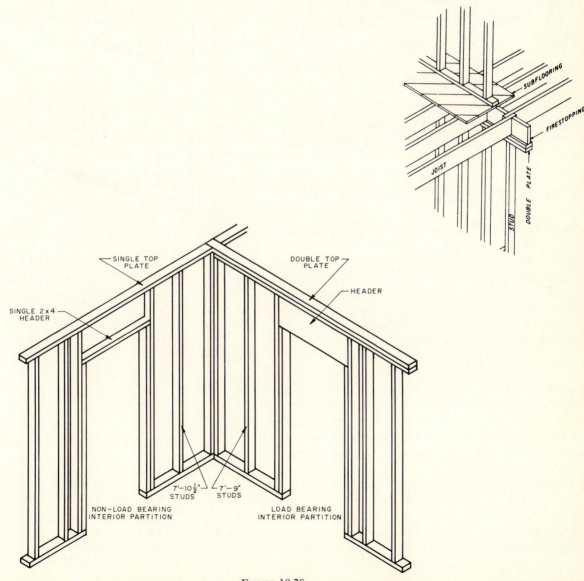

FIGURE 10:28
Framing load-bearing and non-load-bearing interior partitions.

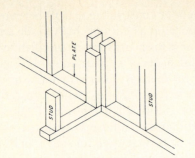

FIGURE 10:29

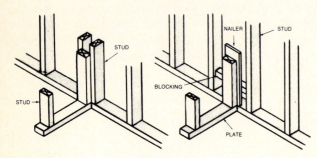

FIGURE 10:30

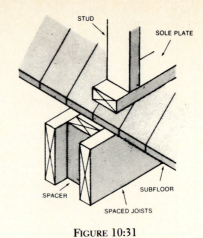

FIGURE 10:31

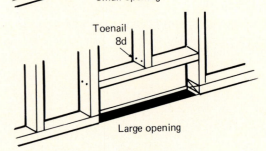

FIGURE 10:32

tween the closest two joists (Figure 10:31). Bearing partitions that run perpendicular to the joists can be strengthened by nailing solid bridging between the joists.

OTHER FRAMING PROCEDURES

Various incidental framings are necessary for installation of plumbing fixtures, heat ducts, and baseboard moldings. These can be installed during the initial framing operation, provided the exact locations have been dictated by the plans. It is possible to delay these procedures until they become absolutely necessary, but that could result in greater expenditures of time, labor, and materials than if done initially.

It is advisable to frame forced-air heating duct openings during the initial framing rather than wait until the heating contractor is ready to install the ductwork (Figure 10:32). The same advice holds true for the insertion of wall backing, which provides a firm support and a prominent nailing surface for the proper installation of sinks, toilets, vanities, and cabinets. The location and the correct height of each item requiring backing must be obtained from the plans (Figure 10:33).

Unused short pieces of 2 × 4 framing can be used to good advantage. Placed vertically in the corners of the rooms or horizontally on top of the sole plate, they not only enlarge the available nailing surface but also provide extra support for the baseboard moldings (Figure 10:34).

Pieces, about $14\frac{1}{2}''$ long, can be nailed between the upright studs, spaced 16″ on center, at the junction of

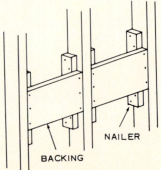

FIGURE 10:33

floor and ceiling in balloon framing. They serve as fire stops that prevent the spread of fire and smoke through the air spaces in the wall (Figure 10:35).

A bathtub requires some form of blocking to provide it with additional support. The blocking can consist of either one long 2 × 4 nailed horizontally to the upright studs or several pieces of similar length, each nailed vertically to an upright stud (Figure 10:36).

A stack wall or plumbing partition (usually a 6″ to 8″ partition) houses the main plumbing stack which connects the sewer, at its lower end, and vents through the roof, at its upper end. Most of the pipes leading to the various bathroom fixtures are housed in this same partition (Figure 10:37).

STEEL STUDS

Steel studs can be used, instead of wooden 2 × 4s, for nonload-bearing partitions that will be eventually covered with gypsum wallboard panels. They are

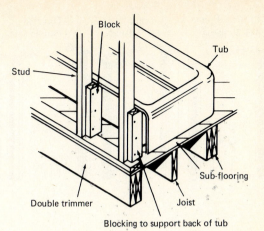

FIGURE 10:36

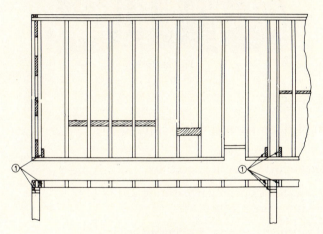

FIGURE 10:34

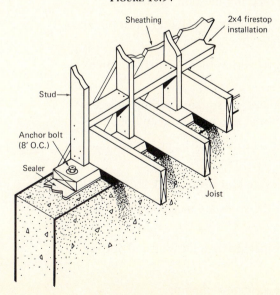

FIGURE 10:35

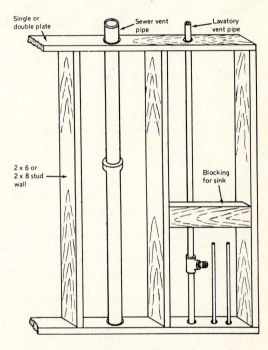

FIGURE 10:37
6″ or 8″ partition housing plumbing vent and water pipes.

available in inch widths of $1\frac{5}{8}$, $2\frac{1}{2}$, $3\frac{5}{8}$, 4, and 6 and lengths of 8′ to 16′. The complementing steel runners come in the stud width chosen and 12′ lengths (Figure 10:38).

The advantages that steel studs have over wooden studs are the following:

1. Less expensive and weigh less
2. No shrinkage, warping, or twisting to contend with
3. Total amount of materials needed is less
4. Noncombustible and more effective in providing adequate sound control
5. Prepunched holes make running electrical wires an easy task

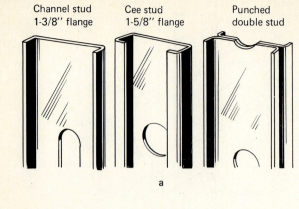

Channel stud 1-3/8" flange   Cee stud 1-5/8" flange   Punched double stud

a

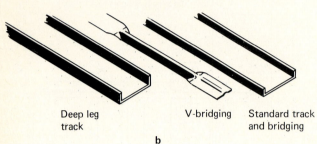

Deep leg track   V-bridging   Standard track and bridging

b

FIGURE 10:38
(A) USG steel stud. (B) USG steel runner.

Light steel studs can be used in load-bearing walls. They are available in inch widths of $3\frac{1}{2}$, 4, $5\frac{1}{2}$, 6, $7\frac{1}{2}$, $9\frac{1}{4}$, $9\frac{1}{2}$, $11\frac{1}{2}$, and $13\frac{1}{2}$ and in lengths to 28'. The runners used to complement the chosen stud width come with a $1\frac{1}{4}$" lip, and in 10' lengths.

Each stud must be $\frac{3}{8}$" less in length than the actual distance between the floor or ceiling runners to compensate for any floor or ceiling movement. Once they are properly sized, the studs are inserted into their respective runners and then twisted into position to achieve a friction fit. All stud flanges should point in the same direction (Figure 10:39).

FIGURE 10:39

QUESTIONS

1. How does balloon framing differ from platform framing with respect to the manner in which the floor joists are nailed?

2. Explain where headers must be located and why they are necessary.

3. Explain how you would rough frame a window opening (measuring 2'8" × 2'10") and a door opening whose dimensions are 2'6" × 6'8".

4. Show how you would lay out the plates, using a 16" o.c. spacing, to accommodate both the door and window mentioned in Question 3 above.

5. List three (3) important pieces of information that a carpenter must obtain from the architectural plans.

6. List, in the proper sequence of steps, the procedures required to frame an exterior wall.

7. Explain how an exterior wall should be braced, be squared with its adjacent wall, and then be checked to make certain that it is both level and plumb.

8. Explain the functional differences between a bearing and a non-bearing partition with respect to location and framing considerations.

9. What is the purpose of a fire stop? Where should it be located?

10. Explain why the subfloor is installed after the studs are nailed in place, with balloon framing.

11. Describe trimmer and cripple studs with respect to their location and purpose intended.

12. Draw and label the structural members of a window opening.

11

CEILINGS

Joists are the basic structural components for all ceilings. Those positioned between upper and lower living areas serve a dual purpose. They act as ceiling joists for the lower level and as the floor joists for the upper level. The size, spacing, and span length of these joists is much more critical than that of ceiling joists which have only attic space above them. Table 11:1 presents the allowable maximum spans for ceiling joist for both 12″ and 16″ o.c. spacings.

Whenever possible, joists should be positioned to run across the width, rather than the length, of the structure. This practice reduces the cost of the materials and the construction time.

To be structurally sound, every joist must be supported at each end. This can be accomplished in a variety of ways.

1. The joist end can rest on a partition (Figure 11:1)

TABLE 11:1
Allowable Spans for 10 lb./sq.ft. Live Load
(Drywall Ceiling)

Joist Size	Spacing	Maximum Allowable Span
2 × 4	16″	10′–0″
	24″	8′–1″
2 × 6	16″	15′–7″
	24″	12′–9″
2 × 8	16″	20′–7″
	24″	16′–10″
2 × 10	16″	26′–3″
	24″	21′6″

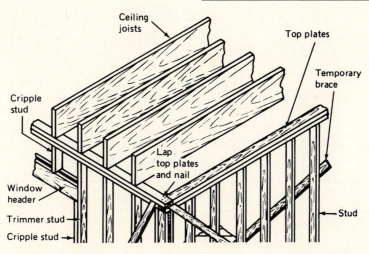

FIGURE 11:1

139

2. It can rest on a support beam or header located below the joist (Figure 11:2)

3. It can be notched to rest on a ledger board whenever the support beam is flush with the joist (Figure 11:3)

4. It can be supported by metal framing anchors or joist hangers (Figure 11:4)

5. Metal straps can be used to anchor the joist end whenever the support beam is located above the joist (Figure 11:5)

It is not imperative that a joist be positioned directly above a wall stud when it rests on a load bearing wall, provided the wall has a double top plate (Figure 11:6). It is imperative, however, that they be aligned with the rafters so that they can be nailed together, even though the joists are nailed in place prior to the placement of the rafters. (Figure 11:7). The outer edge of each joist must be cut or notched to conform to the slope of the roof. (Figure 11:8). In low-pitched hip roof patterns, it is advisable to cut short joists that will run perpendicular and be nailed to the regular joists to avoid problems with the roof's slope (Figure 11:9).

For nailing purposes, joists must be positioned so that the corners of the room have an adequate nailing surface. Nailing strips can be inserted to accomplish the same effect (Figure 11:10).

To estimate the number of joists required to ade-

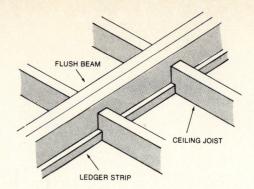

FIGURE 11:3

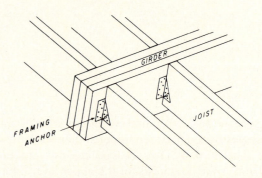

FIGURE 11:4

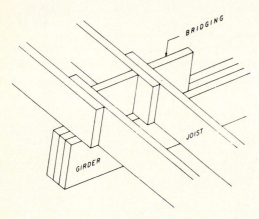

FIGURE 11:2

quately span a ceiling area, the following formula is used: Number of joists = [wall length (width) × $\frac{3}{4}$] + 1; for example, [16' (wall length) × $\frac{3}{4}$] + 1 = 12 + 1 = 13 joists needed.

Residences have either flat or cathedral-type ceilings. The ceiling joists for the cathedral type also serve as the rafters for the structure and thus must follow the roof line. The joists can be covered to produce a smooth surface or can be constructed so that they are left exposed (Figure 11:11).

The joists for a flat ceiling can be covered with either wallboard, plaster, or ceiling tile. Normally, the ceiling area is covered before the wall areas are covered.

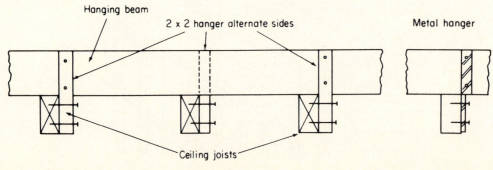

FIGURE 11:5

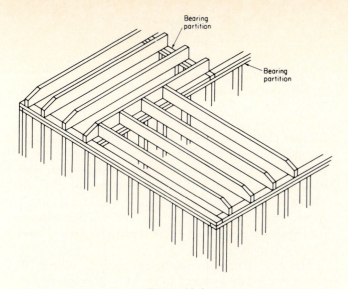

FIGURE 11:6

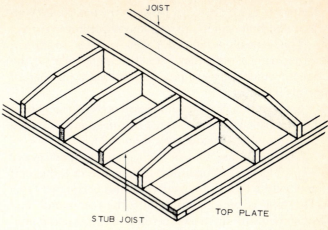

FIGURE 11:9

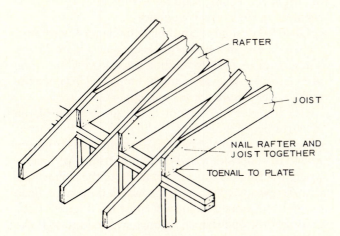

FIGURE 11:7

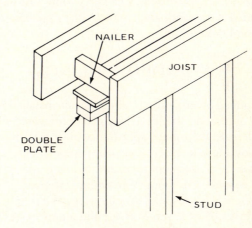

FIGURE 11:10

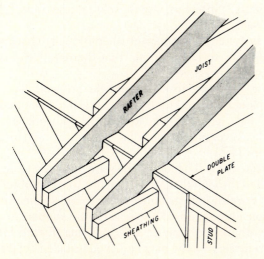

FIGURE 11:8
Joists are notched to conform to shape of rafters.

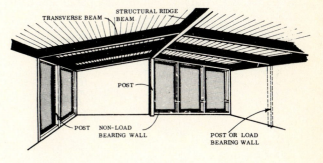

FIGURE 11:11

WALLBOARD

Gypsum wallboard is composed of a fire-resistant gypsum core which is encased on either side with a layer of tough paper—one the finished side, the other the backing side. It is available in thicknesses of $\frac{3}{8}''$, $\frac{1}{2}''$, and $\frac{5}{8}''$; in 4′ widths; and in lengths of 8′, 12′, and 16′. The panels are tapered on the surface face. This makes it simpler to cover the joints between the two panels with tape and joint compound for a smooth, uniform surface.

The gypsum wallboard panels can be positioned either parallel with the joists or perpendicular to the joists, whichever will produce the least number of joints (Figure 11:12).

Each section of wallboard placed on the ceiling must be well nailed. All nails should be driven into the wallboard to form a small dimple (Figure 11:13).

The procedure to seal each joint requires several steps. The joint compound is first applied over the joint area with a trowel, then covered with perforated tape. The tape is pressed firmly into the compound, and smoothed with the trowel before the excess compound is removed. Once dry, two additional separate applications of compound are required, usually one day apart, to achieve a smooth effect. The taped joint area should be sanded with sandpaper prior to the application of more joint compound.

SUSPENDED CEILINGS

A dropped or suspended ceiling may be installed for a variety of reasons: to hide a damaged or uneven ceiling; to modernize the room's appearance and/or to make it more soundproof; to conceal wires, pipes, and joists, especially in a basement area; and to lower the ceiling height to make the room more energy efficient.

The panels for suspended ceilings may be purchased in two sizes, 2′ × 2′ or 2′ × 4′, and an assortment of patterns.

A detailed sketch of the ceiling area must be made on graph paper, with the location of each panel and light

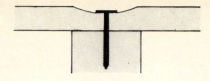

FIGURE 11:13

fixtures clearly marked, before any installation can be started (Figure 11:14). The sketch will also enable you to balance the border so that it will be equal on both sides of the room. Table 11:2 shows how to calculate the number and size of border panels needed for a sample room, using either standard panel size. The accompanying Figure 11:14 shows the ceiling plans for 2′ × 2′ and 2′ × 4′ panels.

The final desired ceiling height must be at least 4″ below any existing object to enable you to maneuver the panels into position. The installation of a dropped or suspended ceiling requires that a series of steps be followed in the correct sequence.

Once the desired ceiling height has been determined, a chalk line is snapped along each wall (Figure 11:15). These lines are used to position the wall moldings. (Figure 11:16). A line level should be employed between the moldings to ensure a level ceiling.

Use the sketch to layout the exact location of the main runners (Figure 11:17). The most effective means of accomplishing this task is to run string lines from one wall to the opposite wall. Where the string lines cross is where the runners must be attached to the ceiling joist with screw eyes and wire, provided they have been properly positioned on the wall moldings (Figure 11:18). Odd-sized or odd-shaped ceilings will require that the border panels and supporting hardware be altered or reduced in size to achieve the desired ceiling symmetry.

Next, the tabs at both ends of the cross tees are inserted, at 2′ intervals, into the main runner slots (Figure 11:19).

It is extremely important to continually check your dimensions to make certain that the required spacing for each panel is correct.

The panels are inserted once the gridwork has been completed (Figure 11:20). Each panel must be tilted upward and slightly on edge to clear the gridwork before it can be lowered into position on the exposed flanges (Figure 11:21).

Attractive ceiling light panels are designed to be easily tilted through the grid system of a suspended ceiling. They are easy to install since they require no measuring, no hole cutting, nor any special fixture support. It conveniently fits into any 2′ × 2′ opening (Figure 11:22). The wire serving the panel must be installed before the gridwork is assembled.

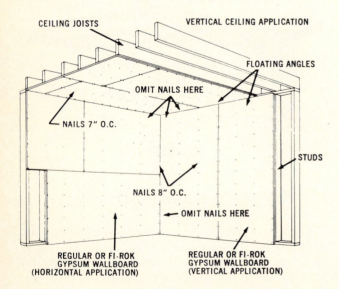

FIGURE 11:12

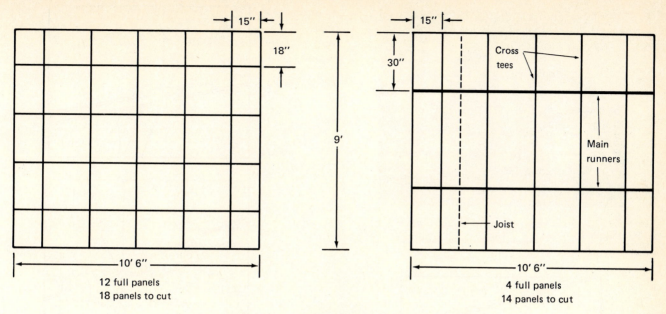

12 full panels
18 panels to cut

4 full panels
14 panels to cut

FIGURE 11:14
(A) 2 × 2 panels, 30 required. (B) 2 × 4 panels, 18 required. (Graph
scale: 1 square = 1 sq. ft.)

TABLE 11:2
Calculating Number of Ceiling Panels for a Room
9′ × 10′ 6″

Directions	2 × 2 Panels	2 × 4 Panels
Short Wall (9′)		
1. Change feet to inches	9′ × 12″ = 108″	Same
2. Divide by panel length	108 ÷ 24″ = 4 lengths, 12″ remainder	108 ÷ 48″ = 2 lengths, 12″ remainder
3. Remainder + length ÷ 2	12 + 24 = 36	12 + 48 = 60″
equals border panel length	36 ÷ 2 = 18″	60 ÷ 2 = 30″
4. Number of panels for one	Two 18″ panels and 3 full panels	Two 30″ panels and 1 full panel

run = 2 boarder panels + full
length panels (one less than
number in (2) above
Note: Above calculations depend on panel length running parallel to short wall.

Long wall		
1. (same as for short wall)	10′6″ × 12″ = 126″	
2.	126 ÷ 24″ = 5 lengths, 6″ remainder	
3.	6″ + 24″ = 30″, 30 ÷ 2 = 15″	
4.	Two 15″ panels and 4 full panels	

Note: Calculations are the same for both sizes; i.e., common 2′ width.

143

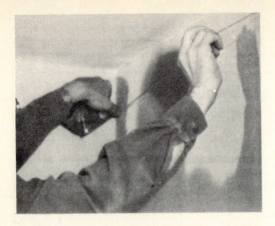

FIGURE 11:15
Snap a connecting chalk line on the fourth wall.

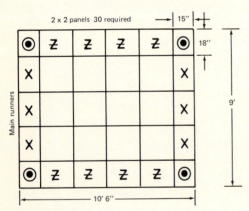

FIGURE 11:16
Nail the wall molding around the perimeter of the room with the top
edge of the molding lined up with the chalk line.

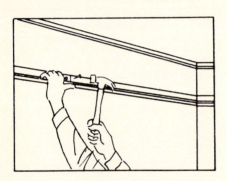

2 x 2 panels 30 required — 15"

18"

Main runners

9'

10' 6"

12 full panels

18 panels to cut

◉ 4 15" x 18"—corner panels

X 6 15" x 24"—place next to shortwall (9')

Z 8 18" x 24"—place next to longwall (10' 6")

FIGURE 11:17
2 x 2 panels, 30 required.
12 full panels
18 panels to cut
4 15" x 18"—corner panels.
6 15" x 24"—place next to short wall (9')
8 18" x 24"—place next to longwall (10'6")

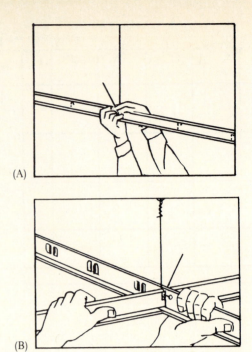

(A)

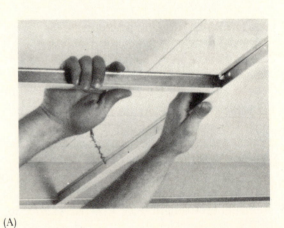

(B)

FIGURE 11:18
(A) Position main tees with hanger wires. (B) Install cross-tees
between the main tees.

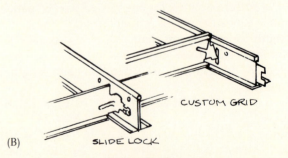

(A)

(B)   SLIDE LOCK   CUSTOM GRID

FIGURE 11:19
(A) Attach 4' cross tees to main runner across the room at 2'
intervals. (B) two different attachment types.

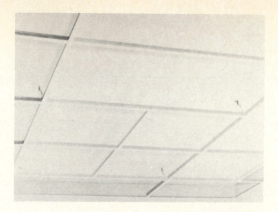

FIGURE 11.20
For 2′ × 2′ panels, attach 2′ cross tees at the midpoints of 4′ cross tees.

FIGURE 11.21

FIGURE 11.22

The border panels must be measured individually and cut with a very sharp utility knife. The knife is used to cut along a straight edge; make certain that the panel's good surface is facing upward (Figure 11:23).

CEILING TILE

To cover any ceiling area, old or new, no building product is as versatile as the ceiling tile. It can be purchased in three stock sizes (12″ × 12″, 24″ × 24″, and

FIGURE 11.23
Cut border panel to size.

24″ × 48″, in a variety of compositions, and in two different thicknesses; it can be installed over furring strips, sheet rock, plaster, or plywood; it can be attached to any smooth ceiling surface by either staples or an adhesive designed specifically for ceiling tile. The tools required are hammer, chalk line, staple gun, and a sharp knife. Selection of the most suitable tile is based on size and appearance, sound absorption, resistance to fire, cost, light reflection, maintenance requirements, and ease of installation.

The number of 12″ × 12″ (1 sq. ft.) tiles required to cover a ceiling area is easily estimated. Use the following formula: Number of tiles = ceiling area (sq.ft.) × 1 + 10%.

A diagram of the ceiling area must be drawn to scale on graph paper before starting the installation. The diagram will help determine the exact location of the furring strips, the size of the border tile, and a good approximation of the number of tiles required (Figure 11:24).

Table 11:3 shows how to calculate the size of border panels using either 12″ × 12″ and 12″ × 24″ tiles for a room 9′ 6″ wide (end wall) by 13′ 4″ long (side wall). The table complements Figure 11:24.

For a rectangular ceiling, locate the midpoint of each wall and snap two chalk lines (perpendicular to each other) on the existing ceiling or joists to mark the center of the room (Figure 11:25). These two lines are essential to guarantee that all the furring strips and all the ceiling tiles run parallel to them.

The first and last furring strips (1 × 3 wooden strips) are nailed with 8d common nails, flush against the wall at right angles to the joists. (Figure 11:26).

It is advisable, during the entire procedure of installing the furring strips, that the nails *not* be driven completely into the furring strips. The exact locations of the second-to-last and next-to-last furring strips depend upon the width of the border tiles; the furring strips are to be nailed so that the stapling edge of these border

TABLE 11:3

Calculating Size of Border Panels for a Room 9′6″ × 13′4″

Directions	12 × 12 Tiles	22 × 24 Tiles
End wall		
1. Add 12″ to the inches portion of the dimension	9′ 6″ +12″ 8′ 18″	Same as for 12″ tiles
2. Divide this measurement by 2 to obtain size of border tiles (side walls)	18 ÷ 2 = 9″	
3. Total number of full tiles plus the 2 border tiles	8 full tiles Two 9″ border tiles	
Side wall Repeat procedure above		
1.	13′ 4″ +12″ 12′ 16″	13′ 4″ − 12′ 0″ = 1′ 4″ = 16″ +24″ 40″
2.	16 ÷ 2 = 8″	40 ÷ 2 = 20″
3.	12 full tiles Two 8″ border tiles	5 full tiles Two 20″ border tiles

Note: When 12 × 24 tiles are used, the length of the tile must run parallel to the furring strips. Also, note how the border tiles are determined for a span involving an odd number of feet.

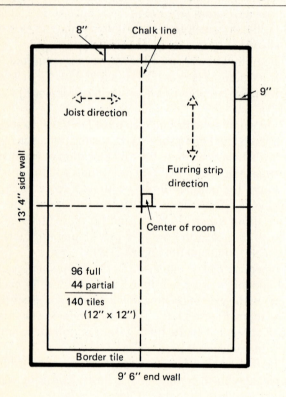

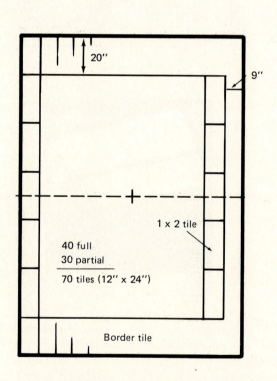

FIGURE 11.24

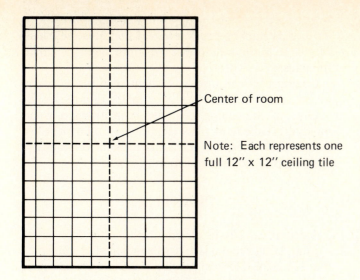

FIGURE 11.25
Use the graph paper above to lay out your room. Note: Each graph
square represents one full 12 × 12 ceiling tile.

tiles will be centered on the strip. The balance of the
furring strips are positioned on 12″ centers for either
12″ × 12″ or 12″ × 24″ tiles. Employing a wooden
spacer strip (Figure 11:27) between the furring strips,
during their installation, will guarantee the desired
spacing and will speed-up this portion of the layout.

Prior to the actual installation of the ceiling tile, the
position of each furring strip should be checked for the
proper distance. The lower surface of each strip must
be checked for level, both along its entire length and in
relation to the other strips. Wooden shims or tapered
wedges may be needed to align some of these strips and
make them level (Figure 11:28). Once all the strips have
been leveled, the nails can be driven all the way into the
furring strips.

The first tile to be installed is a border tile in one corner of the room. (Figure 11:29). Cut the border tile to
size from a full tile by first scoring its surface with a
razor knife, and then positioning it over a sharp edge so
that the excess can be broken off (Figure 11:30). Make

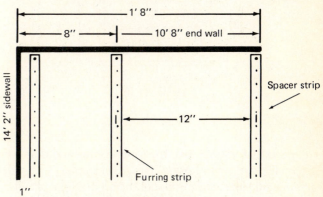

FIGURE 11.27

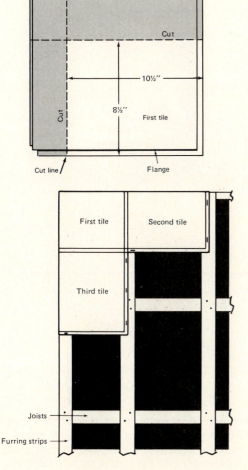

FIGURE 11:26
Furring strips installed perpendicular to joists. Note that first strip is
flush against the wall.

FIGURE 11.28
Your finished ceiling will be only as level as the rows of furring strips.
Use a long straightedge or carpenter's level to check the rows. Correct
any unlevelness in furring strip rows with wood shims.

FIGURE 11:29
Cut the first tile so that the outside edge of the flanges lines up exactly with your chalk lines.

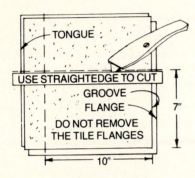

TONGUE

USE STRAIGHTEDGE TO CUT

GROOVE

FLANGE

DO NOT REMOVE THE TILE FLANGES

7"

10"

FIGURE 11.30

certain that you remove the tongue edge and that you keep the side with the wide flange for staples. The wide flange side faces the interior of the room when installed.

This tile is stapled to the furring strip and will require nailing next to the wall. Continue to install border tile in both directions and start to fill in with full size tiles. Make certain that you properly align each tile and keep stapling edges pointed toward the area not yet covered (Figure 11:31).

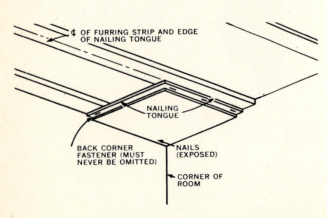

¢ OF FURRING STRIP AND EDGE OF NAILING TONGUE

NAILING TONGUE

BACK CORNER FASTENER (MUST NEVER BE OMITTED)

NAILS (EXPOSED)

CORNER OF ROOM

FIGURE 11.31

The border tile on the opposite wall must have one or both of its stapling edges removed and must be nailed very close to the wall so that the trim molding will hide the nails.

Ceiling tiles can also be cemented in place to any sound, level wallboard or plaster ceiling (Figure 11:32).

Once all the ceiling tiles have been installed, wood molding is nailed into place (Figure 11:33). To make the installation easier, it is advisable to paint or stain the molding prior to its installation.

PLASTER

A ceiling can also be covered with plaster once either gypsum lath or metal lath has been nailed or stapled to the ceiling joist. It is imperative that there are nailers in all the corners of the ceiling and that the joints of the lath be staggered with each course.

The scratch coat, which must be at least $\frac{1}{2}''$ thick, is applied directly to the lath and allowed to dry for at least

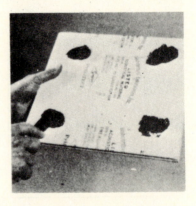

FIGURE 11.32
Cement application for 12 × 12 tiles.

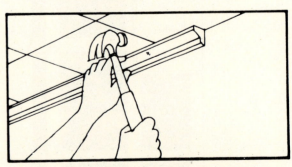

Finish off your ceiling with decorative wall molding. Nail the molding into the wall studs at least every 24". Be sure to avoid nailing into the ceiling tile.

FIGURE 11.33

one day. Then the final or finish coat is applied. This coat can be finished smooth or given a sand texture.

CEILING OPENINGS

Fire regulations specify that provision be made for an entrance into the attic area. This can be accomplished by installing either a small opening in the ceiling of a bedroom closet that is concealed with a plywood cover lid or an attic pulldown stairway, which is usually located in the hallway (Figure 11:34).

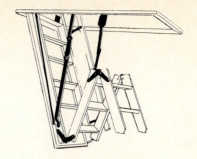

FIGURE 11.34
Wall framing.

QUESTIONS

1. List three (3) different structural methods that can be employed to ensure adequate support for ceiling joists.

2. Explain the procedures used to provide a nailing surface around the entire perimeter of a room's ceiling.

3. How many joists are needed to span the ceiling of a room measuring 16′ × 20′ (assume 16 o.c. spacing)?

4. Estimate the number of 4 × 8 panels, tape, joint compound, and nails needed to cover 4020 square feet of combined wall and ceiling area.

5. List three (3) reasons why a dropped ceiling may be installed.

6. Draw two installation sketches, one using 2 × 2 tiles, the other for 2 × 4 tiles, for a room's ceiling whose dimensions are 12′ × 16′, and calculate the number of tiles needed for each.

7. List three (3) criteria used to select the most appropriate ceiling tile.

8. Installing ceiling tile to furring strips requires that the strips be installed properly. List two (2) items to check before the tiles can be installed.

12

ROOF FRAMING

A roof should be designed to protect the house from the elements, be architecturally attractive, and provide an overhang that shades the interior during the summer but allows the winter sun to shine through and warm the house.

ROOF TYPES

The types of roofs commonly used in residential construction are presented in Figure 12:1. The *flat* roof has only a slight slope and uses the same structural component as both rafter and ceiling joist. The *shed*, or lean-to, roof has one continuous slope and an overhang that is usually greater at the front than at the rear. Its rafters double as ceiling joists.

A modification of the shed roof is the *butterfly* roof which consists of two shed roofs that slope to a low point in the center of the building. Substantial overhangs, both front and rear, are designed to ensure

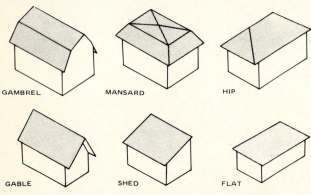

GAMBREL   MANSARD   HIP

GABLE   SHED   FLAT

FIGURE 12:1

both protection and benefit from the sun's rays. This roof type is not used in areas where snowfall is common. The excess weight might be more than the roof could bear.

The roof type most commonly used in residential construction is the *gable* roof. It has two surfaces that slope away from a central ridge and is easy to construct. The roof slope can vary from almost flat to a very steep pitch.

The *hip* roof, popular as the roof type chosen for modern ranch style houses, is constructed with four surfaces slanting from the ridge or peak, giving an illusion that the length of the house is less than it actually is.

To provide additional head room in the second floor, a *gambrel* roof may be used. It has two surfaces, each having a sloping plane. The upper slope forms a 30° angle with the horizontal while the lower slope angle is 60°.

Another form of double-sloped construction that is designed to improve the ceiling height on the second floor or attic is the *mansard* roof. Its upper section has a very slight slope, ranging from 15° to flat, while the lower section approaches the vertical.

To be an effective roof framer, a carpenter must have (1) an understanding of geometry; (2) the ability to read and interpret framing plans (Figure 12:2); (3) the expertise required to fully utilize the information on a rafter or framing table on a carpenter's square (Figure 12:3); and (4) a thorough knowledge of the various roofing terms. In addition, a carpenter must be familiar with the different types of rafters, know where they should be used, and be acquainted with the names and functions of the individual rafter parts.

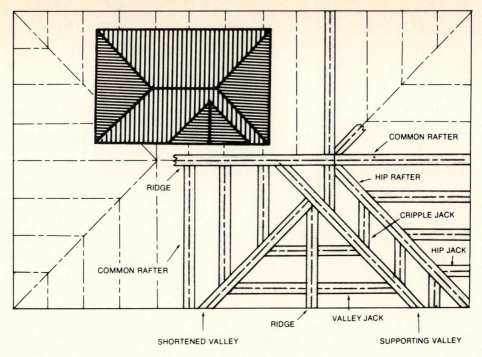

FIGURE 12:2

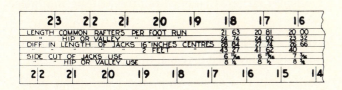

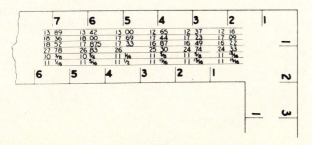

FIGURE 12:3

ROOFING TERMS

The following terms are illustrated in Figure 12:4.

Double Top Plate The support required by most rafter types and one of the principal reference points for many roof term measurements.

Span The horizontal distance between two outside walls.

Run Measurement from the center line of the ridge to the outside of the top plate; equal to one half the total span.

Rise Measurement from the top plate to the center line of the ridge.

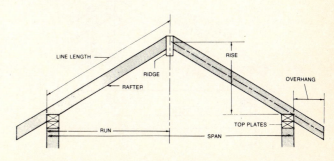

FIGURE 12:4

Measuring Line (Line Length) The diagonal distance from the center line of the ridge to the outside corner of the top plate.

Ridge (Ridge Board) The top edge of the roof where two roof surfaces meet.

Overhang The distance beyond the outside wall or top plate to the end of the rafter.

The following terms are illustrated by Figure 12:5.

Unit of Run The horizontal distance used to describe a roof's slope which is always equal to 12″.

Unit of Rise Expressed as the number of inches a roof rises per foot of run; a value that varies across the total run.

Unit of Span Equal to twice the unit of run, or 24″.

Slope Determined by the formula:

$$\frac{\text{unit of rise}}{\text{unit of run}}$$

It is expressed as, for example, 6″ per 12″ of run, or having a slope of 6 in 12.

The following terms are illustrated by Figure 12.6.

Pitch Determined by the formula:

$$\frac{\text{unit of rise}}{\text{unit of span}}$$

and represents the angle which the roof surface makes with the horizontal.

Comparison of Slope and Pitch A 4 in 12 slope has a $\frac{1}{6}$ pitch, a 12 in 12 slope has a pitch of $\frac{1}{2}$.

RAFTERS

A rafter is the structural framing piece upon which the other roofing materials are placed. A carpenter must be familiar with each type, where it is used, and how it must be cut to be structurally sound and to become an integral part of the entire roof.

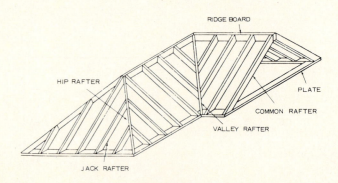

FIGURE 12:6
The rise divided by the span equals the pitch. It is usually expressed as a fraction.

Types of Rafters

Figure 12:7 shows the various rafter types. A description of each type follows:

Common rafter Extends from the plate to the ridge board

Hip rafter Extends from the plate at an outside building corner on a diagonal to the ridge

Valley rafter Extends from the plate at an inside building corner on a diagonal to the ridge, usually doubled to support the weight of roof above

Jack rafter Runs either from the plate to a hip (called a hip jack) or from the ridge to a valley (called a valley jack). They lie in the same plane as common rafters and have the same rise per foot.

Cripple jack rafter Extends between valley or hip rafters, but does not bear on the wall plate

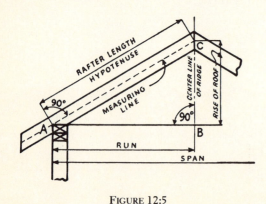

FIGURE 12:5

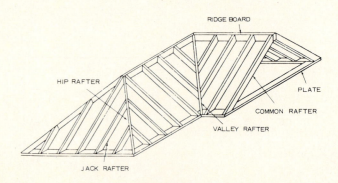

FIGURE 12:7
Rafter types (top view).

Hip jack rafter   Needs a side cut on the top (plumb) cut (Figure 12:8)

Valley jack rafter   Needs a side cut on the bottom (seat) cut (Figure 12:8)

Fly rafter   A rafter that extends beyond the house to provide an overhang (Figure 12:9)

Lookout rafters   Connect common rafters to the fly rafter (Figure 12:9)

Cutting and Sizing Rafters

The cuts required for a common rafter are either vertical (plumb, or heel cuts) or horizontal (seat, or bottom, cuts). The rafter cuts are simple to make, yet must be accurately located to ensure that each rafter is the correct length and is structurally sound.

Figure 12:10, in conjunction with the text that follows, will help you comprehend where and why each cut is made.

A plumb cut is made at the ridge (A), at the tail where it will receive a fascia board (B), at the rear vertical position of the bird's mouth (C), and at the outside corner of the top plate whenever the rafter ends at the plate (D). This particular plumb cut is often referred to as a heel cut.

A seat cut is one that is made where the rafter rests on the top plate as part of the bird's mouth (E). A bottom cut is made at the overhanging end of the rafter as anchorage for the soffit (F).

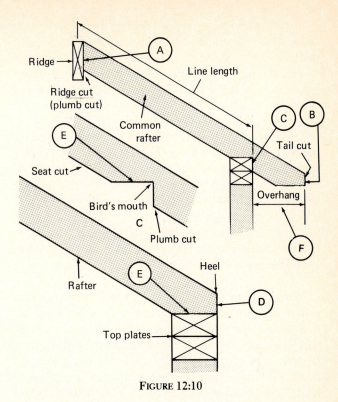

FIGURE 12:10

The bird's mouth is a notch consisting of a horizontal seat cut and a vertical plumb cut. It allows the rafter to fit snugly over the top plate.

The approximate length of a common rafter can be determined with a framing square. The tongue is used to represent the rise while the body of the square simulates the run. The diagonal distance from their respective ends is equivalent to the rafter length without overhang (center of ridge to the outside of the top plate). See Figure 12:11. For the example shown, purchasing 14′ 0″ stock will provide for the desired rafter length plus a 1′ overhang.

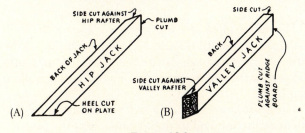

FIGURE 12:8
(A) Hip jack. (B) Valley jack.

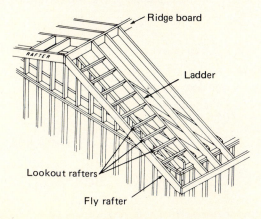

FIGURE 12:9

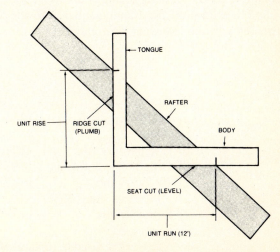

FIGURE 12:11

Most carpenters use the rafter tables on their carpenter's framing square to determine this diagonal distance (rafter length). These tables help the carpenter to quickly find the exact length as well as to locate where cuts must be made for any rafter. It helps avoid the errors inherrent in other methods that may be used to determine rafter length.

The carpenter is aware that the rise per foot of run is always the same for a given pitch. He or she also knows that it is determined by using the following rule: The *rise per foot run* (in inches) is obtained by multiplying the total *rise* (in feet) by 12 and then dividing by the total length of the *run* (in feet).

Table 12:1 shows the rafter length required for a 10′ run for a variety of roof pitches. Figure 12:12 is a representative example for a $\frac{1}{3}$ pitch.

Many carpenters prefer to cut and test their common rafters for suitability as soon as the floor joists on the first floor have been covered. This procedure allows them to check on how well the rafters fit (with respect to desired height, length, bird's mouth, and overhang) while working at ground level, instead of one or more stories above the ground (Figure 12:13). If the two test rafters conform to the plan's specifications, the balance of the rafters can be cut before they are actually needed, using one of the pair as a template (Figure 12:14).

Besides common rafters, a hip roof requires hip, hip jack rafters, and possibly valley and valley jack rafters whenever the roof framing plan specifies a valley.

In most instances, the pitch is the same for all hip roof surfaces. Since the hip rafter joins a common rafter at a 45° angle, its length is related to the diagonal distance for a square whose unit of run is 12″. The unit run for the hip rafter is equivalent to the diagonal, and is thus 17″ (Figure 12:15). To determine the length per foot of run needed for either hip or valley rafters, the second line on the framing square is used.

The other lines on the framing square provide the carpenter with the following information:

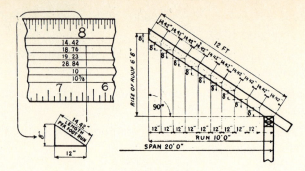

FIGURE 12:12

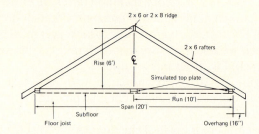

FIGURE 12:13

1. First line: length of common rafter per foot run
2. Second line: length of hip or valley per foot run
3. Third line: difference in length of jacks 16″ centers
4. Fourth line: difference in length of jacks 24″ centers
5. Fifth line: side cut of jack rafter use
6. Sixth line: side cut of hip or valley use

The top end of the hip rafter requires a beveled, V-shaped, miter cut (Figure 12:16) on each side of the rafter's centerline to guarantee that it will fit snugly into the corner formed by the two common rafters (Figure 12:17).

The overhang end of a hip rafter needs the same type of cut as the top end so that it can accept the fascia.

The seat cut of the bird's mouth must be deeper than normal so that the top edge of the hip rafter lies in the

TABLE 12:1

	12 in 12	8 in 12	6 in 12	5 in 12	4 in 12
a) Roof slope	12 in 12	8 in 12	6 in 12	5 in 12	4 in 12
b) Pitch	$\frac{1}{2}$	$\frac{1}{3}$	$\frac{1}{4}$	$\frac{5}{24}$	$\frac{1}{6}$
c) Rise per foot run	12″	8″	6″	5″	4″
*d) Length per foot run	16.97″	14.42″	13.42″	13.00″	12.65″
e) 10′ rafter run (d) times run (10′)	169.7″	144.2″	134.2″	130.0″	126.5″
f) Rafter length (e) ÷ 12	14.14′	12.01′	11.2′	10.83′	10.54′

(*) Found in top line of rafter table under *rise per foot run*

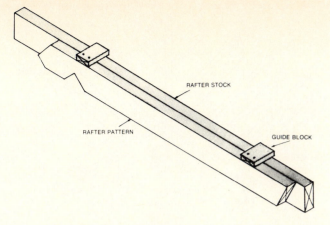

FIGURE 12:14

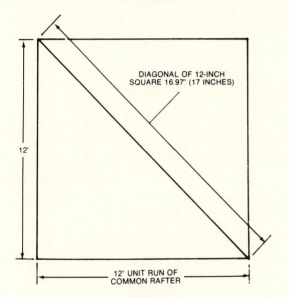

DIAGONAL OF 12-INCH
SQUARE 16.97" (17 INCHES)

12"

12" UNIT RUN OF
COMMON RAFTER

FIGURE 12:15

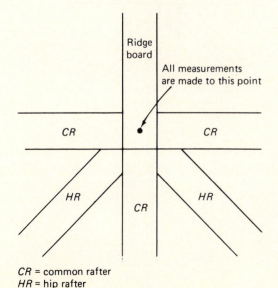

Ridge
board

All measurements
are made to this point

CR   CR

HR   HR

CR

CR = common rafter
HR = hip rafter

FIGURE 12:16

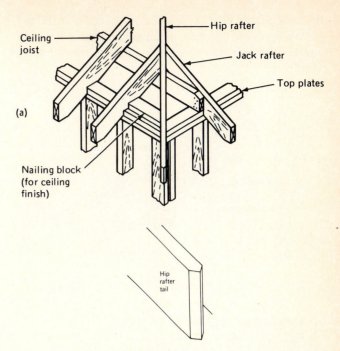

Ceiling
joist

Hip rafter

Jack rafter

Top plates

(a)

Nailing block
(for ceiling
finish)

Hip
rafter
tail

FIGURE 12:17

same plane as the adjoining rafters, making the sheathing procedure run smoothly.

The top edge of each hip jack rafter requires two separate cuts, a plumb cut followed by an angle cut, to enable it to fit flush against the hip rafter (Figure 12:18). When the jack rafters are spaced at equal intervals, the length of each successive one corresponds to a multiple of the number found on either line 3 (16" o.c.) or line 4 (24" o.c.) of the framing square (Figure 12:19).

In most instances, the length of a valley rafter is determined by the same procedure used for a hip rafter. Both ends of a valley rafter, however, require cuts that differ from those given to hip rafters. The top end usually receives a side or cheek cut (Figure 12:20) where it meets the ridge, while the bottom end requires an inverted V-shaped cut to enable it to accept the fascia (Figure 12:21). When both valley rafters meet the ridge, the top end of each one must receive a V-shaped cut similar to that given a hip rafter (Figure 12:22). Note that the top edge of every valley rafter must have its length reduced by one half the 45° thickness of the ridge board.

Valley rafters are often doubled to ensure adequate support and to provide a more substantial bearing surface for the connecting valley jack rafters (Figure 12:23).

The gambrel roof is a modified gable roof constructed with two different lengths of rafters. The upper ones are reasonably flat, are shorter in length than the lower ones, and span from the ridge to the purlin. An upper rafter runs across the length of the roof and is designed to help transmit the weight of the upper roof section to the house's side walls. The lower rafters connect the top plate with the purlin and are longer and form a steeper angle with the horizontal than the upper rafters (Figure 12:24).

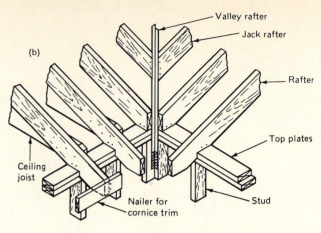

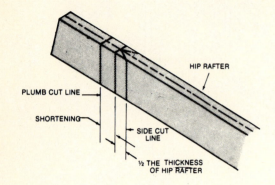

FIGURE 12:18

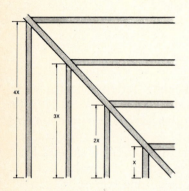

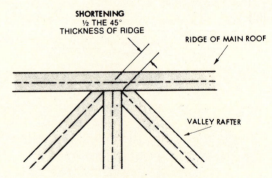

FIGURE 12:21

FIGURE 12:19

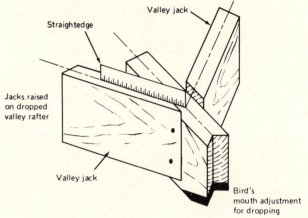

FIGURE 12:22

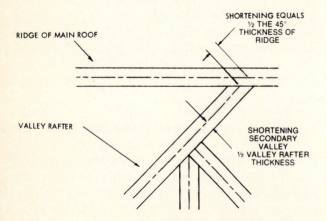

FIGURE 12:20
Valley meeting ridge and valleys meeting each other.

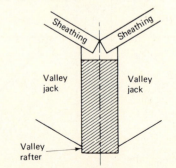

FIGURE 12:23
Valley jack rafters attached to doubled valley rafter.

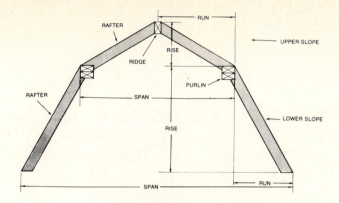

FIGURE 12:24

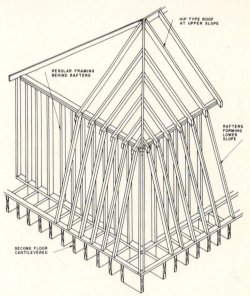

FIGURE 12:25

The lower rafters of a mansard roof are very steep. They extend from a plate placed on top of cantilevered joists to the base of the rafters forming the upper slope, thus completely covering the wall of the upper living level. The upper portion of this roof type is a hip roof that has a slight slope (Figure 12:25).

DORMERS

A dormer will create more living space as well as head room in the second floor of a home with a gable style roof. It is much easier to frame and enclose a dormer while the house is under construction than it is to do it after the roof has been completed.

Since all dormers are, basically, openings in the roof's surface, they must be framed accordingly. The correct procedure is to double each of the side supporting rafters and both the headers, which are located above and below the opening. This enables them to more adequately carry the additional structural weight (Figure 12:26).

Dormers have two basic designs: gable (Figure 12:27) and shed. The roof framing of a shed dormer can start either at the ridge (Figure 12:28) or at a point below the ridge (Figure 12:29) and can extend across a portion of the roof or across its entire length.

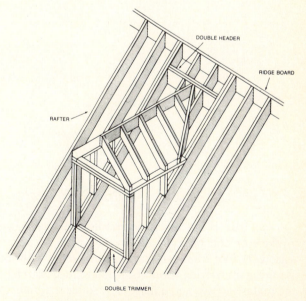

FIGURE 12:26

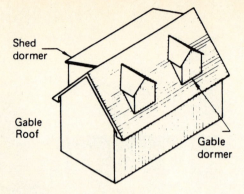

FIGURE 12:27

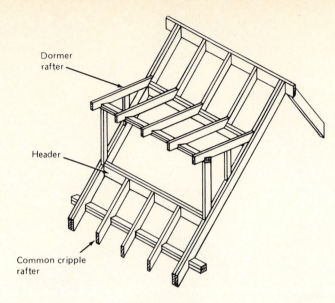

FIGURE 12:29

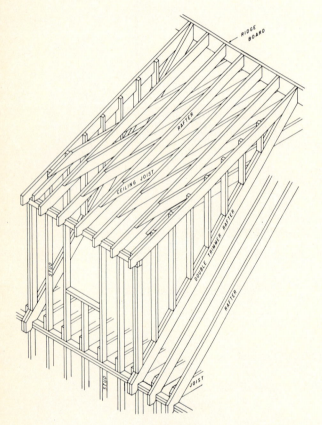

FIGURE 12:28

ROOF TRUSSES

A roof truss is a preassembled frame that may be used in place of conventional rafters and joists for several reasons. A roof truss can span a greater length without requiring underlying supports; it can reduce both labor and material costs; and it makes it possible to frame and sheath the roof area very quickly.

There are three basic types of roof trusses used in residential construction. They are the W-type, the king-post type, and the scissors type (Figure 12:30).

Selecting the best truss to use depends on many factors: the live and dead roof loads, the span to be covered, the weight of the roof, and the slope of the roof (flat roofs have greater stresses than those with steep

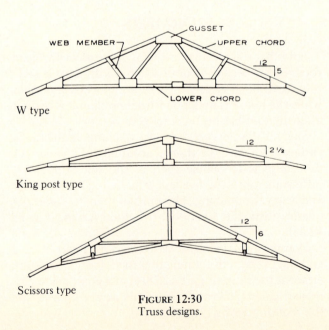

FIGURE 12:30
Truss designs.

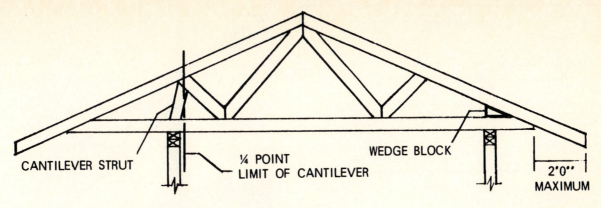

FIGURE 12:31
Cantilever conditions.

pitches). All of these factors, plus the size of the components used to construct the truss, must conform to the local building codes.

When ordering trusses, a builder must provide the manufacturer with specific information related to the following items:

1. The quantity required (Most trusses are placed 2′ 0″ on center.)
2. The distance the trusses will span
3. The truss type
4. The roof slope (e.g., 4 in 12 or $\frac{1}{6}$ pitch)
5. The length which the top chord will extend or overhang the vertical wall and the cantilever distance, if desired (Figure 12:31)
6. The end cut of the top chord; a plumb cut is standard (Figure 12:32)
7. The type of soffit return (Figure 12:33)
8. The type of gable end, the spacing of the studs, and the placement of any openings (Figure 12:34)

The trusses are delivered to the building site on a flatbed truck (Figure 12:35). They can then be erected either by hand or by a crane. The erection method chosen is usually dictated by the span, the wall height, and the contour of the terrain surrounding the building.

When erected by hand, each truss is lifted over the side wall, slid along the top plate, and finally rotated into its upright position with the help of a fork-like lifting pole (Figure 12:36).

Trusses must be securely braced, both during the erection procedure and after they have been completely installed; if one falls, the rest will follow, in domino fashion. The first, the gable-end truss, should be braced from both inside and outside the building

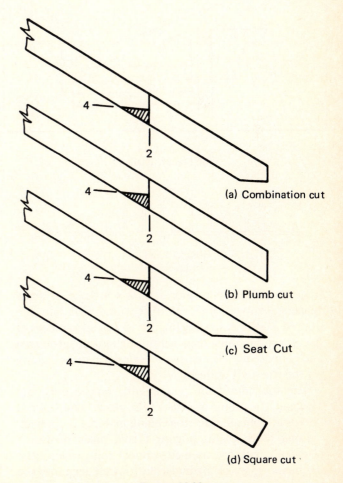

(a) Combination cut

(b) Plumb cut

(c) Seat Cut

(d) Square cut

FIGURE 12:32

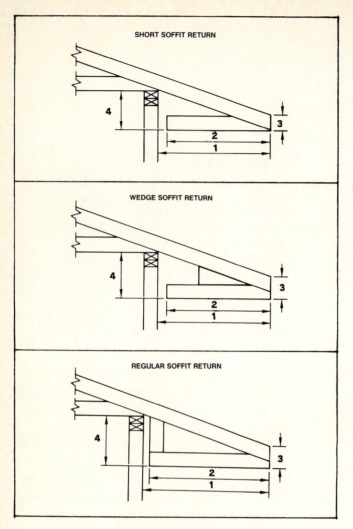

FIGURE 12:33
Soffit return details.

SHORT SOFFIT RETURN

WEDGE SOFFIT RETURN

REGULAR SOFFIT RETURN

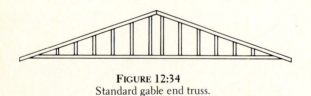

FIGURE 12:34
Standard gable end truss.

FIGURE 12:35

FIGURE 12:36

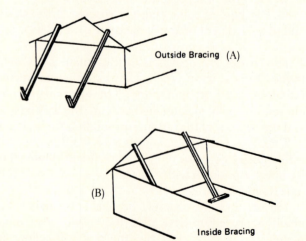

Outside Bracing (A)

(B)

Inside Bracing

FIGURE 12:37
(A) Inside bracing. (B) Outside bracing.

(Figure 12:37). Short pieces of 2 × 4, used as truss spacers, connect adjacent trusses and maintain the desired 2' 0" o.c. spacing (Figure 12:38). Long pieces of 2 × 4, running on a diagonal throughout the attic area and connecting several trusses, are used as cross bracing for the trusses (Figure 12:39).

The top chords are permanently braced once the plywood sheathing is nailed into place. The bottom chords are braced once the ceiling material is applied.

Long tooth plates (Figure 12:40), made of 16- to 20-gauge steel, are used to connect the various truss members. They are hydraulically pressed deep into the lumber.

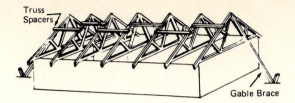

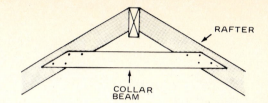

FIGURE 12:38

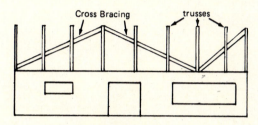

FIGURE 12:39

FIGURE 12:41

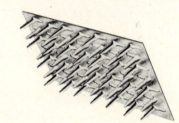

FIGURE 12:40

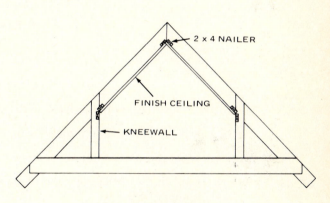

FIGURE 12:42

ADDITIONAL ROOF FRAMING

Collar beams (Figure 12:41) are horizontal framing members that tie certain rafter pairs together. The distance between adjacent collar beams is usually 4′ 0″. They give the roof structure more rigidity.

Knee walls (Figure 12:42) are short vertical stud walls that connect the rafters to the attic or second floor. They may be installed to provide additional roof support, to define the perimeter of an attic room, and/or to create storage space behind them.

FRAMING GABLE ENDS

Notched studs placed between the top plate and the end rafters are used to frame the openings at the ends of gable and gambrel roofs.

Each end stud receives a notched angle cut that conforms to the slope of the rafter. The notch should be as wide as the thickness of the rafter so that the rafter is properly supported on the lip of the notch. The stud should be positioned snugly against the inside of the end rafter so that its edge is flush with the rafter's outer side, and its angled butt end is flush with the rafter's upper edge. (Figure 12:43). The difference in length between any two adjacent end studs will be the same throughout the stud wall provided the spacing between all the studs is equal.

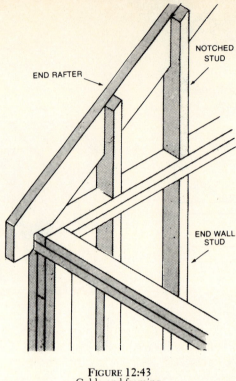

FIGURE 12:43
Gable end framing.

QUESTIONS

1. Compare four (4) roof types with respect to overall design and slope of roof.

2. What is the pitch of a roof having a slope of 6 in 12?

3. What is the basic difference between a hip and a valley rafter?

4. Explain the structural reason why fly and lookout rafters are used on some homes.

5. Draw a common rafter and label each of its typical cuts as being either plumb or seat cuts.

6. Explain why the unit run for a hip rafter is more than it is for a common rafter.

7. What is the primary reason for building either a mansard or a gambrel roof?

8. Explain the framing procedures used to construct any opening in the roof's surface.

9. List three (3) reasons why roof trusses are used instead of conventional rafters and ceiling joists.

10. What is a roof's pitch having a rise of 4′ and a span of 12′?

11. Explain the term *measure line* for a common rafter.

12. Name three (3) types of jack rafters and tell where they are used.

13. Explain where collar beams are installed and their purpose.

14. How are the terms *pitch* and *slope* related to each other?

15. Explain how rafter pairs are nailed to the ridge board.

13

COVERING THE
ROOF FRAME

Applying the finished roof covering to the roof frame is the final operation of a series that must be performed before the roof is completed. The other operations include:

1. Attaching the fascia to the rafter ends. The cornice rake and soffit may also be completed at this time. (Figure 13:1)
2. Completing all the projections that pass through the roof (vent pipes, chimneys)
3. Covering the roof frame with sheathing (Figure 13:2)
4. Installing flashing to all roof joints, around all projections, and where the roof meets a vertical wall
5. Applying the underlayment and drip edge
6. Connecting the gutters and leaders

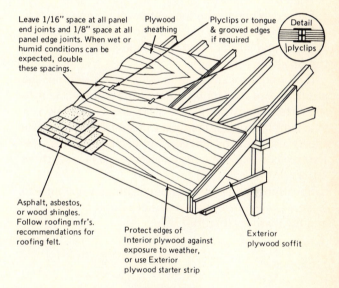

Leave 1/16" space at all panel end joints and 1/8" space at all panel edge joints. When wet or humid conditions can be expected, double these spacings.

Plywood sheathing

Plyclips or tongue & grooved edges if required

Detail plyclips

Asphalt, asbestos, or wood shingles. Follow roofing mfr's. recommendations for roofing felt.

Protect edges of Interior plywood against exposure to weather, or use Exterior plywood starter strip

Exterior plywood soffit

FIGURE 13:2
The roof is built after the exterior walls are sheathed.

It is extremely important for the carpenter to realize the dangers inherent in working on a roof and to take the proper steps to avoid these dangers. Wearing rubber-soled footwear, using a scaffold system instead of ladders to convey supplies to the roof, as well as using roof brackets (Figure 13:3) to stand on, are all measures designed to make the work safer.

SHEATHING THE ROOF

Once the fascia board has been attached to the rafter ends, and all projections through the roof have been completed, the roof frame can be covered with sheathing.

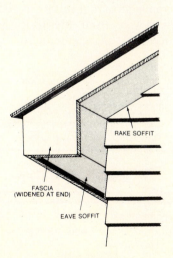

RAKE SOFFIT

FASCIA (WIDENED AT END)

EAVE SOFFIT

FIGURE 13:1
A typical horizontal cornice seals the area below the rafter overhang.

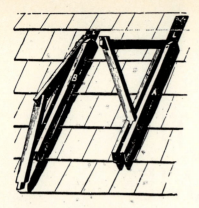

FIGURE 13:3
Roof brackets. (Courtesy of American Plywood Association)

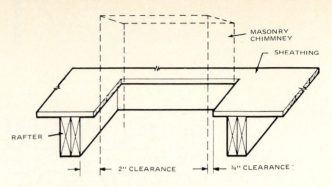

FIGURE 13:5

Roof sheathing materials include plywood sheets, waferboard, solid wood (installed either closed or spaced), plank, or fiberboard decking. The type selected is related to the style of the roof and its slope, local climatic conditions, the top roof covering, and the cost.

Roof sheathing has a dual purpose: to make the roof stronger and sturdier and to provide a nailing base for the finished roof material.

If plywood or waferboard is used, each sheet must be positioned so that its face grain runs perpendicular to the rafters or trusses (Figure 13:4), and so that any vertical joint falls on the centerline of a rafter or truss. The sheets also must be staggered in each successive row. Every piece must be large enough to span across three rafters or trusses.

Sheathing should be cut around chimney openings to leave a ¾" clearance gap on all sides between the wood and the masonry (Figure 13:5).

Heavy tongue-and-groove boards or special fiberboard panels are often used for roofs with inclines ranging from a very slight slope to completely flat, and for those in post-and-beam construction. Because boards and panels serve three functions, they are referred to as decking rather than sheathing. Decking provides a surface on which to attach styrofoam insulation (positioned on top or beneath the decking; Figure 13:6), offers a finished surface, and serves as the covering for the roof's frame.

The types of solid boards that can be used for roof sheathing are common (having square edges and ends), shiplap, and tongue-and-groove, which can be edge-matched, end-matched, or both (Figure 13:7). Boards are nailed at right angles to the rafters or trusses. Care must be exercised to stagger the joints in each successive row (Figure 13:8). End-matched boards allow the possibility of joining boards between rafters or trusses, whereas all other types require that joints be made directly over rafters or trusses.

FLASHING

Flashing is applied to a roof surface as additional protection to prevent water from entering the living area. It is applied around the perimeter of roof projections,

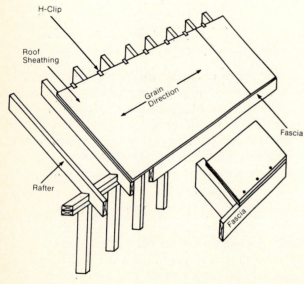

FIGURE 13:4

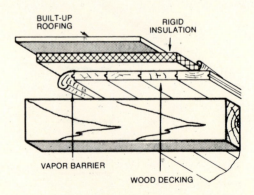

FIGURE 13:6
Detail showing application of vapor barrier and insulation on a plank decking located directly over heated space.

ROOF DECKS

Tongued and grooved
phenolic-bonded plywood

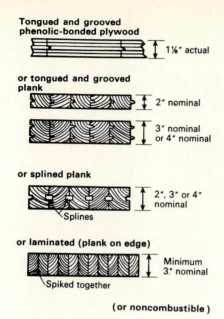

1⅛″ actual

or tongued and grooved
plank

2″ nominal

3″ nominal
or 4″ nominal

or splined plank

2″, 3″ or 4″
nominal

Splines

or laminated (plank on edge)

Minimum
3″ nominal

Spiked together

(or noncombustible)

FIGURE 13:7

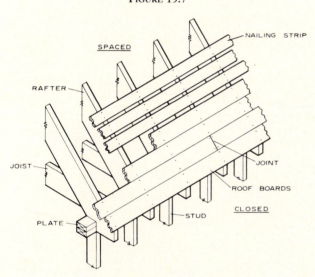

SPACED

NAILING STRIP

RAFTER

JOIST

JOINT

ROOF BOARDS

CLOSED

PLATE

STUD

FIGURE 13:8

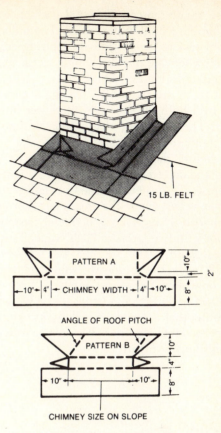

15 LB. FELT

PATTERN A

10″

2″

10″ | 4″ | CHIMNEY WIDTH | 4″ | 10″

8″

ANGLE OF ROOF PITCH

PATTERN B

10″

4″

10″ | 10″

8″

CHIMNEY SIZE ON SLOPE

FIGURE 13:9

Metal flashing, known as counterflashing, may be either one long continuous piece or a series of short pieces (Figure 13:10). Either type must be set 1½″ deep into mortar joints, which have had their mortar removed, before they can be re-cemented in place. The finished roof surface is then cemented, with plastic cement, to the roll flashing.

where two roof surfaces intersect (e.g., a valley), where the roof and a vertical wall intersect, and in the eaves area whenever there is a possibility of winter ice formation. The materials used to flash these areas can be a metal (e.g., galvanized or stainless steel, aluminum, or copper), a heavy mineral-surfaced roll roofing, or plastic.

Two types of flashing are used around a chimney. The finished roof surface must be brought up to the lower face of the chimney. Then, the 90-lb. mineral-surfaced roll roofing is cut to size and embedded in a thick layer of plastic roof cement that is applied to both the chimney and to the finished roof surface (Figure 13:9).

FIGURE 13:10

Some builders prefer to install prefabricated chimneys that have built-in insulation. The individual sections are easy to handle and assemble. Special fittings are used when the prefab chimney is carried through walls and roofs (Figure 13:11).

Vent pipes are flashed by placing a metal flashing sleeve over both the pipe and the lower finished roof surface. The top of the sleeve is nailed into position; then, plastic roof cement is applied to the base of the sleeve before the finished roof surface is cut to fit around and on top of the metal sleeve (Figure 13:12).

There are two methods used to flash a roof valley. The *open valley*, where the area is left exposed, can be flashed with a single layer of metal flashing (Figure 13:13) or with two layers of 90-lb. mineral-surfaced roll roofing (Figure 13:14). Note that the first layer of roofing is 18″ wide, the second 36″ wide, and that both are nailed only along their exterior edges.

A *closed valley* employs one layer of 36″ wide, 55-lb. roll roofing that is nailed only along its exterior edges. This method relies on the valley being covered by overlapping shingles (Figure 13:15).

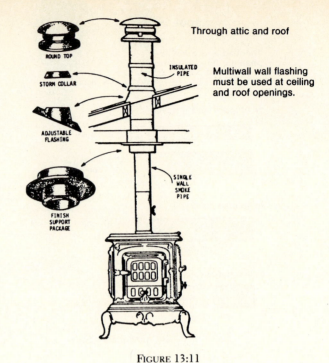

FIGURE 13:11

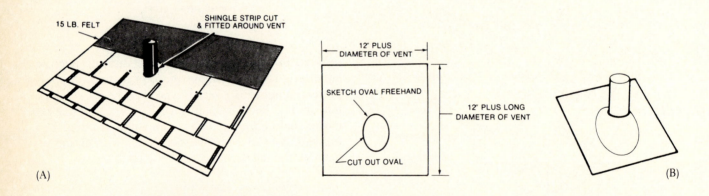

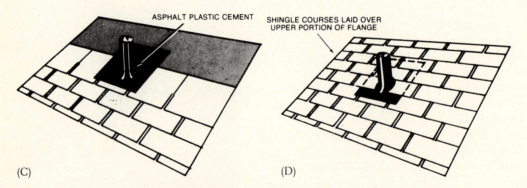

FIGURE 13:12
(A) Lay shingles to the pipe. Cut one to fit around the pipe. (B) Slide a metal flashing sleeve over the pipe. (C) Apply plastic roof cement to the sleeve making a watertight seal around the pipe. (D) Lay shingles in normal manner cutting to fit around pipe.

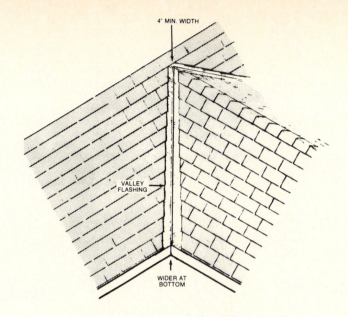

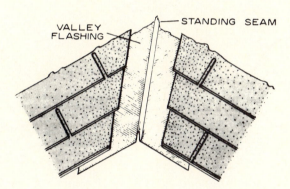

FIGURE 13:13

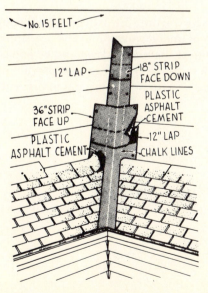

FIGURE 13:14

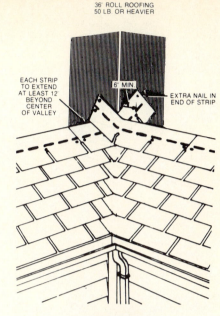

(A)

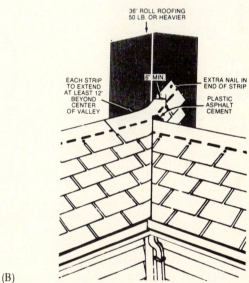

(B)

FIGURE 13:15
(A) Woven valley flashing. (B) Closed valley flashing.

Metal flashing—either stepped or one continuous piece—is used to flash the intersection of the roof and a vertical wall. First, the underlayment is placed over the sheathing and continued up the vertical wall at least 4″. A continuous piece of flashing can then be applied directly over the underlayment, or a stepped metal flashing can be applied over the end of each course of shingles (Figure 13:16). Once the flashing is positioned and covered with the final roof surface, the siding can be applied to the vertical wall. The siding should cover a portion of the flashing but leave at least the bottom inch exposed.

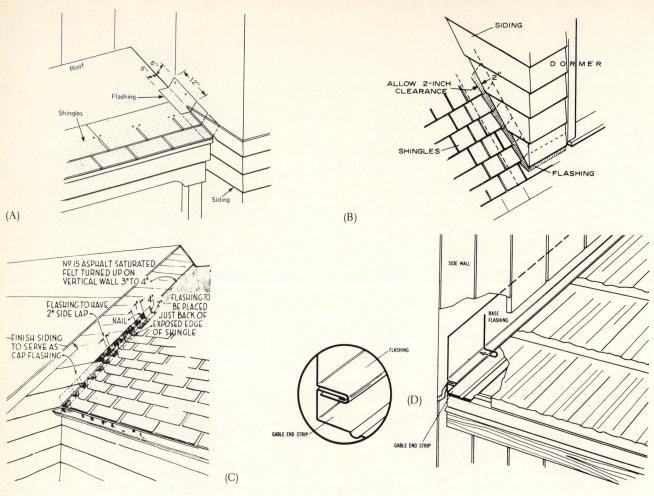

FIGURE 13:16
Methods used to flash junction between roof and vertical wall. (A)
Step flashing at intersecting surface. (B) Step flashing at dormer walls.
(C) Flashing extends 4″ up wall, 2″ onto roof. (D) Gable end strip
flashing.

Some form of roll roofing is used to flash the eaves. It is applied over the underlayment, and, depending upon which type of finished roof material is used, either nailed or cemented into place.

In areas where the roof surface rests against a vertical wall, metal flashing must be nailed to the wall's sheathing so that it will be beneath the final exterior covering. Its other edge must be nailed on top of the finished roof surface (Figure 13:17). Placing plastic roof cement on the finished roof surface, before the flashing is nailed to it, should ensure a water-tight bond.

UNDERLAYMENT AND DRIP EDGE

The type of finished roof surface dictates not only what type of underlayment material should be used to cover the sheathing, but also how it should be applied. At the eaves, the underlayment is installed over the drip edge, while at the rake of the roof, it is placed beneath

the drip edge (Figure 13:18). In general, the top edge of each layer of underlayment must be overlapped at least 2″ by the next succeeding layer and be end-lapped 4″. The purpose of the drip edge is to provide protection for the edges of the sheathing and to prevent leaks.

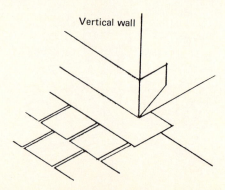

FIGURE 13:17
Flashing first nailed to wall sheathing before being covered with exterior covering.

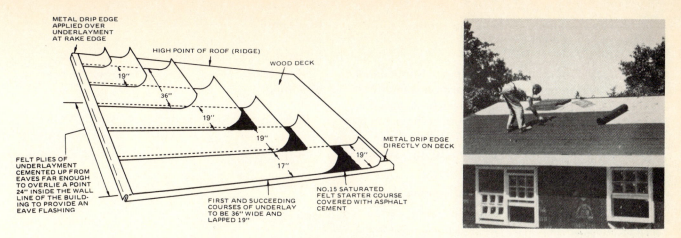

FIGURE 13:18
Underlayment is lapped on the edges and ends. It is placed below the rake drip edge.

ROOF DRAINAGE SYSTEM

Galvanized iron or aluminum gutters collect rainwater from the roof edges and transmit it downward through metal leaders, which direct it away from the foundation. The entire drainage system can be installed before or after the finished roof surface has been applied. The components of a roof drainage system are presented in Figure 13:19.

FINISHED ROOF COVERINGS

Selection of the finished roof covering is based on several factors. They include

1. Cost—both the initial and maintenance costs
2. Durability—the expected length of service
3. Local building code requirements—related to weather factors, roof pitch, and house design

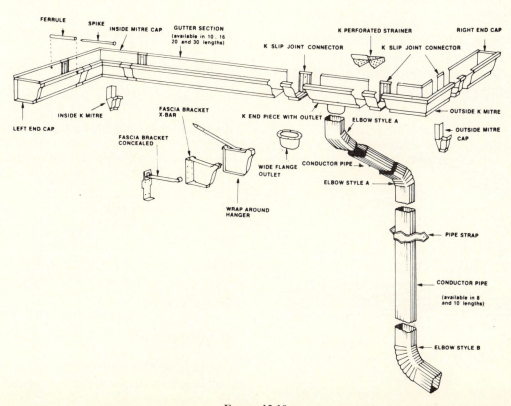

FIGURE 13:19
Components of a roof drainage system.

4. Personal preference—how the finished surface appears when completed.

Each of the four most common types of finished roof coverings used in residential construction is discussed in this section. Special attention is given to the specifics governing how the roof surface must be prepared as well as how the covering should be installed.

The following terms are related to the ensuing presentations:

Starter course An initial layer of roofing material placed at the eaves

Expansion gap The space left between wood shingles or shakes to allow for their expansion

Exposure The distance in inches between the edges of two adjacent courses of finished material

Coverage A term used to imply the amount of weather protection achieved by overlapping the finished roof material. It can range from one thickness (single coverage) to three (triple coverage). See Figure 13:20.

Square The amount of finished roof material required to cover 100 sq. ft. of roof area

Roll Roofing

Materials available

1. SIS—55 lb. or 70 lb., double coverage (51 sq. ft.), 36″ wide, 17″ exposure (Figure 13:21)
2. No. 90—90 lb., single coverage (100 sq. ft.), 36″ wide, 34″ exposure

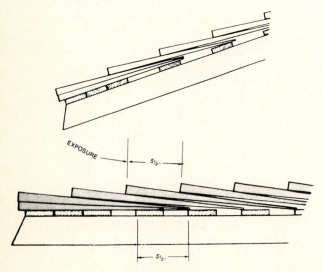

Figure 13:20
(A) Double coverage. (B) Triple coverage. The spacing of open sheathing should be the same as the exposure.

Roof criteria

Flat roofs or those having a very slight slope

Sheathing

Plywood or closed boards

Underlayment

15-lb. felt

Drip edge required

Installation

1. Overhang the drip edge $\frac{3}{4}$″.
2. Lap sides and ends 6″ and nail along entire edge.
3. Roofing is nailed to sheathing with roofing nails, in areas lacking the mineral surface.
4. Plastic roof cement is applied to the nonmineral surface of each layer so that the lower edge of the next course will adhere to it and become waterproof.

Asphalt Shingles

Materials available

Minimum weight of 235 lb. per square

1. Two- or three-tab hexagonal shingles (Figure 13:22)
2. Three-tab square butt shingles (Figure 13:22)

Roof criteria

Roofs with slopes ranging from 2 in 12 to steeper than 4 in 12

Sheathing

Plywood or closed boards

Underlayment

15-lb. felt

Drip edge required

Eave flashing

One layer of 90-lb. mineral-surfaced or 55-lb. smooth roll roofing is used; placed over the underlayment; extends beyond drip edge $\frac{3}{8}$″ to $\frac{1}{2}$″.

1	2		3	4		5		6	7
PRODUCT	Approximate Shipping Weight		Sqs. Per Package	Length	Width	Side or End Lap	Top Lap	Exposure	Underwriters' Listing
	Per Roll	Per Sq.							
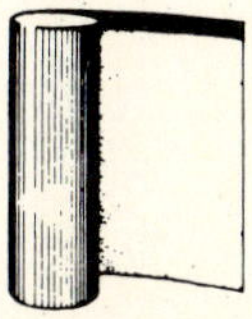 Mineral Surface Roll	75# to 90#	75# to 90#	One	36' 38'	36" 36"	6"	2" 4"	34" 32"	C
			Available in some areas in 9/10 or 3/4 Square rolls.						
 Mineral Surface Roll Double Coverage	55# to 70#	55# to 70#	One Half	36'	36"	6"	19"	17"	C
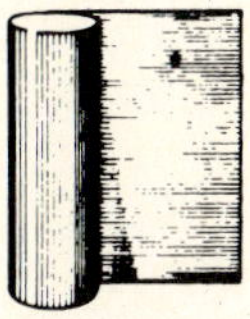 Coated Roll	50# to 65#	50# to 65#	One	36'	36"	6"	2"	34"	None
 Saturated Felt	60# 60# 60#	15# 20# 30#	4 3 2	144' 108' 72'	36" 36" 36"	4" to 6"	2"	34"	None

Asphalt Roofing Manufacturers Association

FIGURE 13:21
(A) Mineral surface roll. (B) Mineral surface roll, double coverage.

1. On 4 in 12, or steeper, it is nailed into place and must extend at least 12" above the inside wall (Figure 13:23).

2. For large overhanging roofs, it is cemented into place, and any joint made must fall below the interior wall.

3. For roofs that are 2 in 12 to 4 in 12, it is cemented into place, and must extend at least 24" above the inside wall (Figure 13:24).

Installation

1. Starter course—row of shingles nailed with tabs pointing up.

2. First regular course—start with a full shingle at the rake and overhang the drip edge by $\frac{3}{4}$" (Figure 13:25).

3. Second regular course—start with a shingle that has one half of one of its tabs removed.

4. Balance of course—alternate directions given for first and second courses.

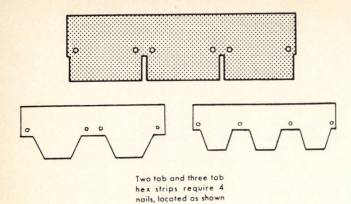

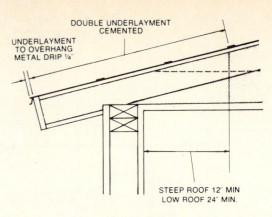

Two tab and three tab
hex strips require 4
nails, located as shown
here.

FIGURE 13:22
Nailing patterns for asphalt shingles.

FIGURE 13:23

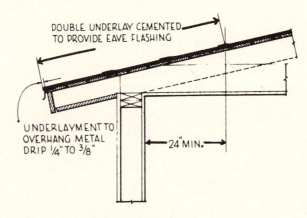

FIGURE 13:24

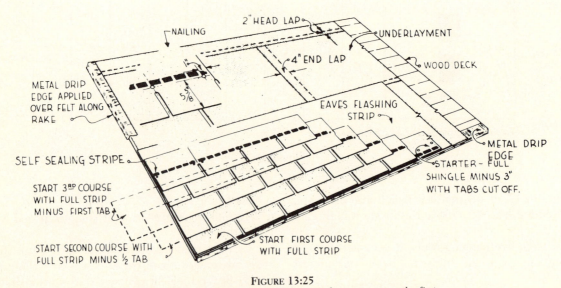

FIGURE 13:25
(A) Lay first course with tabs up. (B) Lay another course over the first
course with the tabs down. Start with a full shingle. (C) Start second
course with a shingle with half a tab cut off. (D) Start the third course
with a full shingle.

5. Use $1\frac{1}{2}$″ galvanized roofing nails ($2\frac{1}{2}$ per square).

6. Exposure is 5″, with double coverage.

7. To cap the ridge or hip, cut shingles into thirds, bend them lengthwise, and nail each exterior edge with one nail; nails hidden by next overlapping piece; 5″ to 6″ exposure (Figure 13:26)

Wood Shingles

Materials available

Number 1 grade red cedar shingles—16″, 18″, or 24″ in length

Roof criteria

Roofs having a slope of 3 in 12, or steeper

Sheathing

Spaced wood (Figure 13:27); or can be solid wood

Underlayment

Not necessary, but may be installed anyway (15-lb. felt)

Drip edge not required

Installation

1. First course—doubled or tripled, and should extend $1\frac{1}{2}$″ beyond the sheathing edge at the eaves and rake.

2. Leave $\frac{1}{4}$″ expansion gap between shingle edges.

3. Joints between rows must be staggered.

4. Use two nails (galvanized steel or aluminim) per shingle; $\frac{3}{4}$″ in from each edge, and $1\frac{1}{2}$″ above the butt line of the next course (Figure 13:27).

5. Shingle exposure decreases with a decrease in roof slope (Table 13:1).

6. The ridge is covered with overlapping shingles (Figure 13:28).

7. Four bundles of 16″ shingles with a 5″ exposure will cover one square (100 sq. ft.).

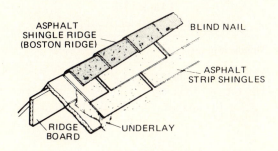

FIGURE 13:26

TABLE 13:1

Shingle length (in.)	Single exposure (in.)		
	5 in 12 slope	4 in 12 slope	3 in 12 slope
16	5	$4\frac{1}{2}$	$3\frac{3}{4}$
18	$5\frac{1}{2}$	5	$4\frac{1}{4}$
24	$7\frac{1}{2}$	$6\frac{3}{4}$	$5\frac{3}{4}$

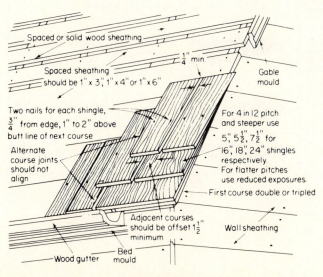

FIGURE 13:27

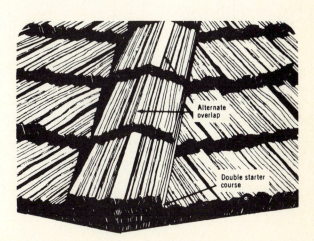

FIGURE 13:28

Wood Shakes

Materials available

Number 1 grade red cedar in three types: straight-split or barn shingles (18″ or 24″ long), hand split and resawn (18″, 24″, or 32″ long), and taper-split (24″ long)—Figure 13:29

Roof criteria

Roofs having a slope of 3 in 12, or steeper

Sheathing

Spaced wood where the on-center distance must be the same as the shingle exposure; solid sheathing for 3 in 12 roofs or those exposed to wind-driven snow and 0° F temperatures.

Underlayment

An 18″ strip of 30-lb. asphalt-saturated felt is used to cover the top of each layer of shakes and the sheathing above it. The bottom edge of this strip is positioned two exposure distances above the shingle butt. Example: 24″ shakes with $7\frac{1}{2}$″ exposure would have the bottom edge of the felt 15″ above the butts, or the top 9″ of each shake would be covered.

Drip edge not required

Eave flashing

Copper flashing must not be used with wood shakes.

1. For 3 in 12 to 4 in 12 slopes, two layers of 30-lb. felt are cemented together with plastic cement.
2. For 4 in 12 slopes, or steeper, nail the two layers of felt together and make certain that they extend at least 24″ beyond the inside wall.

Installation

1. Place a 36″-wide, 30-lb. asphalt felt starting strip at the eaves; use the same material at the ridge and hip.
2. Double the starter shingle course and extend it $1\frac{1}{2}$″ beyond the sheathing edge.
3. Leave a $\frac{1}{4}$″ to $\frac{3}{8}$″ expansion gap between shakes.
4. Stagger the joints between the rows.
5. Use threaded galvanized steel or aluminum nails; two per shake.
6. Coverage can be double or triple (Table 13:2).

TABLE 13:2
Exposures for Wooden Roof Shakes

Double exposure	Triple exposure
13″ for 32″ shakes	10″ for 32″ shakes
10″ for 24″ shakes	$7\frac{1}{2}$″ for 24″ shakes
$7\frac{1}{2}$″ for 18″ shakes	$5\frac{1}{2}$″ for 18″ shakes

Grade	Length and Thickness	20″ Pack		18″ Pack		Shipping Weight	Description
		# Courses Per Bdl.	# Bdls. Per Sq.	# Courses Per Bdl.	# Bdls. Per Sq.		
No. 1 HANDSPLIT & RESAWN	15″ Starter-Finish	8/8 10/10	5 4	9/9	5	225 lbs.	 These shakes have split faces and sawn backs. Cedar logs are first cut into desired lengths. Blanks or boards of proper thickness are split and then run diagonally through a bandsaw to produce two tapered shakes from each blank.
	18″ x ½″ to ¾″	10/10	4	9/9	5	220 lbs.	
	18″ x ¾″ to 1¼″	8/8	5	9/9	5	250 lbs.	
	24″ x ⅜″	10/10	4	9/9	5	225 lbs.	
	24″ x ½″ to ¾″	10/10	4	9/9	5	280 lbs.	
	24″ x ¾″ to 1¼″	8/8	5	9/9	5	350 lbs.	
	32″ x ¾″ to 1¼″	6/7	6			450 lbs.	
No. 1 TAPERSPLIT	24″ x ½″ to ⅝″	10/10	4	9/9	5	260 lbs.	 Produced largely by hand, using a sharp-bladed steel froe and a wooden mallet. The natural shingle-like taper is achieved by reversing the block, end-for-end, with each split.
No. 1 STRAIGHT-SPLIT	18″ x ⅜″ True-Edge*	14 Straight	4			120 lbs.	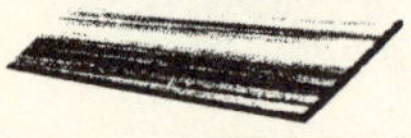 Produced in the same manner as tapersplit shakes except that by splitting from the same end of the block, the shakes acquire the same thickness throughout.
	18″ x ⅜″	19 Straight	5			200 lbs.	
	24″ x ⅜″	16 Straight	5			260 lbs.	

NOTE: * Exclusively sidewall product, with parallel edges.

Courtesy of Red Cedar Shingle and Handsplit Shake Bureau

FIGURE 13:29

QUESTIONS

1. List three (3) preparatory steps that must be completed before the roof receives its final covering.

2. List three (3) structural requirements that must be fulfilled by every sheet of plywood used as roof sheathing.

3. Select one item present on the roof's surface that will require flashing and describe how it should be properly flashed with respect to materials and procedures.

4. Compare the procedures used to apply flashing and roof shingles both to an open valley and to a closed valley.

5. Explain how the installation of underlayment and drip edge differ at the eaves and at the rake of the roof.

6. List three (3) criteria that are used to decide which is the most suitable finished roof material for a specific house.

7. Explain how the installation of eave flashing on a steep roof (greater than 4 in 12) differs from that made on a reasonably flat roof (2 in 12 slope).

8. Discuss the reasons why flashing is used.

9. Explain the difference between an open and a closed valley.

10. Tell where and why underlayment is used on a roof.

14

WINDOWS AND DOORS

The exact location of all exterior doors and windows must be known in advance, as must the manufacturer's rough opening dimensions, so that they can be constructed as an integral part of each exterior wall partition. As a rule, the rough openings for windows should be $\frac{1}{2}''$ larger on each side and $\frac{3}{4}''$ higher than the actual frame, while the opening for each door should be $2\frac{1}{2}''$ larger than the height and width of the frame. The standard height of both, once installed, should be 6′ 8″ above the finished floor.

The centerline for each window and door is obtained by careful examination of the building's floor plan (Figure 14:1). The window and door schedule for the same plan is presented in Figure 14:2. Note that the schedule lists all the pertinent information concerning the specific windows and doors to be installed.

Figure 14:3 shows two different ways to frame a window, with the various structural components properly labeled. Small pieces of $\frac{1}{2}''$ plywood, sheet rock, or lath are positioned between the two 2 × 12s that will serve as the double header, so that the header's thickness will match that of the 2 × 4s used as the wall studs (Figure 14:4).

The use of 2 × 12 headers throughout the structure guarantees a uniform height for all openings, as well as reduces the construction time involved. It alleviates the need to measure, cut, and nail into place the cripples required whenever 2 × 6 or 2 × 8 headers are used.

Table 14:1 gives the standard window heights above the finished floor.

Once the exterior walls have been covered with sheathing, the windows and exterior doors can be installed. Some contractors only install the doors as the means of securing the residence and forego the installation of the windows. They figure that windows could be damaged accidentally or maliciously during the remainder of the construction.

WINDOWS

Energy-conserving windows are installed in all new homes. These windows are equipped with double or triple insulated glass (Figure 14:5) with an insulating barrier of dry air (to prevent condensation) hermetically sealed between the individual panes. They provide homeowners with reduced fuel bills during the winter months, more effective air conditioning during the summer, and could eliminate the need to purchase storm windows.

Table 14:2 lists the glass widths and heights for the common and multiple window types.

Window Types

There is sufficient variety of designs of residential windows so that a homeowner can achieve a specific architectural effect. Some windows slide, others swing, while others are completely stationary with no moving components.

Double-hung

This window type has two sashes that slide up and down within its frame (Figure 14:6). It is a desirable window because it is efficient, economical, and can vary from a simple design consisting of one pane of glass per sash to one having eight panes per sash.

The operation of the sash in the older versions of this window type was controlled by a sash cord connected to a large, concealed window weight, while the newer types rely simply on a friction fit between the sash and

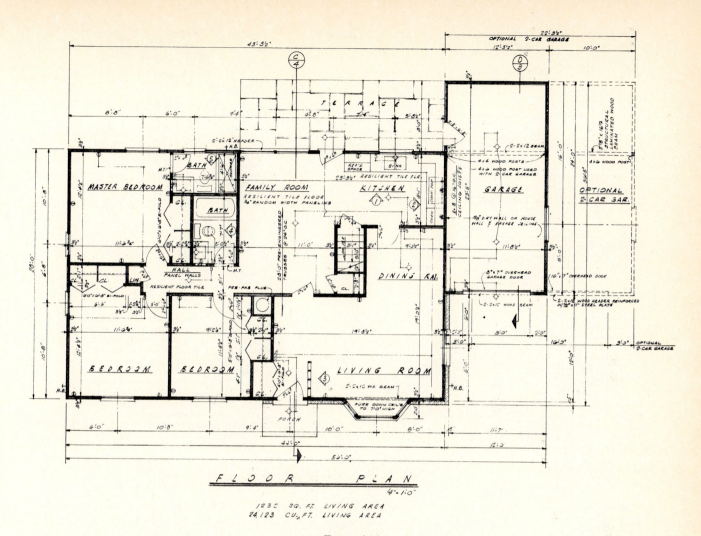

FIGURE 14:1
Sample floor plan.
(Courtesy of Home Planners, Inc.)

Quantity	Code	Description	Location	Anderson #
		Window schedule		
1	A	Bay, 40 × 49 fixed (2), 16 × 24 D.H.	Living room (2)	1846
4	B	40 × 24 D.H. (4 wide, 4 high)	Bedrooms (front, rear)	3846
			Living room (side)	
1	C	Double 40 × 24/36 D.H. (4 wide, 5 high	Playroom	3856
2	D	24 × 24 D.H. (3 wide, 4 high)	Bedrooms (left side)	2446
1	E	24 × 36 casement (3 wide, 3 high)	Bathroom	CX135
1	F	Double casement 20 × 36 (each)	Kitchen	C 235
1	G	40 × 49 fixed	Garage	
		Door schedule		
1	H	3' 0" × 6' 8" × 1¾"	Front door	
2	I	2' 8" × 6' 8" × 1¾"	Rear door, garage door	
1	J	8' 0" × 7' 0"	Overhead garage door	

FIGURE 14:2
Window and door schedule.

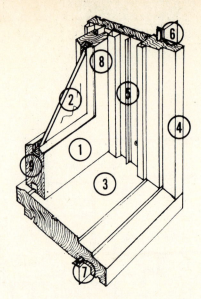

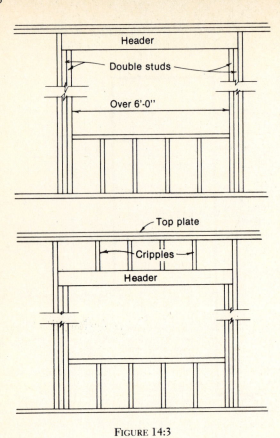

FIGURE 14:3

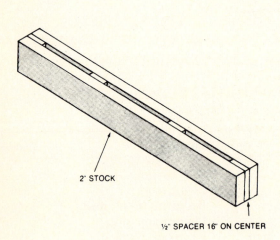

2" STOCK

½" SPACER 16" ON CENTER

FIGURE 14:4

Typical window parts: (1) Sash protected with a patented 4-step polyurea factory finishing process. (2) Select quality welded insulating glass. (3) Wood sill covered with white rigid vinyl (PVC). (4) Wood outer frame member covered with rigid viny (PVC). (5) Rigid vinyl (PVC) jamb liner. (6) Vinyl anchoring flange and windbreak fits into groove of outer frame. (7) Soft vinyl sill windbreak. (8) Rigid vinyl rib on jamb liner fits into polypropylene covered urethane foam weatherstripping on sash. (9) Foam type weatherstripping on bottom and top rails. (A)

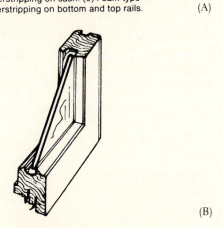

(B)

FIGURE 14:5
(A) Typical window parts. (B) Double-pane insulating glass.

TABLE 14:1

Height, floor to:	Window sill (base)	Window head (top)
Living room	1' 0"	6' 8"
Dining room	2' 6"	6' 8"
Kitchen	3' 6"	6' 8"

its frame. Screens and storm windows for windows containing only single paned glass are installed on the outside frame of the window so that the sash can be easily opened and closed.

Horizontal Gliding

This type of window is equipped with one or more movable sashes that slide horizontally within the frame. These movable sashes can be easily lifted out of their tracks for easy cleaning (Figure 14:7).

TABLE 14:2
Common Window Sizes (Without Frame)

Window type	Glass widths (inches)	Glass heights (inches)
Casement	13, 16, 20, 24, 32, 44	31, 36, 43, 55, 67
Awning	19, 31, 36, 43, 55	16, 20, 32, 36, 67
Double hung	16, 20, 24, 28, 32, 36, 40	14, 16, 20, 22, 24, 28, 60, 68
Gliding	28, 40, 52, 64	29, 35, 41, 53
Basement/utility	27	10, 14, 18
Bow	74, 97, 120, 143, 163	31, 36, 43, 55, 67
Bay (box type)	50, 74	31, 36, 43, 55
(casement type)	62, 69, 86, 93, 110, 117	31, 36, 43, 55, 67
(double-hung type)	84, 91, 96, 103, 111	49, 53, 61
Awning combinations (with a narrow mullion strip between components)	19, 31, 36, 43, 68, 77, 91, 115, 118, 140, 175	16, 19, 36, 43, 57, 67, 68
Casement/awning combinations	19, 31, 36, 43, 56, 68, 80	55, 67, 79, 91, 103, 116

Note. Add 5 inches to the stated glass size to obtain the rough opening dimension for a specific window.

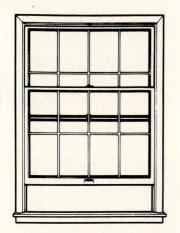

FIGURE 14:6

FIGURE 14:7
Gliding windows.

Casement

A crank handle controls the operation of a hinged casement window. Sash locks are needed to ensure a weather-tight seal between the sash and the frame when the window is closed. (Figure 14:8). Because of the design, casement windows are usually not as thick as the residential wall's thickness and thus will require pine extension jambs to extend the interior frame to the finished wall. This will allow the window to be trimmed properly with molding (Figure 14:9). Casement windows are ideal for areas where the operation of a double hung window is impractical; e.g. over a kitchen sink.

Because of the swinging nature of casement windows, screens must be positioned on the inside of the window frame. Casement windows are frequently used in conjunction with large stationary picture windows to provide ventilation.

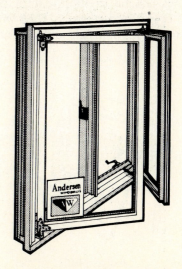

FIGURE 14:8

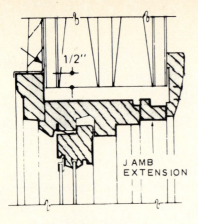

FIGURE 14:9

Awning

Most awning windows are equipped with sliding hinges at the top so that, as they are cranked open, the top rail moves downward as the base of the frame swings outward (Figure 14:10). They are popular in bedroom areas because they provide privacy, adequate ventilation, and the opportunity to rearrange furniture without interference from a window positioned too low on the wall. Screens are mounted on the inside of the frame.

Like the casement window, they are frequently placed in conjunction with fixed glass panels. A possible objection to the use of either awning or casement windows is that the location may interfere with the flow of traffic on walkways or patios situated directly below them.

Hopper

This window style is the direct opposite of the awning window. It is hinged at the base and swings into the room, thus possibly presenting difficulties in the operation of drapes and curtains. Their primary use, therefore, is to provide both light and ventilation for the basement area.

FIGURE 14:10
Awning window.

Bays and Bows

Both of these types are designed to project out on angles from the exterior wall. The side units of bay windows make either a 30° or a 45° angle with the house wall and with the fixed front panel(s), which run(s) parallel to the exterior wall (Figure 14:11). A bow window differs from a bay in that it usually contains more units, each one making a more gradual angle with the exterior wall. (Figure 14:12)

FIGURE 14:11

FIGURE 14:12

Both of these types are equipped with casement units for ventilation purposes, as well as the stationary unit(s). Most of these windows require additional support beneath them, extension jambs to finish their interior, and usually some type of roof structure for protection.

Jalousie Window

This window style has numerous glass slats stacked horizontally above each other in a hinged metal frame. All the slats operate simultaneously whenever the window is cranked open or closed (Figure 14:13). Because of its design, this window type is not very energy efficient, and thus, in most northern climates, it is limited to unheated sunporches.

Window Greenhouse

This window style provides the plant enthusiast with a place to grow and display a variety of house plants. Units have insulated glass, plant shelves, support brackets, extension jambs, and a protective roof. They range in width from 3′ to 6′ and from 3′ to 4′ feet in height. Units without a means of ventilation should be avoided, since plants need a change of air frequently for ideal growth (Figure 14:14).

Skylights

These windows are available as either fixed, moveable, or motorized units to be installed in the roof area. The frame for a skylight requires a double header at each end to support the shortened rafters (Figure 14:15). The roof shingles below the rough opening must be nailed into place before the unit can be installed.

Once the unit has been properly positioned and nailed into place, the balance of the roof shingles are installed. Those on each side of the unit must be cut $\frac{3}{4}''$ from the frame to ensure proper drainage (Figure 14:16).

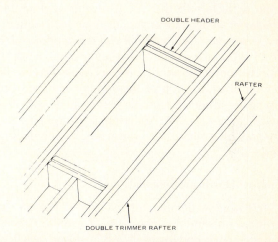

FIGURE 14:15

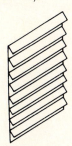

FIGURE 14:13

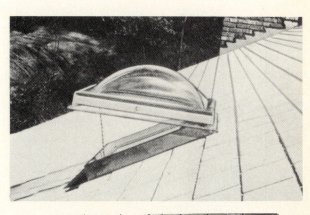

FIGURE 14:14

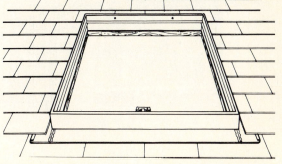

FIGURE 14:16

Installation

The installation of most windows should proceed smoothly, provided the rough opening for each one is large enough and is level and plumb. Since most windows are inserted into the rough opening (Figure 14:17) from the outside, a carpenter will require several people to assist in the initial installation. Great care must be exercised in unpacking the window, carrying it to its location, guiding it into its opening, and then holding it in position while the carpenter plumbs and levels it within its frame.

To make certain that the head (top) of the window frame is flush with the 2 × 12 header (Figure 14:18) shingles or wedge blocks are positioned at frequent intervals between the rough sill and the window sill (Figure 14:19). These wedges may then need to be adjusted until the top of the window is level. Once this has been accomplished, the top of the window is nailed firmly into place, either through the perma shield with 2″ galvanized roofing nails (Andersen), or through the exterior wooden frame with galvanized casing nails (Figure 14:20).

Window Hardware Installation

Most window manufacturers install all the necessary hardware on windows before they leave the factory. A carpenter is needed to install sash locks and sash lifts on some types of double-hung windows.

There are two types of sash locks for double-hung windows. Either one can be used to make certain that the two sash are properly locked. The sash lift design may vary among window manufacturers, but the basic function is the same: to enable any individual to raise and lower the bottom sash (Figure 14:21).

Whenever a carpenter is faced with the installation of a large casement type window, the task becomes simpler and safer, and the window frame lighter to manage, if

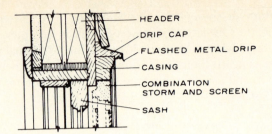

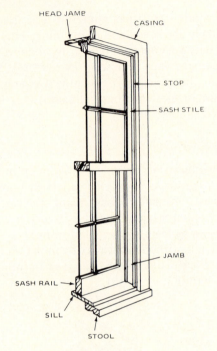

FIGURE 14:18
Double hung window in frame construction.

FIGURE 14:17
A layer of builders' felt is nailed around the rough opening before the window unit is set in place.

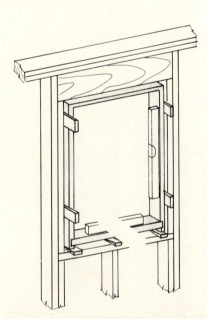

FIGURE 14:19
Wooden wedges are used to level the window unit in its rough opening.

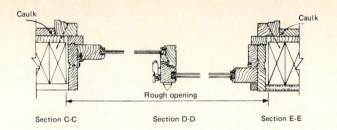

Section C-C   Section D-D   Section E-E

FIGURE 14:20
This type of window unit is installed by nailing the wood molding to
the studs forming the rough opening.

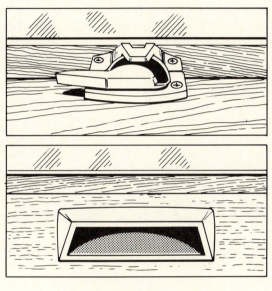

FIGURE 14:21
(Courtesy of Andersen Windowalls)

FIGURE 14:22
(Courtesy of Andersen Windowalls)

he or she understands fully how to remove and, subsequently, reconnect each casement section. This involves disconnecting the necessary hardware before placing the window frame in its rough opening (Figure 14:22).

DOORS

Whenever possible, it is advisable to install factory assembled doors ("prehung") rather than expend the time and energy needed to build the jambs, install the hardware, and finally miter, cut, and nail all the required trim moldings in their respective positions.

Exterior Doors

All exterior doors have a standard thickness of $1\frac{3}{4}''$. They are available in two basic types: the panel, or stile and rail door, and the flush door (Figure 14:23). The stiles and rails of most panel doors are made of solid stock, while the panels may be plywood, hardboard, or solid stock. A sash door is a specialized panel door with one or more of its wood panels replaced by glass. Figure 14:24 shows a representative sampling of exterior door designs.

Flush doors are constructed with face panels of either plywood, metal, hardboard, or plastic laminates that are bonded to a solid core of wood, flakeboard, or polyurethane. Most interior flush doors, however, have hollow, rather than solid cores and are thus much lighter in weight. The hollow core system consists of cardboard strips positioned between the wooden stiles and rails in a lattice-like arrangement so that the plywood surface (skin) can be glued to the cardboard's edges (Figure 14:25).

For exterior door installation, both the header and the floor joists must be notched, and the subfloor cut, so that the top of the door sill is level with the finished floor (Figure 14:26).

The rough opening height, measured from the top of the subfloor to the base of the header, is determined by adding $2\frac{1}{2}''$ to the door height (6' 8" is the standard door

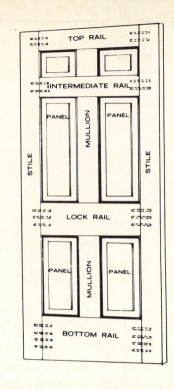

(A)

Glue–block core door

Mat—formed composition-core door

(B)

hollow—ladder core door

hollow—mesh or cellular core door

FIGURE 14:23
(A) Parts of a panel door. (B) Types of wood flush doors.

height) and $3\frac{1}{2}''$ to the door's width (3' 0" is standard entrance door width). All exterior doors must swing inward, and are purchased as either left-hand or right-hand swing. This determination is made by standing outside the house. A left hand swing door has its hinges on the left side (Figure 14:27).

Doors can be purchased as either complete prehung units (Figure 14:28) with the appropriate jambs, door, and hardware, or as separate components requiring that the jambs be installed before the door can be hung (Figure 14:29).

A single entrance door should be at least 3' 0" wide, while a rear door should be either 2' 6" or 2' 8" wide. Most exterior doors are equipped with a sill whose interior surface should be level with the finished floor. Wedges and cleats are usually needed beneath the sill to attain this desired elevation, and also to adequately support the sill.

These shims and wedges should be placed in the vicinity of where the hinges and the lock strike plate will be (Figure 14:30). One of the wedges, on the hinge jamb, should be positioned 7" down from the top; another 11" up from the bottom; and a third one located equidistant between the first two. The lock plate for the door is usually positioned 36" above the sill while an antiburglar deadbolt should be placed 60" above the sill.

After this has been accomplished, a level is used to check the door jamb on the hinge side for plumb and straightness. Wood shingle shims are then inserted between the studs and the top and side jambs to wedge the door frame in its rough opening (Figure 14:31).

Once the top and side jambs are level, finishing nails are driven through the jamb, the wedge, and into the stud. The final installation procedure involves driving casement nails through the exterior casing into the underlying studs.

Even though many exterior prehung doors are purchased with only two hinges, the addition of a third hinge midway between the other two, helps prevent warping, provides for additional support, is safer for its users, and should extend the life of the door.

Sliding Glass Doors

The procedure for the installation of a sliding glass door frame is the same as that discussed for an exterior door, except that for sliding glass doors, special attention must be given to making certain that the sill is well supported and that it is level and straight in the rough opening. If not, the heavy glass door(s) will not slide back and forth properly.

Most sliding glass door units have at least one sliding panel and one stationary panel. Manufacturer's catalogs and installation instructions use an "X" notation to de-

Door designs B2 thru B10 have DEEP SCULPTURED PANELS — BOTH SIDES! Illustrations are actual PHOTOS of designs assembled by Iroquois. All doors are glazed with TEMPERED (safety) glass.

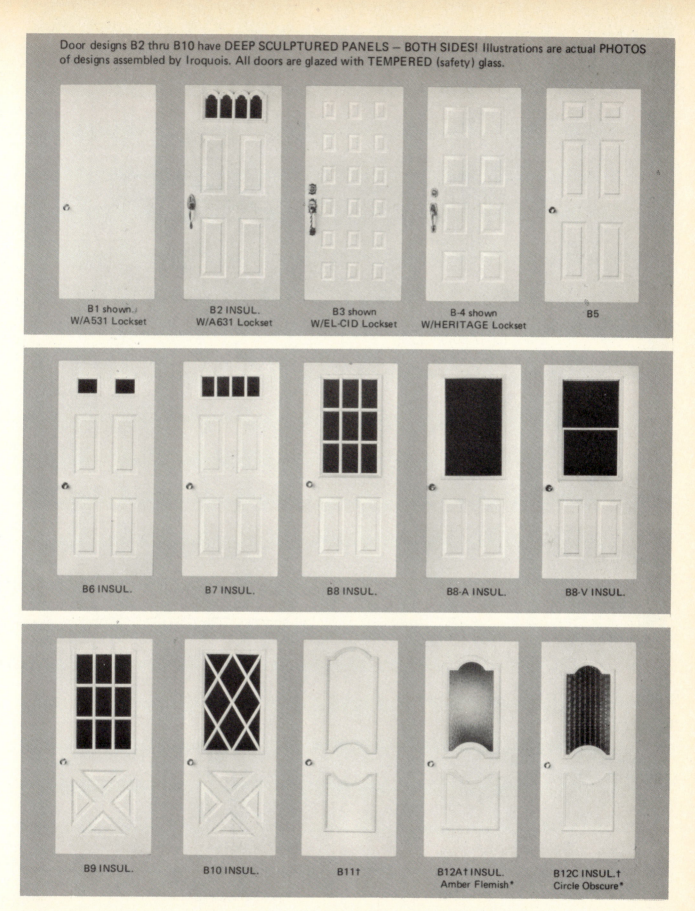

FIGURE 14:24

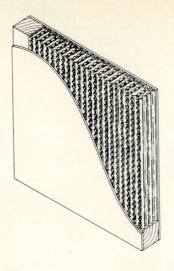

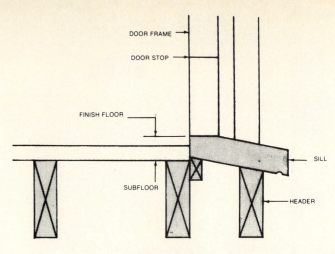

FIGURE 14:26
The floor joists must be cut to receive the door sill.

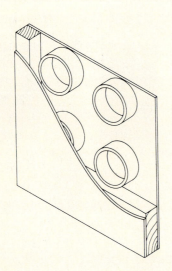

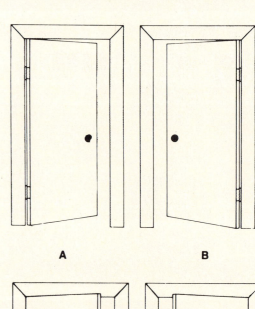

A   **B**

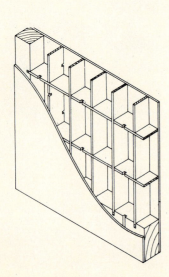

FIGURE 14:25

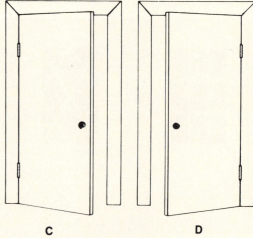

C   **D**

FIGURE 14:27
The "hand" of a door refers to which way it opens and the location of the lock. (A) left hand door; (B) right hand door; (C) left hand reverse door; (D) right hand reverse door.

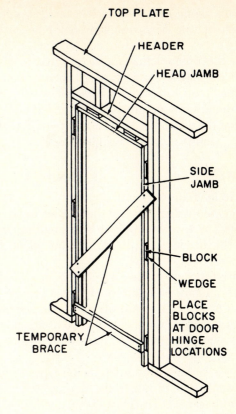

TOP PLATE

HEADER

HEAD JAMB

SIDE JAMB

BLOCK

WEDGE

PLACE BLOCKS AT DOOR HINGE LOCATIONS

TEMPORARY BRACE

FIGURE 14:30

FIGURE 14:28
Prehung door.

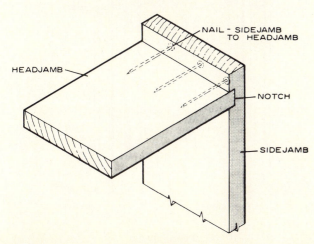

NAIL - SIDEJAMB TO HEADJAMB

HEADJAMB

NOTCH

SIDEJAMB

FIGURE 14:29

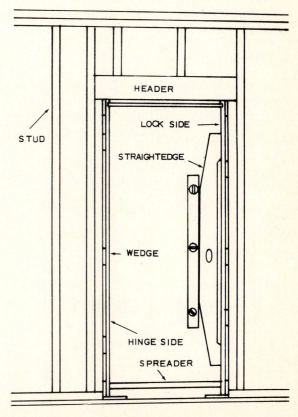

HEADER

LOCK SIDE

STRAIGHTEDGE

STUD

WEDGE

HINGE SIDE

SPREADER

FIGURE 14:31

187

note each sliding panel and an "O" for each stationary panel, illustrating views from outside the house (Figure 14:32).

Both the unit's actual width (referred to as "unit dim." in the diagrams) and its rough opening dimension ("rgh.opg.") are listed above the unit, while the two height dimensions are presented to the left of the unit. Note that the difference between the two width dimensions is $\frac{3}{4}''$ and that of the height as only $\frac{3}{8}''$.

Interior Doors

The rough framing for interior doors is very similar to that for exterior doors, except that the header is usually a double 2 × 4, with cripple studs filling the space between the header and the double top plate (Figure 14:33).

Most interior doors have a thickness of $1\frac{3}{8}''$, a height of 6′ 8″, and are complemented with a prehung jamb which is 6′ $8\frac{1}{2}''$ long to provide a $\frac{1}{2}''$ clearance beneath the door. Most bedroom entrance doors are at least 2′ 6″ wide, while bathroom and single closet doors should be at least 2′ 0″ wide.

There is sufficient variation in the types of interior door installations to satisfy the desires of any homeowner with respect to usable wall space and appearance. The two designs most commonly used for interior doors are the plain flush door and the louver door, which may be constructed with a partial or full louver (Figure 14:34).

Pocket Doors

Most entrance doorways are equipped with flush, hollow-core doors that swing into the room from the hallway. Small bedrooms with limited wall space may require the installation of a pocket door, so that all the in-

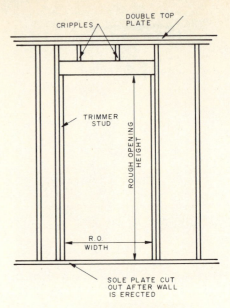

FIGURE 14:33
Framing for a typical door opening.

Full Louvered   Louvered with raised
                       bottom panel

FIGURE 14:34
Two types of interior doors.

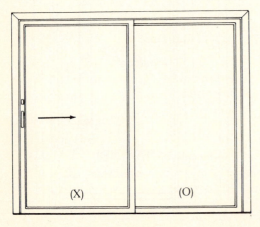

FIGURE 14:32

terior walls can be used to their full potential (Figure 14:35).

Pocket doors can be preassembled at the factory or assembled at the site. They are designed to disappear into the wall. The rough opening for a pocket door frame has to be large enough to house both the pocket and the finished door opening.

Bypass Doors

Standard $1\frac{3}{8}''$ thick interior doors are used in the installation of bypassing door sets. Bypass door units are an efficient means of enclosing a large closet or storage area without the space problems associated with a swinging door.

The height of the rough opening should be $1\frac{1}{2}''$ more than the door height, which allows $\frac{3}{8}''$ clearance beneath the door. The rough opening width is determined by using the following formula: the combined width of the doors minus a $1''$ overlap on each pair of doors. For example, three $32''$ doors = $96'' - 2''$, overlap = $94''$ rough opening (track width). Each installation hardware package comes complete with all the necessary components for the track length desired—$47''$, $59''$, $71''$, $94''$, $95''$, or $118''$ tracks are available (Figure 14:36).

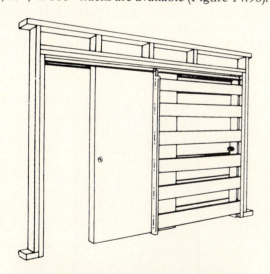

FIGURE 14:35
Structural details of a pocket type sliding door unit. (Ideal Company)

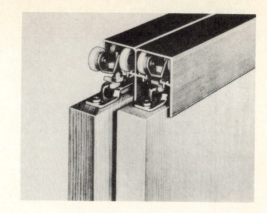

FIGURE 14:36

Bi-fold Doors

This door style is popular for closets, wardrobes, and as room dividers. A bi-fold unit may include one pair of doors, two pairs, or more than two pairs. The pairs are hinged together, glide along an overhead track, and are supported by pivot brackets attached to the side jambs (Figure 14:37).

Cafe Doors

This style is a modification of a bi-fold door whose sole purpose is to serve as a decorative room divider; its length and design do not completely fill the door opening.

Accordion Folding Doors

These doors may be constructed of narrow wooden, metal, or plastic slats that are connected by means of concealed hinges. They glide along an overhead track with an action similar to that of a one directional drapery operated by a draw cord.

Complete Door and Hardware Assembly

There are situations when it is not appropriate to install a prehung door. When this happens, the carpenter must follow a definite series of steps to guarantee a proper installation.

1. The head and side jambs must be positioned, shimmed, and leveled in the rough opening. Jambs are usually nominal $1''$ thick and have a width equal to the overall finished wall thickness ($4\frac{5}{8}''$ drywall with $\frac{1}{2}''$ walls; $5\frac{1}{4}''$ plaster).
2. The door is then positioned between the jambs and held in place with shims and wedges (Figure 14:38).

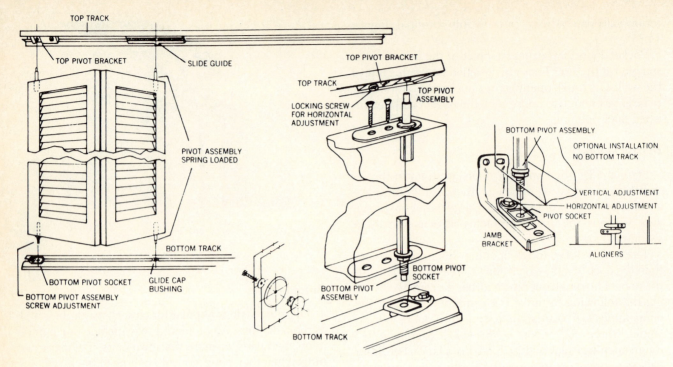

FIGURE 14:37

3. Once in position, the location of the hinges is marked, both on the door and on the jamb (Figure 14:39). At the same time, the correct clearances must be checked so that the necessary adjustments, either to the door or to the jamb, can be made. Table 14:3 is designed to aid you in selecting the most suitable hinge size for the task at hand.

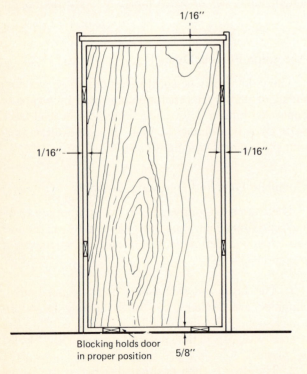

Blocking holds door
in proper position

FIGURE 14:38

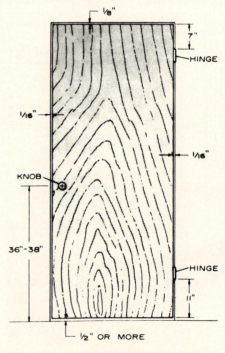

FIGURE 14:39
Hinge placement and door clearance.

TABLE 14:3
How to Choose Hinges for Doors

Thickness of door	Width of door	Suggested hinge height
$\frac{3}{4}$″ cabinet doors up to $1\frac{1}{8}$″	Up to 24″	$2\frac{1}{2}$″
$\frac{7}{8}$″, $1\frac{1}{8}$″ screen or combination door	Up to 36″	3″
$1\frac{3}{8}$″	Up to 32″	$3\frac{1}{2}$″
	Over 32″	4″
$1\frac{3}{4}$″	Up to 36″	$4\frac{1}{2}$″
	Over 36″ to 48″	5″
2″	Up to 42″	5″ or 6″

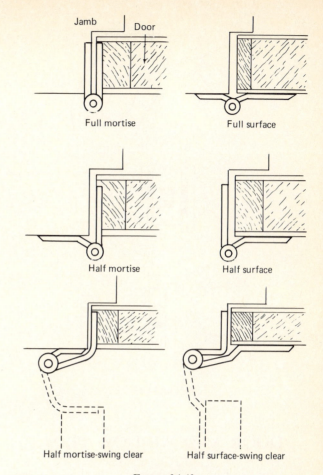

FIGURE 14:41
The six common ways of mounting hinges. The hinges are especially designed for the method of mounting.

Figure 14:40 shows the various parts of a butt hinge, while Figure 14:41 shows the six methods used to mount hinges.

4. The hinge outline is scribed, both on the door edge and the jamb; then the mortise (gain) is cut out with a chisel or router, making certain that the hinge is flush with the wood (Figure 14:42).

5. The door is then hung by its hinges and checked for proper fit, jamb clearance, and ease of movement. If all the above checks are found to be suitable, the door is then marked for the knob and the lock locations, and the jamb marked for the strike plate (Figure 14:43).

6. The hole for the door knob is drilled first, and then the hole for the lock set is drilled. It is imperative that the correct size drill bit be used for each hole. The knob and lock set are installed (Figure 14:43).

7. The strike plate is then cut out with a chisel and installed.

8. The door is then rechecked for final fit, clearance, and ease of movement.

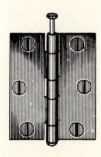

FIGURE 14:40
Parts of a butt hinge.

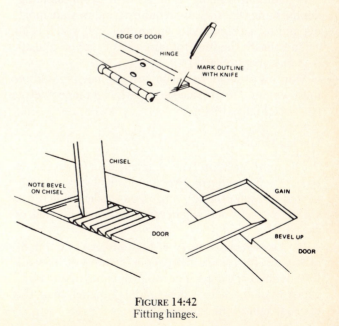

FIGURE 14:42
Fitting hinges.

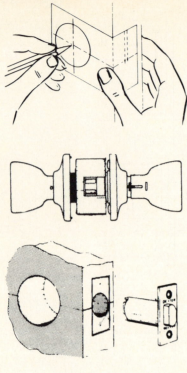

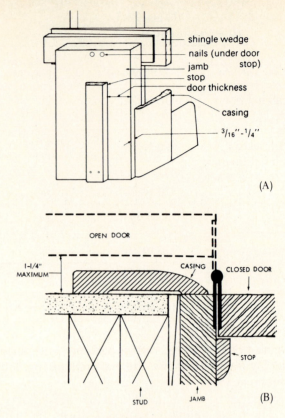

shingle wedge
nails (under door stop)
jamb
stop
door thickness
casing
3/16″ - 1/4″

(A)

OPEN DOOR

1-1/4″ MAXIMUM

CASING

CLOSED DOOR

STOP

STUD   JAMB   (B)

FIGURE 14:44
(A) Side view; (B) top view.

FIGURE 14:43

DOOR AND WINDOW TRIM

The finished trim is added around the interior perimeter of all doors and windows once the walls have been completed. Casing molding is used to trim each door and hide the shims and wedges employed to square the door jambs (Figure 14:44).

In addition to being trimmed with casing molding, double hung windows also require a window stool molding and its accompanying apron molding, which is located beneath the stool for purposes of support and appearance (Figure 14:45).

Muntins (wooden or plastic inserts) may be used on most windows to provide them with certain geometric patterns (Figure 14:46).

Garage Doors

The most common type of garage door is the roll up type, which consists of several individual sections, connected to each other by hinges, and rollers which run within a steel track whenever the door is opened or closed.

Most garage doors are 6′ 8″ high and can be either 8′ or 9′ wide, for a single garage door opening, or 16′ wide for a double car garage. The rough opening should be 3″ wider than the actual door and $1\frac{1}{2}$″ higher than the door's height to the finished floor.

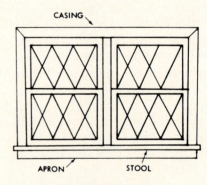

CASING
APRON   STOOL

FIGURE 14:45

FIGURE 14:46
Muntins.

The installation of a roll up type garage door is done in a series of steps.

1. Nail a 2 × 4 or 2 × 6 to the inside edge of the studs outlining the door frame (Figure 14:47). This will act as the support jamb for the track (back jamb).

2. Temporarily nail the exterior door stop to this jamb.

3. Place the various sections of the door in the opening and connect them with hinges.

4. Attach the roller holders to the door sections and then insert the rollers in their holders.

5. Place the vertical track sections on the rollers and attach them to their respective 2 × 4 jambs.

6. Connect the horizontal sections of the track, at one end, to the vertical sections, and the other end to a firm brace that extends downward from a ceiling joist.

7. Raise the door and prop it in the open position.

8. Attach the door's counterbalancing mechanism (Figure 14:48) while the door is open.

9. Close the door and make the adjustments necessary for a smooth operation.

10. Remove the exterior door stops, reposition them to ensure a snug fit, and then renail.

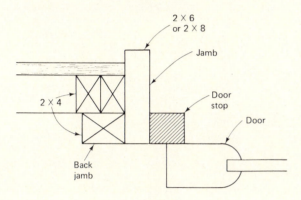

FIGURE 14:47

FIGURE 14:48

QUESTIONS

1. State the generalized rules used to determine the rough openings for doors and windows when their actual frame size is known.

2. List three (3) advantages that windows with double thick glass have over those with single panes.

3. Explain the reason why extension jambs must be used with the installation of certain window types.

4. List the operations associated with the installation of a door that are eliminated whenever a prehung door is chosen.

5. Explain where wood shingles should be placed to help square and plumb the door jambs and how the ensuing nailing procedure is accomplished.

6. Compare bypass and bi-fold doors with respect to installation, operation, and the accessibility of the materials concealed by the doors.

7. List the steps involved in the assembly and installation of an overhead garage door.

8. State the generalized recommendation used to determine the rough opening for window framing.

9. Compare the amount of ventilation possible from double hung, casement, and gliding windows.

10. Explain what the terminology *double insulated glass* means.

11. State the purpose of installing a drip cap above every exterior door and window.

12. State the generalized statement associated with the rough framing of doors.

13. Explain the notation XO for a sliding glass door.

15

EXTERIOR COVERINGS

The final decision as to which one of many suitable materials should be chosen for a home's exterior is usually determined once the following considerations or questions have been carefully examined and answered:

1. Will a sheathing layer be necessary?
2. Are the materials readily available?
3. Will the material selected be energy efficient?
4. What environmental and geographical factors will affect the home?
5. Is the material maintenance free or will it require treatment periodically (such as painting, staining or sealing)?
6. What will be the overall cost of the entire installation (materials and labor)?

It is possible that the most appropriate material may be totally rejected solely on the basis of appearance and personal appeal.

SHEATHING OR SUBSIDING LAYER

4 × 8 sheets of either $\frac{3}{8}''$ or $\frac{1}{2}''$ C/D plywood or $\frac{1}{2}''$ nailable fiberboard are used as the underlying layer for the final exterior covering. Both types are nailed directly to the upright studs and plates. Staggering the vertical joints in each succeeding row alleviates the need for wall braces (Figure 15:1). Each piece should be thoroughly nailed around its entire perimeter and thus may require the insertion of nailing strips between the studs. The sheathing layer is covered with building paper before the final exterior covering is applied (Figure 15:1).

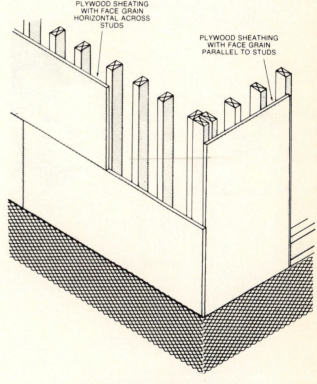

FIGURE 15:1

For masonry veneer walls in houses with one and a half to two stories, sheathing nailed to a balloon frame is preferred to that of a platform frame to reduce the possibility of expansion cracks between floors. There is little shrinkage throughout the length of the upright studs in a balloon frame. The floor headers in a platform frame, on the other hand, do experience sufficient shrinkage to cause the resultant cracks. Very little damage, however, should be expected in a one-story house.

The sheathing must be rigid enough to support the rust proof metal ties which are employed to anchor the bricks to the frame, even though there is a $\frac{1}{2}$″ to 1″ air space between the bricks and the sheathing (Figure 15:2).

Strip or open sheathing involves the use of 1 × 3 or 1 × 4 wood strips strategically positioned on the upright studs for the nailing of shingles (Figure 15:3). This practice is usually only recommended for warm climate areas.

Whenever texture one-eleven, or another of its plywood relatives (Figure 15:4) having at least a thickness of $\frac{5}{8}$″, is used as the exterior covering, the installation of sheathing is not necessary. In most instances, covering the studs once, rather than twice, should reduce the overall cost of time and materials. Texture one-eleven should not be used, however, unless built-in diagonal bracing is installed in each of the four exterior walls (Figure 15:5).

Applying a full 4′ × 8′ sheet of plywood sheathing in each corner of the house is an alternate method of pro-

viding the necessary diagonal bracing (Figure 15:6). It is a common practice used whenever fiberboard, plastic foam, or gypsum sheathing are used to cover the balance of the exterior upright studs.

REVERSE BOARD AND BATTEN
(303 SIDING)

Deep, wide grooves cut into brushed, rough sawn, coarse or scratch sanded or natural textured surfaces. Grooves 1/4″ deep, 1″ to 1½″ wide, spaced 8″, 12″ or 16″ o.c. with panel thickness of 5/8″. Also available in 3/8″ or 1/2″ thickness panels with 3/32″ deep grooves. Provides deep, sharp shadow lines. Long edges shiplapped for continuous pattern. Finish with exterior pigmented stain or leave natural. Available in redwood, cedar, Douglas fir, lauan, southern pine and other species.

CHANNEL GROOVE
(303 SIDING)

Shallow grooves typically 1/16″ deep, 3/8″ wide, cut into faces of 3/8″ thick panels, 4″ o.c. Other groove spacings available. Shiplapped for continuous patterns. Available in similar surface patterns and textures as Texture 1-11. Finish with exterior pigmented stain. Available in redwood, Douglas fir, cedar, lauan, southern pine and other species.

BRUSHED
(303 SIDING)

Brushed or relief-grain surfaces accent the natural grain pattern to create striking textured surfaces. Available in 3/8″ or 5/8″ panels. For finishing, follow individual manufacturer's recommendations. Available in redwood, Douglas fir, cedar, Sitka spruce, lauan and white fir.

ROUGH SAWN AND KERFED
(303 SIDING)

Rough sawn surface with narrow grooves providing a distinctive effect. Long edges shiplapped for continuous pattern. Grooves are typically 4″ o.c. Also available with grooves in multiples of 2″ o.c. May be variable. Especially suited for exterior pigmented stain. Available in Douglas fir, cedar, redwood, lauan, southern pine and other species.

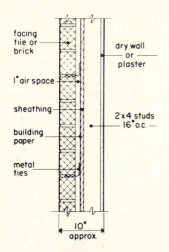

facing tile or brick

1″ air space

sheathing

building paper

metal ties

dry wall or plaster

2 x 4 studs 16″ o.c.

10″ approx.

FIGURE 15:2

FIGURE 15:3

FIGURE 15:4
Surface texture and patterns of selected plywood siding panels. (Courtesy of American Plywood Association) (Continued)

FINE-LINE
(303 SIDING)

Fine grooves cut into the surface to provide a distinctive striped effect; reduces surface checking and provides additional durability of finish. Finish with exterior pigmented stain or acrylic latex emulsion finishes. Available factory-primed. Shallow grooves about 1/4" o.c., 1/32" wide. Also available combined with Texture 1-11 or channel grooving spaced 2", 4" or 8" and reverse board and batten. Available in several species.

STRIATED
(303 SIDING)

Fine striations of random width, closely spaced grooves forming a vertical pattern. The striations conceal nailheads, checking and grain raise, and also conceal butt joints. Finish with exterior acrylic latex paint system or pigmented stain.

FIGURE 15:4 (CONTINUED)

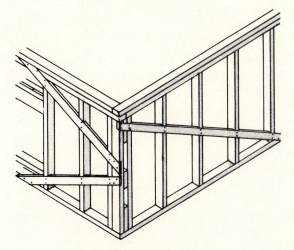

FIGURE 15:5
Set let in braces flush and nail to each stud.

FIGURE 15:6
Plywood panels provide adequate bracing at the corners of the building.

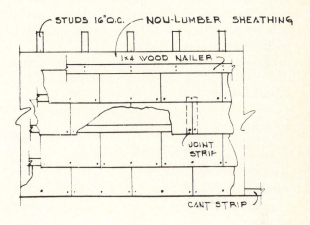

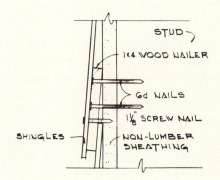

FIGURE 15:7
How to install vertical siding over plywood and fiberboard and other sheathing that will not hold nails.

Since none of these three materials are capable of holding nails, the exterior covering selected to cover them must be nailed either directly to the studs or to 1 × 2 or 1 × 3 wooden nailer strips that have been applied directly over the sheathing and serve as the nailing surface for the exterior covering material (Figure 15:7).

AVAILABILITY OF MATERIALS

This consideration may have a definite bearing on the installation cost of the home's exterior. A brick home built in an area where bricks are manufactured should cost less than a similar one far removed from the factory, because shipping and labor charges for delivery would be less.

In some locales, a specific exterior covering may not only necessitate a special order but may also take months to obtain.

ENERGY EFFICIENCY

The exterior covering for any home, regardless of its geographical location, must be structurally sound and installed properly so that the effects of air infiltration and heat loss transmission are kept to a minimum.

It is possible for you to select the proper underlayment or sheathing, suitable thickness of insulation, and interior wall covering not only to complement the chosen exterior covering but also to achieve the desired degree of energy efficiency for a home in a specific locale.

Table 15:1 lists the building materials commonly used for wall construction and their respective R-values. The R-value is a number which suggests a material's ability to resist the transmission of heat. Any material that acts as a good insulator has a high R-value. By adding all the R-values for the various components chosen, the carpenter will be able to determine whether or not the home will match the expected energy requirements for its environmental area.

An example of the various wall components, installed in a cold climate home, and their respective R-values, is presented below.

Outside air	0.17
Shingles, double course	0.87
15-lb. felt paper	0.07
$\frac{3}{8}$" plywood	0.47
3" insulation	11.00
$\frac{1}{2}$" gypsum wallboard	0.45
Inside air	0.68
Total	13.71

The complete wall structure, exterior and interior, for a home constructed in a cold climate should have a minimum R-value range of 12 to 14, a range of 9 to 10 for a home in a moderate climate, and a range of 7 to 8 for a home located in a warm climate. A map showing the recommended R-values for the various parts of the United States is presented in Figure 15:8.

The placement of the home on the lot may have some bearing on which exterior covering would be the most energy efficient, especially when a portion or the entire home will rely on solar energy as a source of heat.

Limiting the number or reducing the size of the windows on the north and east sides of the home will definitely help in the conservation of energy. Also, it could possibly affect the selection of the exterior covering because there will be less cutting required and thus less waste.

If most of the prevailing winds originate from these two directions, the use of smaller or fewer windows has

TABLE 15:1
The R-values for Certain Building Materials

Material	R-value
$\frac{1}{4}$" plywood	0.31
$\frac{3}{8}$" plywood	0.47
$\frac{1}{2}$" plywood	0.62
$\frac{5}{8}$" plywood	0.78
$\frac{25}{38}$" insulated fiberboard	2.06
15 lb. felt (builder's paper)	0.07
Texture one-eleven ($\frac{5}{8}$")	0.58
$\frac{3}{8}$" bevel siding	0.46
$\frac{1}{2}$" bevel siding × 8", lapped $7\frac{1}{4}$"	0.81
16" shingles, $7\frac{1}{4}$" exposure	0.87
Asbestos-cement shingles	0.03
4" brick veneer	0.44
8" concrete wall	0.64
8" concrete blocks (cinders)	1.72
(sand and gravel)	1.11
1" stucco	0.20
1" nominal board siding	0.74
2"–$2\frac{1}{2}$" insulation	7.00
3"–4" insulation	11.00
5"–7" insulation	19.00
$\frac{1}{4}$" paneling	0.25
$\frac{1}{2}$" gypsum wallboard	0.45
$\frac{3}{4}$" lath and plaster	0.40
air space between layers	0.97
outside air	0.17
inside air	0.68
1" styrofoam insulation	5.00
aluminum siding with reflector foil	2.50

the added advantage of helping to reduce air infiltration and heat loss through those exterior walls.

MAINTENANCE REQUIREMENTS

Homes covered with brick, stone, stucco, asbestos shingles, hardboard, aluminum or vinyl siding, and texture one-eleven plywood should remain maintenance-free forever.

Cedar and redwood products are widely used as exterior coverings because of their natural appearance and their resistance to decay. They may be left alone to weather naturally or be covered with a clear wood preservative, stain, paint, or linseed oil to help them more effectively resist the ills of weather.

Homes covered with wood shingles or siding, however, should receive a coat of paint or stain every five to seven years for both protection and appearance.

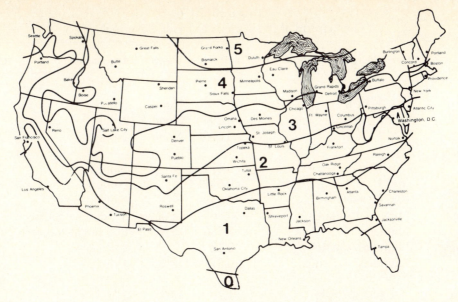

FIGURE 15:8
Heating zone map shows recommended R-values in different parts of
the United States.

Recommended R-Values

Heating Zone	Attic Floors	Exterior Walls	Ceilings Over Unheated Crawl Space or Basement
1	R-26	R-Value of full wall	R-11
2	R-26	insulation, which is	R-13
3	R-30	3½" thick, will depend	R-19
4	R-33	on material used.	R-22
5	R-38	Range is R-11 to R-13.	R-22

R-Values Chart

	Batts or Blankets		Loose Fill (Poured In)		
	glass fiber	rock wool	glass fiber	rock wool	cellulosic fiber
R-11	3½"-4"	3"	5"	4"	3"
R-13	4"	4½"	6"	4½"	3½"
R-19	6"-6½"	5¼"	8"-9"	6"-7"	5"
R-22	6½"	6"	10"	7"-8"	6"
R-26	8"	8½"	12"	9"	7"-7½"
R-30	9½"-10½"	9"	13"-14"	10"-11"	8"
R-33	11"	10"	15"	11"-12"	9"
R-38	12"-13"	10½"	17"-18"	13"-14"	10"-11"

OVERALL COST

The overall cost of any exterior covering is determined by adding the cost of the materials to the expected cost of labor to install the materials.

To calculate the cost of materials, it is first necessary to determine the total square-foot area of the exterior (excluding any major parts that won't be covered). This value is reduced by the total combined area of windows and doors, and the resultant value is then increased by a percentage to allow for anticipated waste.

Table 15:2 is a presentation of the materials commonly used as exterior coverings, their respective sizes, the cost per square foot (related to 1 × 8 T/G Cedar at $1.00/ft²), the anticipated waste percentage, and the amount of overlap required.

A rough approximation of the labor costs for the installation of a specific exterior covering is obtained by using the following formula:

Total labor costs = total cost of materials + 20%

Contractors, specializing in the installation of exterior coverings, estimate a contract on a square foot basis. The price usually includes both labor and materials, plus anticipated profit. Covering a home with brick veneer will definitely be more time-consuming than covering the house with 4′ × 8′ panels of texture one-eleven, and therefore should be expected to cost more.

PREPARATION PROCEDURES

Story Pole

Before a home can be effectively covered with any type of horizontal siding or shingles, it will be necessary to measure the height of each wall. The distance measured should extend from 1″ below the top of the foundation to the soffit area (Figure 15:9).

To determine the number of courses of shingles or siding that will be needed, the overall height measurement for the wall is divided by the expected exposed height of the material (material width minus minimum required lap). This is done to guarantee an equal spacing between each and every course (Figure 15:10). It

TABLE 15:2
Comparison of the Various Exterior Coverings

Item	How sold, size	Cost/sq.ft.	Waste	Lap amount	Comments
Sheathing					
Nailable fiberboard	$\frac{1}{2}'' \times 4' \times 8'$	$0.286	None	None	
Plywood C/D	$\frac{3}{8}'' \times 4' \times 8'$	0.294	None	None	
	$\frac{1}{2}'' \times 4' \times 8'$	0.366	None	None	
Finished exterior covering					
Texture 111	$\frac{5}{8}'' \times 4' \times 8'$	0.617		None	
	$\times 4' \times 9'$	0.631		None	
	$\times 4' \times 10'$	0.659		None	
Shingles, #1 grade	1 box = 1 square	0.563	5%	4", double course	
Barn shakes (16")	4 bundles = 1 square	0.586	5%	$8\frac{1}{2}''$, single course	
Undercourse shingles	2 bundles/square	0.12		4"	
Asbestos shingles	3 bundles = 1 square	0.575	8%	$1\frac{1}{2}''$	12" × 24" standard size
Bevel siding (cedar)	$\frac{1}{2}'' \times 6''$	1.26	33%	1"	
	$\frac{1}{2}'' \times 8''$	1.14	33%	$1\frac{1}{4}''$	
	$\frac{3}{4}'' \times 10''$	1.32	29%	$1\frac{1}{2}''$	
Cedar T/G	$1'' \times 6''$	1.07	16%	None	
	$1'' \times 8''$	1.00	14%	None	
Pine #2 T/G	$1'' \times 6''$	1.13	16%	None	
	$1'' \times 8''$	1.12	14%	None	
Shiplap	$1'' \times 8''$	0.752	16%	$\frac{3}{8}''$	
	$1'' \times 10''$	0.733	14%	$\frac{3}{8}''$	
Masonite (tempered)	$\frac{1}{8}'' \times 4' \times 8'$	0.240		None, when used vertically, battens cover seams	
	$\frac{1}{4}'' \times 4' \times 8'$	0.286		$1-1\frac{1}{2}''$, when used horizontally	
Aluminum siding	double 4" or 8"	0.800		1"	
Vinyl siding	double 4" or 8"	0.690		$1\frac{1}{2}''$	pieces cut $\frac{1}{4}''$ short for expansion
Brick	$2\frac{1}{4}'' \times 3\frac{3}{4}'' \times 8''$	used 1.00 new 2.00			

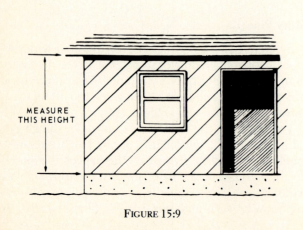

FIGURE 15:9

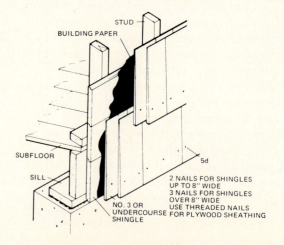

FIGURE 15:10

may be necessary to adjust the height of each course slightly in order to reduce the amount of cutting and notching around windows and doors.

After all the calculations have been made, a 1 × 2 inch story pole is marked where the base of each course should be located on the wall. The story pole is then placed on each outside and inside corner, and next to each window and door frame, so that the sheathing can be marked accordingly with a pencil or a nail driven into the sheathing (Figure 15:11).

Flashing

Windows and doors that are not protected from driving rains by a wide roof overhang should have metal flashing applied over their drip caps before the exterior covering is installed (Figure 15:12).

Corners

It is advisable to install all inside and most outside corners before the exterior covering is applied. Inside corners can be constructed from wood and nailed in place, or commercially available metal corners can be purchased and inserted (Figure 15:13). Which of these two should be employed may be dictated by the exterior covering selected and/or personal preference.

Whether or not the outside corners are applied before or while the exterior covering is being installed is usually dependent upon which material has been chosen. Outside corners can be constructed using two pieces of trim lumber placed at right angles to each

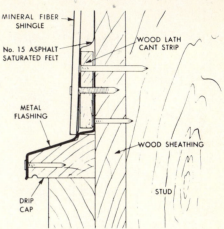

FIGURE 15:12

FIGURE 15:11

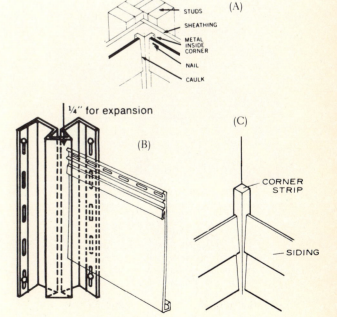

FIGURE 15:13
Inside corners; metal (A,B), wood (C).

other, or they can be purchased as integral components of a specific siding material; e.g., aluminum siding. Either type is installed in advance of the exterior covering material (Figure 15:14).

Other commercially available outside metal corners are attached at the same time that the exterior covering is applied (Figure 15:15).

Spacer Strip

To achieve the desired pitch for the first course of shingles or siding, a spacer strip must be nailed to the

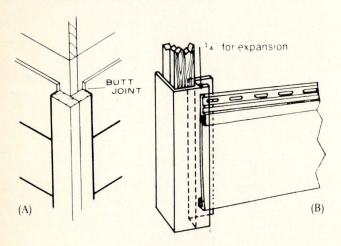

FIGURE 15:14
Outside corners; wood (A), metal (B).

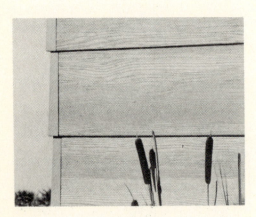

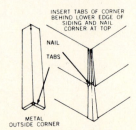

FIGURE 15:15

plate resting on the foundation (Figure 15:16). Its thickness must be the same as the thinner edge of the exterior covering material.

Nail Selection and Nailing Techniques

All nonmasonry exterior coverings should be fastened to underlying support members with some form of noncorrosive nails; aluminum and galvanized (zinc-coated) steel are the two most commonly used. Some exterior materials require face nailing while others, such as tongue-and-groove siding, are designed so that the nails are completely hidden from view (Figure 15:17).

Prior to the actual installation, the carpenter must know the correct nail and its proper length, the nailing technique used for that specific covering, and the quantity required to adequately fasten the material. This quantity may be expressed in either actual number of nails or pounds of nails required per square foot.

INSTALLATION INFORMATION

For simplicity and ease of comparison, the important aspects and techniques associated with the installation of each type of exterior covering will be presented in tabular form. This information will be listed in five categories, identified by numbers 1 to 5. These categories are

1. Sizes available
2. Type and length of nails to use and the nailing technique
3. Treatment around window and door casings
4. How the joints are handled
5. Special treatments and techniques required

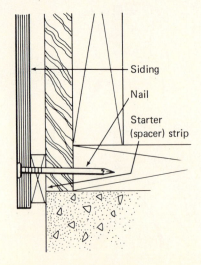

FIGURE 15:16

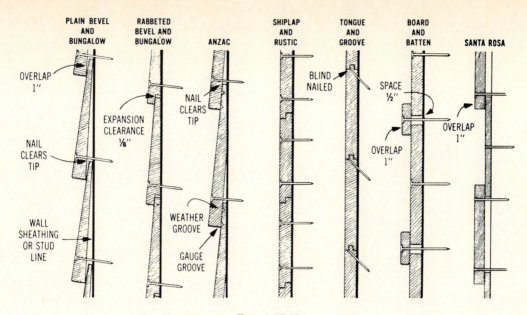

FIGURE 15:17
Nailing details for different kinds of siding.

Siding Types

Aluminum

1. Horizontal—double 4″, double 5″, or 8″; vertical —12″ or 16″ widths (with or without laminated backing)
2. ($1\frac{1}{2}$″) Aluminum driven into panel slots at right angles to sheathing
3. J-bar is used around windows and doors; also trim strips (Figure 15:18).
4. The joints are overlapped $1\frac{1}{2}$″.
5. A #8 wire should connect the siding to the cold water service to prevent an electrical hazard. Corners are inserted before the installation starts. Panels are easily cut with a power saw or tin snips. A $\frac{3}{4}$″ furring strip (air space) beneath the siding changes the R-value from 2.5 to 5.5.

Vinyl

1. The sizes are the same as for aluminum.
2. ($1\frac{1}{2}$″) Aluminum driven into slots, but left loose for expansion purposes.
3. Undersill trim and cap are used around doors and windows.
4. The joints are overlapped $1\frac{1}{2}$″.
5. All horizontal cuts must be $\frac{1}{4}$″ shorter than the actual length required, for expansion purposes.

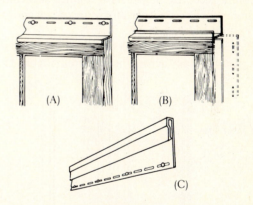

FIGURE 15:18
(A) Cap molding; (B) "J" channel; (C) undersill trim.

Cedar

1. Beveled $\frac{1}{2}$″ × 6″, $\frac{1}{2}$ × 8″, $\frac{3}{4}$″ × 10″, applied horizontally
2. 6d galvanized for $\frac{1}{2}$″, 8d galvanized for $\frac{3}{4}$″; face nailed to underlying studs $\frac{1}{2}$″ above the butt edge or 1″ above butt edge to miss the piece underneath. This allows for expansion and prevents splitting.
3. Siding is cut to fit tightly against casings and is caulked.
4. Joints are butted so that both ends can be nailed to the stud beneath. It may require drilling to avoid splitting.
5. A spacer strip is required (Figure 15:19).

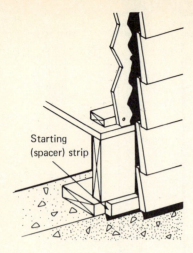

FIGURE 15:19

Tongue and Groove Cedar and Knotty Pine

1. 1 × 6 or 1 × 8 applied vertically.
2. 8d galvanized toe-nailed through the tongue into solid sheathing at 2′ intervals.
3. Pieces are cut to fit tightly against casings and then caulked.
4. Joints are butted and may require face nailing. The joints must be staggered for appearance and structural reasons.
5. The siding may be nailed to 1 × 2 furring strips running horizontally across the wall area (Figure 15:20).

Shiplap

1. 1 × 8 or 1 × 10 applied vertically
2. 8d galvanized face nailed at 2′ intervals to underlying sheathing
3. Pieces are cut to fit tightly against casings and then caulked.
4. The joints, if any, are staggered and face nailed.
5. Many of the overlapping grooves may require sealing to resist the effects of driving rains (Figure 15:21).

Shingle Types

Cedar Shingles

(Number 1 grade or dimension shingles)

1. 16″ long in widths ranging from 3″ to 14″
2. In double coursing, the underlayment shingles are stapled to the sheathing; then the top course is positioned ½″ below the undercourse and face nailed with 1½″ shingle nails (Figure 15:22).
3. The shingles are cut to fit tightly against the casings and then caulked.
4. Each shingle edge should butt tightly with the one adjacent to it.
5. It is best to align the butt edge with the tops of windows and doors whenever possible. A straight edge is used to maintain a level course. A spacer strip is needed for the undercourse.

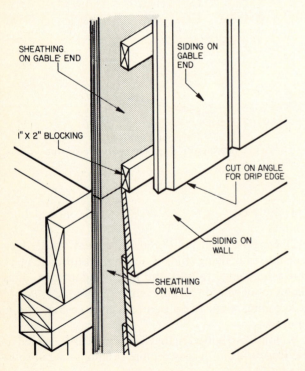

FIGURE 15:20
Blocking strips are used to attach siding on the gable end.

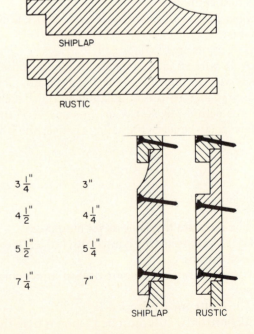

FIGURE 15:21

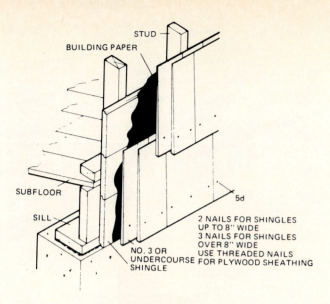

FIGURE 15:22

Cedar Barn Shakes

1. 16″ long with varying widths and uniform thicknesses

2. Same as the number 1 shingles above

3. They may require planning or cutting with a knife to fit tightly.

4. The edges, even though they are already planed, may require trimming to fit snug against the adjacent shake.

5. Whenever wood corners have not been installed, the outside shake must be overlapped with the other end shake. This procedure is staggered with each course (Figure 15:23).

Hand Split Cedar Shakes

1. 24″ long, various widths, and thickness tapered from $\frac{1}{8}$″ to 1″

2.–5. Same as for barn shakes.

Asbestos Shingles

1. 12″ wide and 24″ long

2. They are face-nailed, using at least three specially designed nails per full shingle, into a solid sheathing surface.

3. They are designed to be cut close to the casings, and the space remaining is then liberally caulked.

4. A felt paper joint strip is positioned at each joint.

5. Horizontal nailing strips may be needed if the sheathing is too soft for the nails to hold. A spacer strip is needed. Metal corners are added during the installation. A special cutter is used to both cut and punch holes in the shingles (Figure 15:24).

Panel Types

Texture one-eleven (Plywood)

1. $\frac{5}{8}$″ × 4′ × 8′, 9′ or 10′ long, applied vertically

2. 8d galvanized face nailed into underlying studs or solid sheathing at 2′ intervals.

3. Panels are cut to fit tightly against the casings and then caulked.

4. There is no problem, provided the joints butt tight together over the entire vertical height. Wooden batten strips can be used to cover each joint (Figure 15:25).

5. All joint edges should be sealed to repel driving rains. Pieces left from door and window cutouts may be used elsewhere on the wall provided the vertical grooves line up (Small pieces with horizontal joints will require solid blocking beneath for nailing purposes).

FIGURE 15:23

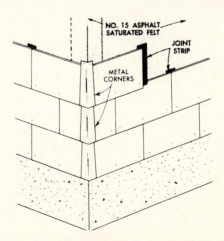

FIGURE 15:24

STANDARD BOARD AND BATTEN: Drive one 8d nail midway between edges of the underboard, at each bearing. Then apply batten
strips and nail with one 10d nail at each bearing so that shank passes
through space between underboards.

FIGURE 15:25
Batten strips.

Masonite (Hardboard)

1. $\frac{7}{16}$″ × 4′ × 8′, 9′, or 10′ long applied vertically
(Thinner pieces are used as horizontal siding.)

2. **and** 3. Same as Texture one-eleven

4. Batten strips are used to cover all vertical joints.
For appearance, these strips are usually nailed every 2′
rather than just over the joints.

5. Cut panels with a fine tooth saw blade. To avoid
damage to the edges, nails should be at least $1\frac{1}{2}$″ from
the edge.

Masonry Types

Brick or Stone

1. Both are applied as a veneer to the face of a solid
wall.

2. A mixture of one part mortar cement to three
parts sand is used to cement the components together.
Metal ties, nailed to the studs, secure the veneer to the
framework and should be used every 15 vertical inches
and every 32 horizontal inches.

3. The veneer adjoins all window and door casings
which may be set flush with the exterior surface or be
recessed. Lintels (Figure 15:26) must be used over all
casings and are supported by the components lining
each side of the casing.

4. Weep holes placed every 4′ in the bottom course
of the veneer allow for the escape of any water buildup
(Figure 15:27).

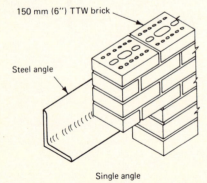

FIGURE 15:26

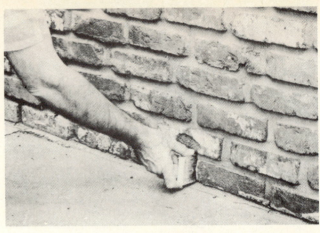

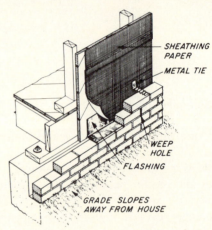

FIGURE 15:27
Weep hole.

5. Excess mortar left on the veneer's surface can be
cleaned up with muriatic acid. There must be a wide
foundation base for any masonry veneer.

Stucco

1. A base or scratch coat, consisting of three parts
sand and one part mortar cement, is applied and forced
through the underlying galvanized metal lath, which is
kept away from the solid sheathing by $\frac{1}{4}$″ wooden spacer
strips (Figure 15:28).

2. 6d galvanized nails are driven through the metal
lath into the solid sheathing.

3. Wood or metal moldings are used around all openings and edges to ensure the proper desired thickness.

4. Stucco is best applied to homes with balloon framing to reduce the possibility of shrinkage and the
resulting cracks in its surface, which is characteristic of
stucco applied to homes with platform framing.

5. The scratch coat can be made smooth with a wet
sponge, given a distinctive pattern, painted, or even
covered with a second finish coat.

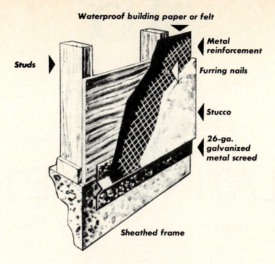

Waterproof building paper or felt

Metal reinforcement

Furring nails

Studs

Stucco

26-ga. galvanized metal screed

Sheathed frame

FIGURE 15:28
Frame construction with sheathing.

QUESTIONS

1. List four (4) questions that must be examined and answered before a final decision can be made on which exterior cover to choose.

2. Explain why it is better to use balloon framing when it is known in advance that the exterior surface will be some form of masonry.

3. What can be done to make a new home more energy efficient?

4. How many courses of shingles, with a 10″ exposure, will be needed to treat an 8′ wall?

5. Explain when and how unprotected doors and windows should be covered with flashing.

6. Select an exterior cover that requires a spacer or starter strip. Tell where the strip is installed and the reason for its use.

7. Describe a lintel and tell where it is used and explain why it is needed.

8. Explain how excess mortar can be removed from a masonry surface.

9. Is there a siding type that can be applied directly to the exterior studs?

10. Explain the importance of using a story pole to locate the height of each row of horizontal siding.

11. Why must exterior brick veneer walls be constructed with weep holes?

16

INSULATING THE HOME

A new home can be insulated once its exterior framework has been completely covered and finished and the rough plumbing, rough wiring, and heating lines have been installed.

Insulation is defined as any material having the ability to restrict the flow of heat. All building materials have this capacity. The insulating effectiveness of a material is related to its density and porosity (the more porous and less dense it is, the better are its insulating properties).

INSULATION CHARACTERISTICS AND CAPABILITIES

Commercially available insulating materials include glass, organic, or mineral fibers, and glass or polystyrene foam. The material should be fire-, moisture-, and vermin-proof, and should have insulating abilities that are effective for the life expectancy of the home.

A well-constructed home that is properly insulated should prove to be both economical and efficient with respect to heating fuel savings during the winter months and reduced electrical charges for air conditioning during the summer months.

When installed, an insulating material should be able to effectively control the three types of heat transfer: radiation, conduction and convection (Figure 16:1) and should provide protection from interior damage due to condensation.

The builder must do everything possible to prevent the movement of warm moist air from the interior to a cooler exterior surface because the water vapor in the air will condense at that point and form water droplets that could result in damage to the home's interior.

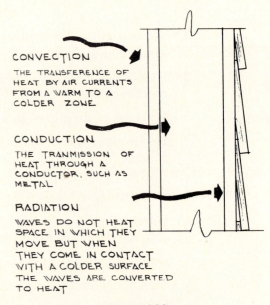

CONVECTION
THE TRANSFERENCE OF
HEAT BY AIR CURRENTS
FROM A WARM TO A
COLDER ZONE

CONDUCTION
THE TRANMISSION OF
HEAT THROUGH A
CONDUCTOR, SUCH AS
METAL

RADIATION
WAVES DO NOT HEAT
SPACE IN WHICH THEY
MOVE BUT WHEN
THEY COME IN CONTACT
WITH A COLDER SURFACE
THE WAVES ARE CONVERTED
TO HEAT

FIGURE 16:1
Three types of heat transfer.

To ensure that this situation does not occur, the builder must thoroughly insulate the home, install a vapor barrier on the warm side of the entire living area, and provide adequate ventilation in both nonliving and living areas to guarantee an adequate movement of air throughout these areas.

Insulation is rated by R value and not by inches of thickness of the material. The R value is a number that suggests a material's ability to resist the transmission of heat. A good insulating material will have a high R value. The more air pockets there are in material and the smaller each pocket is, the better (or higher) its R value is.

208

The heating zones within the United States (Figure 16:2) have been used to determine the optimum recommended R values for homes situated in specific geographical locations (Table 16:1).

It should be apparent from an examination of the map and the table that the colder northern areas require more insulation and, thus, higher R values, than the warmer southern areas.

Table 16:2 can then be used by the builder to determine the most appropriate insulating material to use to achieve the prescribed R values for the home being constructed.

All forms of insulation require a vapor barrier to prevent them from absorbing moisture. Some commercial forms are sold with a built-in vapor barrier, while other forms require installation before they can be used (Figure 16:3). It is also important to realize that insulation, when compressed instead of being loose fitting, loses its efficiency and effectiveness.

CONSIDERATIONS IN PLANNING INSULATION

To properly insulate a home, it is necessary to first calculate the total area (in square feet) involved for each of the following areas: walls, ceilings, and floors. Since each of these areas may require a different insulation type, or different R-value, to do the job properly, you may decide to use batts for one area and rigid board in another area. You must also decide how you will install the vapor barrier.

Whenever the insulation, in the form of batts or blankets, is to be placed between studs, joists, or rafters, the total area is multiplied by 0.90 (if the framing members are 16″ o.c.) or by 0.94 (if they are 24″ o.c.). For example: 1,500 sq. ft. of ceiling area with joists spaced 16″ o.c.

would require 1,350 sq. ft. of insulation. Since R-30 insulation covers the framing members when installed, the total amount of insulation needed equals the total square foot area.

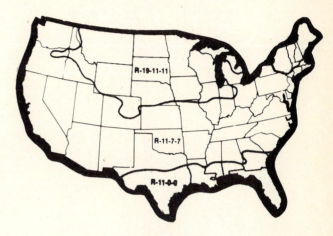

FIGURE 16:2
U.S. heating zones. Federal Energy Administration, *Home Energy Savers Workbook*, November 1976.

TABLE 16:1
Recommended R-Values

Heating zone	Attic floors	Exterior walls	Ceilings over unheated crawl space or basement
1	R-26	R-value of full wall	R-11
2	R-26	insulation, which is	R-13
3	R-30	3½″ thick, will de-	R-19
4	R-33	pend on material	R-22
5	R-38	used. Range is R-11 to R-13.	R-22

TABLE 16:2
R-Values for Common Insulation Materials

	Batts or blankets		Loose and blown fill				
R-Value	Glass fiber	Rock wool	Glass fiber	Rock wool	Cellulose fiber	Vermiculite	Perlite
R-11	3½″	3″	5″	4″	3″	5″	4″
R-13	4″	3½″	6″	4½″	3½″	6″	5″
R-19	6″	5″	8½″	6½″	5″	9″	7″
R-22	7″	6″	10″	7½″	6″	10½″	8″
R-26	8″	7″	12″	9″	7″	12½″	9½″
R-30	9½″	8″	13½″	10″	8″	14″	11″
R-33	10½″	9″	15″	11″	9″	15½″	12″
R-38	12″	10½″	17″	13″	10″	18″	14″

*Source: *DOE Insulation Fact Sheet*

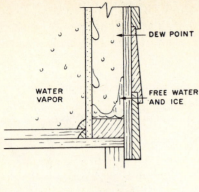

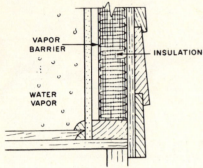

FIGURE 16:3

Loose fill insulation, on the other hand, is sold in bags containing a specific number of cubic feet (volume). Table 16:3 shows the relationship between the coverage, R-value, and thickness in inches, for one 12-lb bag.

Once you have decided upon the most appropriate types of insulation to use in the home, it is advisable to consult with several suppliers to obtain the best price for equivalent insulation. Hiring an insulation contractor to perform the entire operation may be even less expensive, especially when you consider the time it will take you to install the insulation.

TABLE 16:3
Coverage and R-Value Chart
(Per 12 lb. Bag)

Thickness (inches)	1	2	3	4	$4\frac{1}{2}$	8	9
R-value	2.4	4.8	7.2	9.6	11.0	19.0	22.0
Maximum sq. ft./bag coverage*	33	17	11	8	$7\frac{1}{2}$	4	$3\frac{1}{4}$
Minimum wt./sq. ft. (lbs.)	$\frac{1}{3}$	$\frac{2}{3}$	1	$1\frac{1}{3}$	$1\frac{1}{2}$	$2\frac{2}{3}$	3

Each bag of zonolite attic insulation contains approximately 3 cubic feet

*Coverage includes joists 16″ o.c.

Table 16:4 is a comparison of the common types of insulation with respect to the sizes that are available, the locations within the house where they are used, and whether or not their installation requires a vapor barrier.

Figure 16:4 should help to inform you where to place the insulation in a typical residential home.

VENTILATION

Even though the installation of a vapor barrier on the warm side of ceilings and walls is done to minimize the damage effects of condensation, the home must be designed with adequate ventilation provisions to allow any trapped condensation to escape.

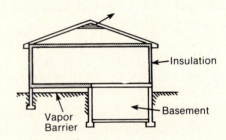

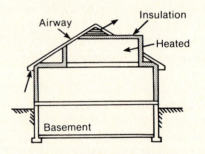

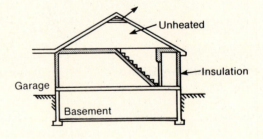

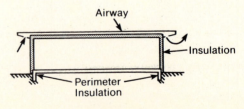

FIGURE 16:4
Where to insulate.

TABLE 16:4
Comparison of the Common Types of Insulation

Type	Sizes available	Where used	Composition	Vapor barrier	Comments
Batts	Widths of 15″ and 23″ Lengths of 4′ and 8′ Thickness of $3\frac{1}{2}$″–7″ in $\frac{1}{2}$″ graduations	Walls Ceilings Floor joists	Glass wool or rock wool	Can be purchased with or without	Easy to install
Blankets	Widths of 15″ and 23″ Length of 57′ Thickness $3\frac{1}{2}$″–7″	Same as batts			
Loose fill	Packaged in bags	Ceiling joists	Glass fiber Vermiculite Perlite Rock wool Cellulosic fiber	Required	Installation time half that of batts
Blown in		Attic floors	Rock wool Glass fiber Cellulosic fiber	Required	Must add 1″ more than necessary (due to settling)
Foamed in		Joists Walls	Urea formaldehyde resin		Has higher R-value than blown in materials
Rigid foam boards	Widths of 2′ and 4′ Length of 8′ Thickness of $\frac{1}{2}$″ to 4″	Foundation walls	Styrofoam Urethane	None required	Must be covered with gypsum wallboard for fire safety

Attic areas that are well insulated must be designed with excellent cross ventilation to prevent the accumulation of moisture. This may be accomplished by the installation of any one of the following: ridge vent, roof vent, triangular- or rectangular-shaped gable vents (Figure 16:5), or a combination of soffit vents and any of the vent types previously mentioned (Figure 16:6).

Table 16:5 enables you to determine the square inches of ventilation required for an attic area of a specific size. It is based on the premise that 1 sq. ft. of unobstructed ventilation opening is adequate for every 300 sq. ft. of attic floor area.

Concrete block vents (Figure 16:7) are designed to be incorporated into masonry walls to properly ventilate crawl space areas. A minimum of four, one per wall, are required to ensure adequate ventilation.

The effectiveness of a home's insulation value is enhanced by adequate ventilation. Heat build up in an unvented attic can reach temperatures exceeding 160°F causing the living areas to become hotter than if the attic was properly vented (Figure 16:8).

Improperly ventilated attic areas collect moisture during the winter months. This moisture condenses as it cools, forming water droplets that first saturate the insulation before they eventually leak into the walls and ceilings (Figure 16:9).

Homes located in areas experiencing snow fall can be easily checked for proper attic ventilation. They are identified by having snow remain on their roofs. These homes have effectively isolated the attic area from the living area by providing adequate air movement and, thus, prevented any condensation buildup in the attic area.

HOW TO INSULATE SPECIFIC AREAS

Walls

Batts or blankets are placed between the studs with the vapor barrier surface facing the interior of a room. The kraft paper flange is stapled, at 6″ to 8″ intervals, to the studs (Figure 16:10).

Installing blankets may be faster than installing batts. The length of the wall between the top and bottom

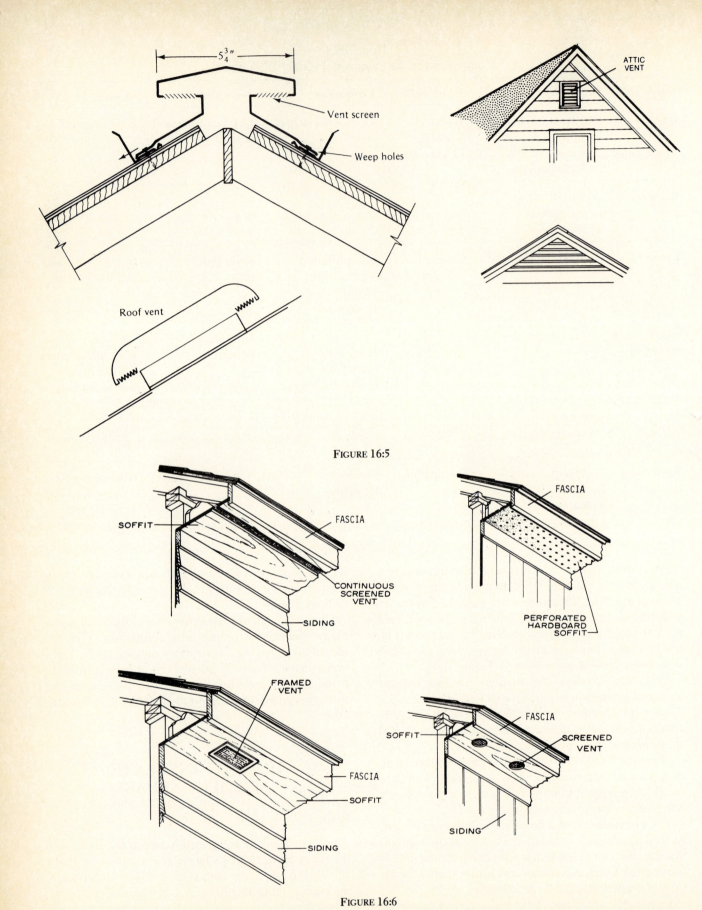

FIGURE 16:5

FIGURE 16:6

TABLE 16:5
Free Area Ventilation Guide
Square inches of ventilation required for attic areas

Width (in feet)	20	22	24	26	28	30	32	34	36	38	40	42
20	192	211	230	250	269	288	307	326	346	365	384	403
22	211	232	253	275	296	317	338	359	380	401	422	444
24	230	253	276	300	323	346	369	392	415	438	461	484
26	250	275	300	324	349	374	399	424	449	474	499	524
28	269	296	323	349	376	403	430	457	484	511	538	564
30	288	317	346	374	403	432	461	490	518	547	576	605
32	307	338	369	399	430	461	492	522	553	584	614	645
34	326	359	392	424	457	490	522	555	588	620	653	685
36	346	380	415	449	484	518	553	588	622	657	691	726
38	365	401	438	474	511	547	584	620	657	693	730	766
40	384	422	461	499	538	576	614	653	691	730	768	806
42	403	444	484	524	564	605	645	685	726	766	806	847
44	422	465	507	549	591	634	676	718	760	803	845	887
46	442	486	530	574	618	662	707	751	795	839	883	927
48	461	507	553	599	645	691	737	783	829	876	922	968
50	480	528	576	624	672	720	768	816	864	912	960	1008

(Length (in feet) is the vertical axis.)

Using length and width dimensions of each rectangular or square attic space, find one dimension on vertical column, the other dimension on horizontal column. These will intersect at the number of square inches of ventilation required to provide 1/300th.

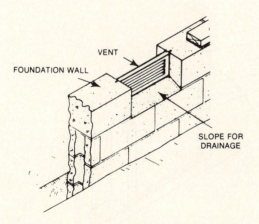

FIGURE 16:7
Cement block vents are designed to be mortared into the same space as an 8 × 16″ cement block. At least four should be used to vent crawl space.

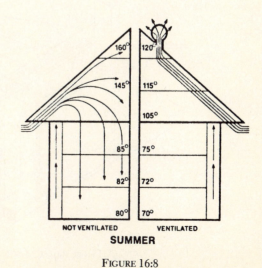

SUMMER

FIGURE 16:8

NOT VENTILATED   VENTILATED
WINTER

FIGURE 16:9

plates can be measured, marked off on the subfloor, and all the required pieces can be cut to size in one operation. Each length should be 2″ to 3″ longer than needed to ensure a snug fit at both ends.

Unfaced fiberglass batts or blankets are designed to fit snugly between the studs without staples. When installed, they must be covered with a polyethylene plastic vapor barrier that must be stapled to the studs (Figure 16:11).

Masonry foundation walls can be insulated by either of two methods. An adhesive can be used to apply rigid styrofoam directly to the masonry walls, which, in turn, can be covered with wallboard or paneling (Figure 16:12). The second method employs a 2 × 4 stud wall positioned just inside the foundation walls so that $3\frac{1}{2}″$ batts or blankets can be inserted between the studs before the wallboard or paneling is nailed into place (Figure 16:13).

Ceilings

Whenever the area above the ceiling will not be heated (e.g. an attic area), the insulation is placed between the ceiling joist. Batts or blankets should be installed with the vapor barrier facing the heated room below. They can be installed from below before the ceiling material is put into place (Figure 16:14) or afterwards, if the attic area is accessible (Figure 16:15).

Whenever either loose fill or blown-in insulation is to be used, a vapor barrier of polyethylene plastic, asphalt coated paper, or aluminum foil must be installed. It is attached to the underside of the joists before the ceiling material is installed. For large areas, it is imperative that the edges of the vapor barrier be well overlapped, and that it be carefully cut and properly fitted around electrical boxes, heat ducts, and water pipes (Figure 16:16).

A ceiling beneath an area that is to be heated (e.g., bedrooms over an unheated garage) is usually insulated with either blatts or blankets, which are installed with the vapor barrier facing upward. To hold them in place, several methods can be employed until the wallboard is installed. The batts or blankets selected can be "pressure fitted" so that no stapling is required; wire mesh can be stapled to the underside of the joists; or "lightning rods," flexible lengths of wire rods that are bent in a reverse U-shape, can be used (Figure 16:17).

It is important that the insulation completely cover the *entire* area of a ceiling, with special care given to the junction with the outside walls. In attic areas, the insulation should extend at least to the outer face of the top plate but should not rest against the roof sheathing or extend into the soffit area; you must not block free air movement through the vents into the attic area (Figure 16:18). In homes where the ceiling joists extend beyond the exterior walls, the insulation must not only fill the

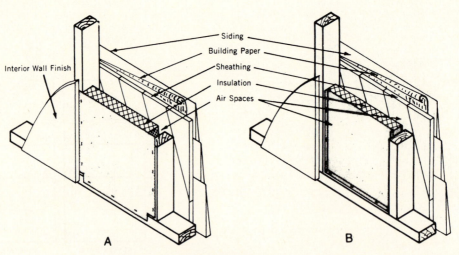

FIGURE 16:10
(A) Flange stapled to edge of stud.   (B) Insulation recessed to ensure
an air space.

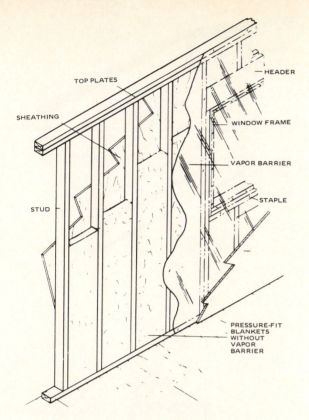

FIGURE 16:11
Insulation that is press fit between studs must have a vapor barrier stapled over it.

FIGURE 16:13

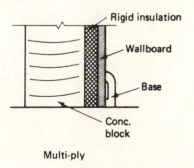

Multi-ply

FIGURE 16:12

FIGURE 16:14
Blanket insulation can be installed in the ceiling before the finished ceiling is nailed to the joists. (Courtesy of Owens Corning Fiberglass)

entire area but also cover the inside surface of the header (Figure 16:19).

Insulation applied to the ceiling in finished attic rooms must be done with care. Blankets or batts are stapled to the rafters with the vapor barrier facing the attic room. The thickness of the insulation used must be *less* than the depth of the rafters to ensure that it does not actually touch the roof sheathing. This would hamper the desired air flow and possibly damage the roof (Figure 16:20).

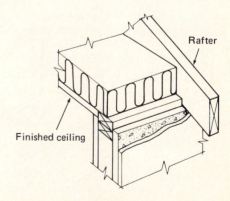

FIGURE 16:15
Blanket insulation can be installed in the ceiling after the finished ceiling is in place. (Courtesy of Owens Corning Fiberglass)

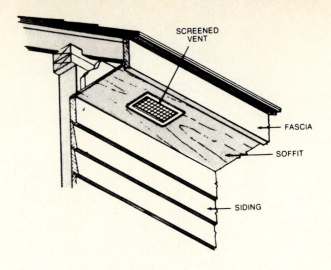

FIGURE 16:16
Polyethylene film vapor barrier—lap joints at least 3″; wrap tightly around outlets, doors, windows, heat ducts, and pipes.

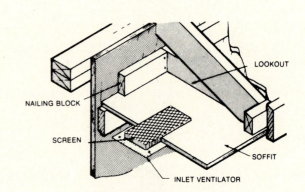

FIGURE 16:18

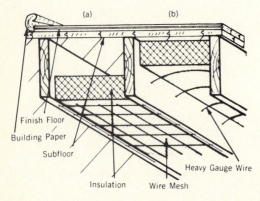

FIGURE 16:17

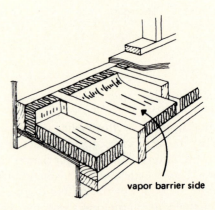

FIGURE 16:19

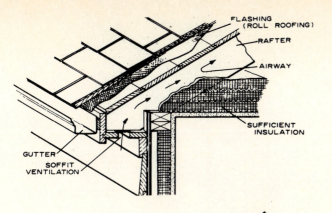

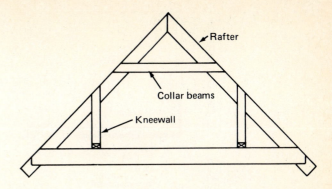

FIGURE 16:21
Insulation is placed between collar beams and studs forming the kneewall.

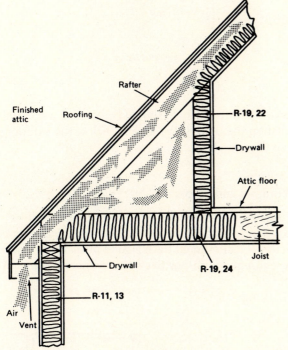

FIGURE 16:20
Extend insulation over and across the top plate of exterior walls—but always be sure not to block vents under the eaves.

summer months. Prior to pouring the concrete, all the rough plumbing pipes and heat lines must be buried beneath the soil line and covered with either gravel or sand to level the surface. Next, a polyethylene plastic vapor barrier is installed over the entire floor area and rigid styrofoam insulation is attached to the inside surface of the perimeter foundation. This must extend downward to the frost line (Figure 16:23).

The slab can also be completely covered with wood flooring or with carpeting; or it can be first covered with rigid insulating panels placed between wood furring

Also, an allowance must be made for proper air flow in the top peak of the roof. Collar ties must be nailed to each set of rafters and then blankets or batts stapled to the underside of the 2 × 4 collar ties (Figure 16:21). Louvers will have to be installed, then, in the roof peak of each gable end of the home to provide the necessary air flow (Figure 16:22).

Floors

There are several means of insulating a concrete slab foundation to eliminate the heat flow that occurs between the exterior and interior of the home during the

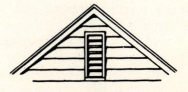

FIGURE 16:22

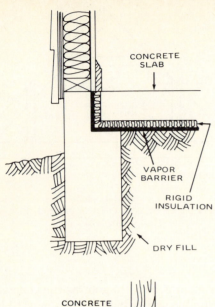

CONCRETE
SLAB

VAPOR
BARRIER

RIGID
INSULATION

DRY FILL

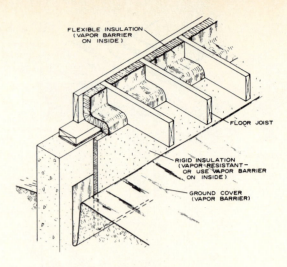

FLEXIBLE INSULATION
(VAPOR BARRIER
ON INSIDE)

FLOOR JOIST

RIGID INSULATION
(VAPOR-RESISTANT-
OR USE VAPOR BARRIER
ON INSIDE)

GROUND COVER
(VAPOR BARRIER)

FIGURE 16:24

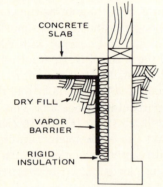

CONCRETE
SLAB

DRY FILL

VAPOR
BARRIER

RIGID
INSULATION

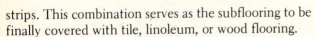

FIGURE 16:23

lower than that inside, should be covered with rigid insulation that extends at least 2′ below the exterior soil line (Figure 16:26).

Windows and Doors

It is imperative that the selection of all windows and exterior doors be done with a keen eye to energy efficiency. Doors should have a good insulation value and be equipped with positive-seal weatherstripping. Win-

strips. This combination serves as the subflooring to be finally covered with tile, linoleum, or wood flooring.

Floors above a crawl space foundation will require the installation of batts or blankets between the joists with the vapor barrier facing the living spaces above. They are held in place with wire mesh, lightning rods, or strips of wood. A polyethylene plastic vapor barrier must be placed on the ground to prevent moisture rising from the ground. A 2″ to 3″ layer of concrete poured on top of the plastic ensures a sealed floor and helps control dust, excessive humidity, and insect invasion (Figure 16:24).

Vents must be positioned on all four foundation walls to provide the proper ventilation that prevents the build up of humidity (Figure 16:25). One square foot of vent area is needed for every 150 sq. ft. of crawl space whenever a vapor barrier is not used. To be totally effective, the soil level inside the crawl space must be higher than the grade outside the foundation. The vertical height in any crawl space area should be at least 28″. Taller foundation walls, where the soil line outside is

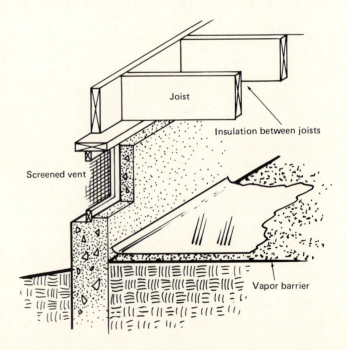

Joist

Insulation between joists

Screened vent

Vapor barrier

FIGURE 16:25

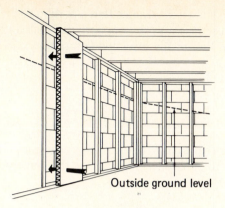

FIGURE 16:26

FIGURE 16:28
Window infiltration.

dows containing double-paned insulating glass should be used throughout the home.

After the windows and exterior doors have been installed, insulation must be cut and tightly fitted around their interior frames. Covering this insulation with strips of plastic (Figure 16:27) should prevent any air infiltration (Figure 16:28). Applying caulk where the exterior frame meets the outside sheathing is also advisable in order to guarantee a sealed exterior (Figure 16:29).

Utilities

Insulation should be carefully fitted behind electrical boxes and plumbing pipes (Figure 16:30). Pipes, so protected, should resist freezing even when the outside temperature is very low. Pipes running through an unheated area (e.g., crawl space or attic area) are most ef-

FIGURE 16:29

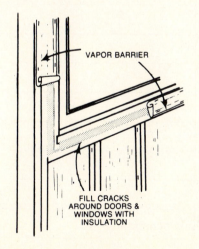

FIGURE 16:27

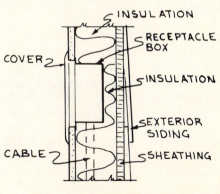

FIGURE 16:30

fectively protected from freezing by being wrapped with commercially available styrofoam insulation. Wrapping each hot water line from heater to fixture with styrofoam will greatly retard the heat loss through the piping and reduce water heating costs.

Garage Area

For most homes, the only wall in the garage to be insulated will be the one adjacent to the interior living area, while garage ceilings are only insulated when there is living space directly above them.

If it is known in advance that a portion of the garage will be used for a workshop, both the walls and ceiling must be thoroughly insulated before they can be covered.

FINAL NOTE

Table 16:6 is a new home energy efficiency check list for you to be familiar with during the construction of every new home.

TABLE 16:6
New Home Energy Efficiency Check List

NOTE: The measures listed below are a set of cost-effective devices and construction practices which should be included in average new homes. Some of these items are replaced by other systems in more advanced home designs.

✓	Insulation:		Minimum R-values	Insulation type
		Walls	R-19	Batt. or blown in
		Ceiling, Roof	R-30 to R-38	Batt, loose fill
		Floor over basement or crawl space	R-19 to R-22	Batt
		Finished basement wall	R-13 to R-19	Batt
		Masonry foundation	R-13	Rigid board, closed cell (suitable
		Insulated slab on grade (perimeter):	R-13	for ground contact)
		Insulated slab on grade (underneath):	R-6	
	Windows:	Double-glazed on south		
		Triple-glazed on north, east and west		
		Consider skylights or clerestories for lighting and direct gain		
		Movable night insulation (R-4 minimum)		
		Thermal breaks on metal windows		
	Doors:	Air-lock entries		
		Solid core in wood door		
		Thermal breaks in metal doors		
		Sweeps and weatherstripping on all doors		
		Storm doors, tightly fitting		
	Others:	Continuous vapor barrier, well sealed at necessary penetrations		
		Sill sealer between top of foundation and sill plate		

QUESTIONS

1. Explain why a vapor barrier should be installed.

2. Should the foil face on insulation batts or blankets face towards or away from the living area they are designed to insulate?

3. Compare the two methods that are used to insulate masonry foundation walls.

4. Explain why a vapor barrier must be attached to the ceiling joists before the ceiling material is installed when loose fill or blown-in insulation will be added to the ceiling material.

5. Explain the reasons why attic insulation should terminate at the top plate and not extend into the overhanging soffit area.

6. Tell how attic rooms should be properly insulated to guarantee adequate air flow throughout the entire area.

7. List three (3) methods used to hold batts or blankets in place when they are installed between joists in floor areas lacking heat below them.

8. To properly insulate a crawl space, certain procedures must be performed. List three (3) of them.

9. What are the two common widths of blanket insulation?

10. State the rule of thumb associated with the proper positioning of insulation that has its own vapor barrier.

11. Explain the various procedures that can be employed to decrease attic condensation.

17

INTERIOR WALL COVERINGS

Even though the framing of walls is one of the first structural operations to be completed in any house building sequence, applying the desired covering to the interior surface of these same walls is one of the last. The interior walls are covered only after the roof, the exterior covering (including the windows and doors), the plumbing, the electrical and heating systems, and all ceiling areas have been completely installed and all exterior walls have been thoroughly insulated.

They can be covered with a variety of materials, some of which can be applied directly to the studs, while others require some type of backing before they can be installed. Walls may be covered with large 4' × 8' sheets (e.g., gypsum wallboard, wood paneling, or hardboard), with individual pieces of tongue-and-groove pine or cedar, with various masonry or ceramic materials, and with plaster.

All the walls and partitions of the house must be completely framed before they can be covered with any interior material (Figure 17:1). Some walls, especially basement walls, require the installation of furring strips (Figure 17:2) before they can be covered. This is best accomplished by installing horizontal 1 × 2 strips 16" o.c. with a vertical strip every 48".

GYPSUM WALLBOARD

A single layer of $\frac{1}{2}$" wallboard, applied directly to the studs in a horizontal position (Figure 17:3), has become a very popular means of covering walls because of the speed of application and the relative ease of concealing the joints.

FIGURE 17:1

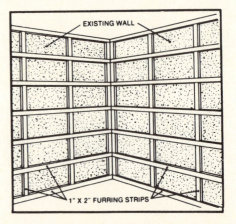

FIGURE 17:2

222

FIGURE 17:3
Make certain to stagger all end joints.

The wall area to be covered is measured accurately before any cuts are performed on the wallboard. For straight cuts, a utility knife is first drawn along a sheet rock square to score the top (finished) side of the wallboard (Figure 17:4). The panel is then positioned to allow the smaller portion to be snapped or flexed along the scored line towards the larger portion. Once accom-

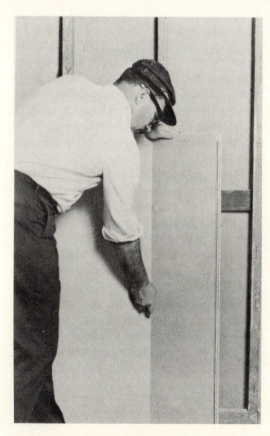

FIGURE 17:4
Steps to cut a wallboard panel.

plished, the back side of the panel is cut with the utility knife to separate the two pieces.

Small cut-outs for electrical boxes as well as other unusual shapes are removed with either a reciprocating saw, a saber saw, or a drywall saw (Figure 17:5).

Wallboard panels can either be installed with special nails or screws (Figure 17:6) or with an adhesive, which has the advantage of eliminating the need to conceal the heads of nails or screws with joint compound (Figure 17:7).

The nail pattern used to minimize nail popping is known as double nailing (Figure 17:8). Figure 17:9 shows the recommended spacing for nails. When a nail is driven into a wallboard surface, the hammer head should make a slight depression or "dimple" in the surface (Figure 17:10). This makes the procedure of hiding the nail holes with joint compound that much simpler.

A power screwdriver is preferred by many builders because of the speed with which a panel can be installed. It is especially ideal whenever the wallboard will be attached to metal studs (Figure 17:11).

Concealing the joints between panels must be performed according to the following steps:

1. Apply a liberal coat of joint compound with a 5″ to 6″ joint knife to both panel edges comprising the joint, making certain to fill the depression between the two panels.

2. Measure the length of reinforcing tape required to cover the entire joint.

3. Center the tape over the joint by hand and then, starting at one end of the tape, firmly run the joint knife across its entire length, making certain that no wrinkles result and the excess compound is removed.

4. Apply a thin coat of compound over the tape and allow it to dry thoroughly.

5. Apply a second coat over the taped area and extend its coverage beyond the edges of the first coat.

FIGURE 17:5

Selector Guide for USG Brand Screws

Fastening Application	Fastener Used
GYPSUM PANELS TO STANDARD METAL FRAMING (1)	
½″ single-layer panels to standard studs, runners, channels	⅞″ Type S Bugle Head
⅝″ single-layer panels to standard studs, runners, channels	1″ Type S Bugle Head
½″ double-layer panels to standard studs, runners, channels	1 5/16″ Type S Bugle Head
⅝″ double-layer panels to standard studs, runners, channels	1⅝″ Type S Bugle Head
1″ coreboard to metal angle runners in solid partitions	1¼″ Type S Bugle Head
½″ panels through coreboard to metal angle runners in solid partitions	1⅞″ Type S Bugle Head
⅝″ panels through coreboard to metal angle runners in solid partitions	2¼″ Type S Bugle Head
GYPSUM PANELS TO 12-GA. (MAX.) METAL FRAMING	
½″ and ⅝″ panels and gypsum sheathing to 20-ga. studs and runners	1″ Type S-12 Bugle Head
USG Self-Furring Metal Lath through gypsum sheathing to 20-ga. studs and runners	1¼″ Type S-12 Bugle Head
½″ and ⅝″ double-layer gypsum panels to 20-ga. studs and runners	1⅝″ Type S-12 Bugle Head
Multi-layer gypsum panels to 20-ga. studs and runners	1⅞″ Type S-12 Bugle Head
WOOD TRIM TO INTERIOR METAL FRAMING	
Wood trim over single-layer panels to standard studs, runners	1⅝″ Type S Trim Head
Wood trim over double-layer panels to standard studs, runners	2¼″ Type S Trim Head

Selector Guide for USG Brand Screws

Fastening Application	Fastener Used
METAL STUDS TO DOOR FRAMES, RUNNERS	
Standard metal studs to runners	⅜″ Type S Pan Head Also available with Hex Washer Head
Standard metal studs to door frame jamb anchor clips 20-ga. studs to runner Other metal-to-metal attachment (12-ga. max.)	⅜″ Type S-12 Pan Head
Standard metal studs to door frame jamb anchor clips (heavier shank assures entry in clips of hard steel)	½″ Type S-12 Pan Head
Strut studs to door frame clips, rails, other attachments in ULTRAWALL partitions	½″ Type S-16 Pan Head Cadmium Plated
TRIM AND ACCESSORIES TO METAL FRAMING	
Door hinges and trim to door frame Aluminum trim to metal framing (screw matches hardware and trim)	⅞″ Finishing Screw Type S-18 Oval Head Cadmium Plated
Metal base splice plates through panels and runner	1¼″ Type S Bugle Head
Batten strips to standard metal studs in demountable partitions	1⅛″ Type S Bugle Head
Aluminum trim to interior metal framing in Demountable and ULTRAWALL partitions	1¼″ Finishing Screw Type S Bugle Head Cadmium Plated
GYPSUM PANELS TO WOOD FRAMING	
⅜″, ½″ and ⅝″ single-layer panels to wood framing	1¼″ Type W Bugle Head
RC-1 RESILIENT CHANNEL TO WOOD FRAMING	
Screw attachment required for ceilings, recommended for partitions	1¼″ Type W, ⅞″ or 1″ Type S Bugle Head (see details above)
For fire-rated construction	1¼″ Type S Bugle Head (see details above)
GYPSUM PANELS TO GYPSUM PANELS	
Multi-layer adhesively laminated gypsum-to-gypsum partitions (not recommended for double-layer ⅜″ panels)	1½″ Type G Bugle Head

Notes: (1) Includes USG Standard Metal Studs, Metal Runners, Metal Angle Runners, Metal Furring Channels, RC-1 Resilient Channels. If channel resiliency makes screw penetration difficult, use screws ⅛″ longer than shown to attach panels to RC-1 channels. For 20-ga. Metal Studs and Runners, always use Type S-12 screws. For steel applications not shown, select a screw length which is at least ⅜″ longer than total thickness of materials to be fastened. USG Brand Screws are manufactured under U.S. Patent Nos. 2,871,752; 3,056,234; 3,125,923; 3,207,023; 3,221,588; 3,204,442; 3,260,100.

FIGURE 17:6

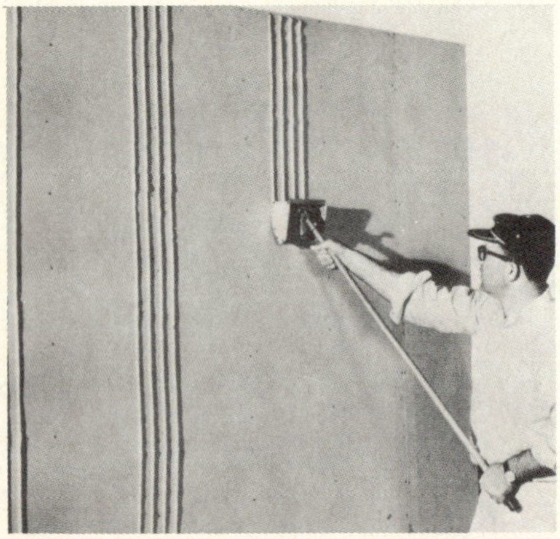

FIGURE 17:7
Spreader application of adhesive.

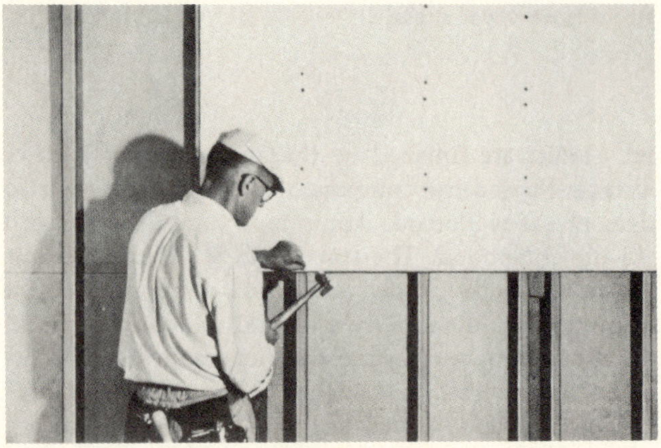

FIGURE 17:8
Double nailing.

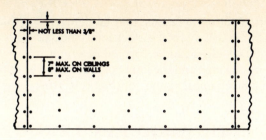

FIGURE 17:9

6. Allow this coat to dry; sand it lightly before applying a third coat, which should be feathered out beyond the edges of the second coat. (Figure 17:12).

Outside metal corners are nailed (Figure 17:13) to the underlying studs and are then covered with joint compound (Figure 17:14). Ceiling joints and inside corners, covered with reinforcing tape that has been folded along its centerline, are treated in the same manner as regular joints (Figure 17:15). The time required to con-

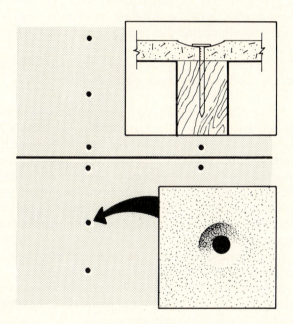

FIGURE 17:10

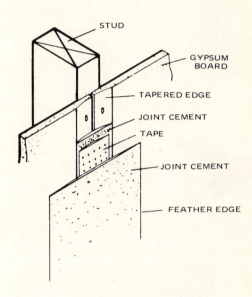

FIGURE 17:12
Steps to cover a wallboard joint.

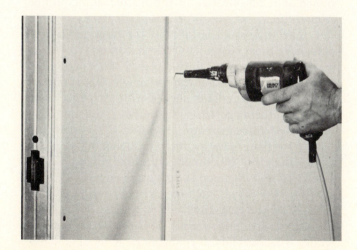

FIGURE 17:11

FIGURE 17:13

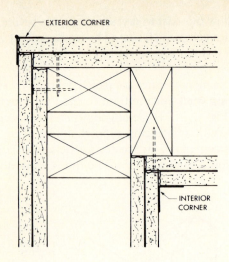

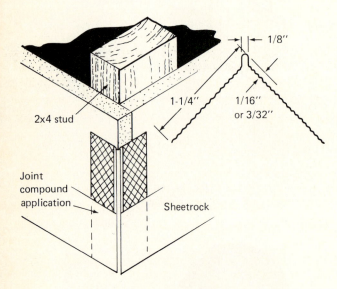

FIGURE 17:14

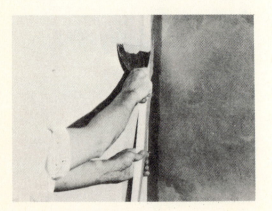

FIGURE 17:15

ceal the tape in these areas may take longer than that required to finish normal surface joints, because only one of the two edges can be covered at a time.

Both the tape and the joint compound can be applied with a mechanical applicator (Figure 17:16).

PLYWOOD PANELING

Prefinished 4′ × 8′ plywood panels can be applied to smooth wallboard backed walls, to furring strips, or to open studs with nails or adhesives (Figure 17:17). Many carpenters prefer to install panels with adhesives instead of using nails, because the installation takes less time and eliminates the task of concealing the nail heads.

To achieve success with the installation of plywood panels, the following suggestions should be used:

1. To estimate the number of panels needed to adequately cover the wall surfaces in any room, begin by adding the lengths of all walls together to determine the room's perimeter. Table 17:1 tells the number of panels required for various room perimeters. It is advisable to draw, on graph paper, the permanent features of each wall (e.g., windows, doors, electrical outlets, and heat ducts) to help you determine where each panel will be located (Figure 17:18).

It may be possible for you to deduct one or more panels from your initial determinations.

2. Make certain that the surfaces to which the panels will be applied are straight and plumb to guarantee a smooth, finished appearance.

3. Most installations are started in one corner of the room, with additional panels being added in succession until the final panel abuts the first one. For rooms having a large picture window or fireplace, the panels should start on one side of the feature and continue around the room.

4. The first panel installed is the most critical. Its outer edge must be plumb with the floor and rest on an underlying stud. A 4′ level is placed vertically on the wall so that a pencil line can be drawn that is plumb. The panel's outer edge is aligned with the pencil mark so that its other edge, facing the corner of the room, can be checked for squareness (Figure 17:19). When not square, the panel edge must be cut to match the irregularities found in the corner of the wall (Figure 17:20). The cut line is obtained by using a divider.

5. Cut the panels, face up, whenever either a hand or table saw are used, and face down, when an electrical circular or saber saw is used. This practice is done to reduce the possibility of splintering.

6. To cut a panel to accommodate a window or door

1. Mechanically tape joints.

2. Wipe down with broad knife.

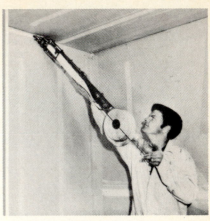

3. Mechanically tape interior corners.

4. Finish sides of angles with corner rollers and corner finisher.

5. Apply second coat of compound over tape using a hand finisher.

6. Spot fastener heads and apply second coat to metal reinforcement.

FIGURE 17:16
Taping joints using an Ames tape applicator and an Ames compound applicator. (Courtesy of U.S. Gypsum)

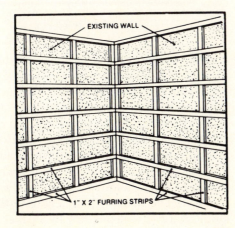

FIGURE 17:17

TABLE 17:1

Perimeter	Number of 4' × 8' panels needed
36'	9
40'	10
44'	11
48'	12
52'	13
56'	14
60'	15
64'	16
68'	17
72'	18
92'	23

FIGURE 17:19

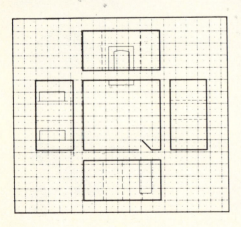

FIGURE 17:18

FIGURE 17:20

opening properly, it will be necessary to transcribe all the critical measurements to the back of the panel (Figure 17:21). It is advisable to drill holes in the corners of the portion to be cut out so that a saber saw blade can be easily inserted and used to perform the cut. Also, make certain that the panel is well supported, especially around the area to be cut out, to avoid any mishaps.

7. Careful planning is a necessity when paneling a foyer area in a split level house where two stairways are involved. The height, from ceiling to the base of the lower stairway, will require two panel lengths. Care must also be exercised, when a grooved panel has been

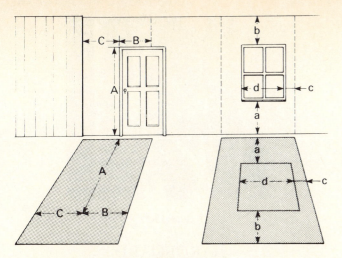

FIGURE 17:21

selected, that the grooves of the top panel align perfectly with those of the bottom panel.

8. Covering the seams and joints at the ceiling and floor and around windows and doors must be done carefully, with the proper molding types, to give the project a finished appearance. The 45° angles should be cut with a power miter box (Figure 17:22) and then trimmed with a coping saw.

9. Electrical outlets must be measured, marked, and then cut out with extreme care so that the panel not only lines up perfectly with the outlets but also fits tightly and is plumb with the adjacent panel (Figure 17:23).

COVERING BASEMENT WALLS

Masonry walls can be insulated with rigid insulation placed between furring strips (Figure 17:24). The furring strips provide a nailing surface for the finished wall material.

FIGURE 17:22
Parts of a power miter saw. (Courtesy of DeWalt)

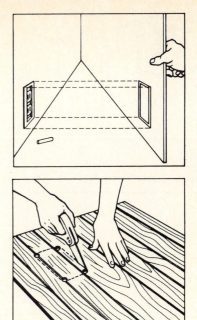

FIGURE 17:23
(Courtesy of Georgia Pacific)

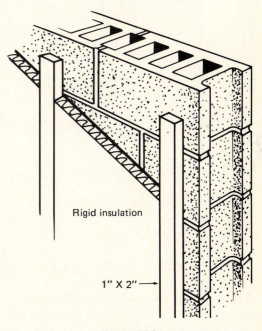

Rigid insulation

1" X 2" ⟶

FIGURE 17:24
Rigid insulation placed between furring strips.

Another method uses an adhesive to install the rigid insulation on the masonry wall and also to glue the finished wall material to the rigid insulation (Figure 17:25).

A third method for covering basement walls involves the framing of stud walls that are positioned just inside the masonry walls (Figure 17:26). Blanket or batt insulation is placed between the studs before the finished wall material is installed with nails, screws, or an adhesive.

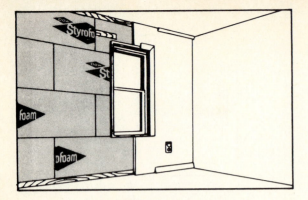

FIGURE 17:25

FIGURE 17:26

SOLID WOOD PANELING

Solid wood panels (tongue and groove) of pine, cedar, and redwood (4″ to 12″ widths, with nominal thickness of 1″) can be applied horizontally or vertically to the interior walls with nails or adhesives. When installed horizontally, the individual tongue and groove pieces are applied directly to the studs or wallboard backing; however, when installed vertically, furring strips are needed to provide each piece with an adequate nailing or adhering surface (Figure 17:27).

Other wooden products that can be used to cover interior wall surfaces to achieve certain dramatic effects include cedar shingles and shakes, pecky cypress, and barn siding with shiplap joints.

FIGURE 17:27

HARDBOARD

These prefinished 4′ × 8′ sheets are $\frac{1}{4}$″ thick and are tough, resistant to moisture, and available in various colors and patterns. They can be installed with adhesives, nails, or screws; some are designed to be placed in metal moldings along their edges. "Pegboard" is a specialized hardboard used to display or store many household items.

CERAMIC TILES

Ceramic tiles are available in an endless array of patterns, glazes, shapes, and colors, and range in size from the small 1″ mosaic squares to the large 12″ squares.

To estimate the number of $4\frac{1}{4}$″ × $4\frac{1}{4}$″ square tiles needed to tile a specific room, the room perimeter must be multiplied by the height, resulting in the number of square feet of wall surface to be tiled. This value must then be multiplied by 8 (the number of tiles necessary to cover one square foot) to obtain the total number of tiles required. It is suggested to purchase extra tiles to replace those that break during the installation or those that become accidentally damaged at a later time.

Covering walls with ceramic tiles is relatively easy, provided the following suggestions and requirements are met:

1. The wall surfaces must be dry, firm, plumb, and free of wax, loose paint, grime, and wallpaper. Bathroom walls, exposed to repeated contact with water, can only be covered with water resistant gypsum wallboard.

2. You should know in advance how many tiles will be required for each horizontal row, the approximate size of the border tiles, and whether or not they will need cutting. Drawing a plumb, vertical line in the center of each wall will enable you to obtain these answers.

3. The adhesive must be applied to the walls with a notched-edge trowel (Figure 17:28), limiting the area covered in one application to 15 sq. ft.

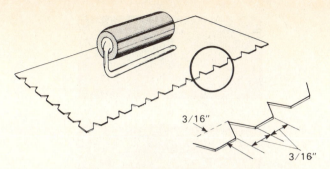

FIGURE 17:28

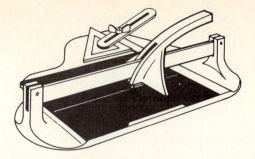

FIGURE 17:30
Glass cutter.

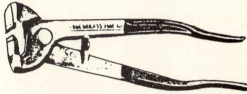

FIGURE 17:31
Picture of tile nippers in use is from the *Franklin Glue Book*, p. 155.

4. Start adding the tile at floor level (exception: bathroom walls). Each one should be turned slightly as it is firmly pressed into the adhesive to ensure an adequate bond. The practice of installing one less tile per horizontal row, as you continue towards the top, enables you to properly check both the vertical and horizontal lines for straightness with a minimum of tiles installed (Figure 17:29).

5. Tiles can be scribed with a glass cutter and then snapped over some rigid surface, or can be cut quickly along the scored line with a tile cutter. (Figure 17:30) Unusual shapes (e.g., around exposed plumbing lines) can be scribed with a glass cutter and then slowly chipped away with a tile nipper (Figure 17:31) or pliers.

6. Once all the tiles have been positioned on a particular wall, checked for straight lines, and the adhesive has had at least 24 hours to cure, the latex grout is applied. It is mixed with water and applied with a rubber-surfaced trowel or sponge (Figure 17:32), using arc-like strokes to force the grout into the spaces between the individual tiles. The excess grout is removed with a dampened sponge as soon as it starts to dry. The wall is

FIGURE 17:29

FIGURE 17:32
From the *Franklin Glue Book*.

rinsed with a damp cloth only after the grout is completely dry.

7. Bathroom walls must be tiled in a special manner. A horizontal line, at least one tile length above the lowest point of the bathtub line, must be drawn with a level across the entire wall. The wall area is marked out with

full tiles so that the border tiles in each corner are approximately the same size. Vertical lines drawn perpendicular to the initial horizontal line will guarantee straight tile lines in both directions (Figure 17:33).

BRICK AND STONE VENEER

Both of these materials make excellent wall coverings that must be attached to a firm wallboard surface with an adhesive. Bricks are installed from the top of the wall downward, whereas the individual stone pieces start at the floor and continue towards the ceiling. This technique gives both a more natural appearance. Half bricks, near the floor, can easily be hidden by a baseboard molding while they are not as easily masked near the ceiling. Smaller pieces of stone placed near the ceiling, however, are more acceptable and appear more realistic than if placed near the floor.

The spacing between individual pieces for both types is $\frac{3}{8}''$ to $\frac{1}{2}''$ and is usually filled with grout or additional adhesive to simulate mortar joints (Figure 17:34).

PLASTER

Plaster has long been recognized as an extremely desirable covering for both walls and ceilings. It is durable, provides structural rigidity, and is adaptable to a variety of surface shapes. Plaster has lost much of its appeal to modern-day contractors, who, because of cost consciousness, have rejected its use in lieu of wall coverings that are faster to apply and to finish.

Very few modern homes are presently built with plastered walls for the following reasons:

1. Qualified plaster masons are becoming difficult to find.
2. The amount of time spent installing the materials is usually more than with other drywall materials.
3. The walls and ceilings must receive at least two separate applications, the initial plaster base plus one or more finished coats.

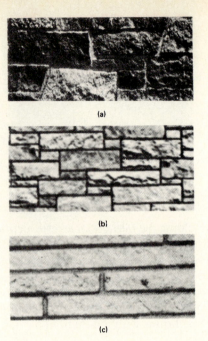

FIGURE 17:34

For those who still prefer plastered walls, regardless of cost, the following steps are required to achieve success:

1. Sheets of gypsum lath must be applied to the studs as the plaster base with metal lath placed in corners for reinforcement purposes (Figure 17:35).
2. Wooden strips, known as plaster grounds and having a thickness equal to the plaster base and plaster combined, must be installed around doors and windows to serve as a guide or leveling surface for maintaining a particular plaster thickness throughout the entire room.

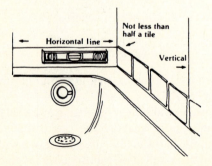

FIGURE 17:33

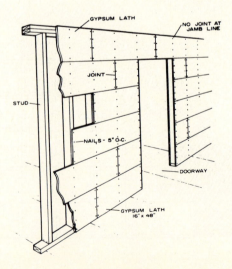

FIGURE 17:35

3. The base coat is applied to the walls, made smooth with a trowel, and then given at least one day to dry.

4. One or more finish coats are then applied directly over the base coat.

HIDING BASEMENT FEATURES

The appearance of the basement area can be improved by hiding certain features. They can be easily covered with a finished wall material once they have been properly framed out. In most instances, the framing procedure is not complex.

Open cellar stairs can be converted into a closed stairwell by simply building a frame beneath the stairs to provide a nailing or gluing surface for the finished material (Figure 17:36).

Metal I-beams and lally columns must be completely boxed in with furring strips or studs so that the finished wall surface can be glued to them (Figure 17:37). This same procedure can also be used to hide heating pipes and ductwork.

Once a framework of studs or furring strips is constructed around a deep set basement window, the casement can easily be covered with the finished wall material (Figure 17:38).

Panel along stairs with pieces left over from door and window cutouts.

Setbacks such as some windows can be made decorative with scraps of paneling, lumber and molding.

FIGURE 17:36

Use narrow pieces of paneling combined with molding to "box in" pipes and to cover posts and beams.

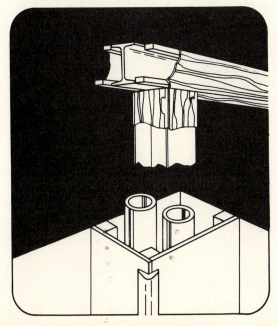

FIGURE 17:37

A closet or other small room can be framed like this and paneled.

FIGURE 17:38

QUESTIONS

1. List the steps involved in taping a joint between two wallboard panels.

2. Compare the techniques used to cut a wood panel using a circular saw to those used when a table or hand saw are employed.

3. Explain the concerns associated with the installation of a grooved wall panel in the foyer of a split level home.

4. Explain the procedures that must be followed to mark and cut out the openings for electrical boxes in any wall panel.

5. Estimate the number of 4 1/4" square ceramic tiles needed to cover 100 square feet of wall surface.

6. What is the minimum length of time that ceramic tiles must be embedded in their adhesive before the latex grout can be applied?

7. List the steps required to both prepare the walls and then apply plaster to them.

18

FLOOR COVERINGS

Once the walls and ceiling areas have been completed throughout the house, the finished floor coverings can be installed. Depending upon personal preference, the type of floor coverings can vary from room to room within the house. Knowing which type of covering will be used in each room, before the actual construction begins, will help the carpenter select the proper materials and techniques for that room. For example, if the living room will definitely be covered with carpeting, there will be no reason to install a wood floor beneath it, since an underlayment plywood floor will be sufficient.

Coverings for floors are divided into four separate categories. *Wood flooring*, in the form of strips, planks, or parquet blocks, has always been popular in living and dining room areas. *Resilient type surfaces*, laid as seamless sheets or as square tiles, have become widely acceptable as coverings for kitchen and bathroom areas because they are practical, easy to apply and keep clean, and they are durable and convenient. *Rigid materials*, such as ceramic, slate, brick, and flagstone, can be used to dress up a drab foyer or accentuate the fixtures in a bathroom. *Carpeting*, sometimes used within and outside the house, is popular as a floor covering material.

SUBFLOORING

All finished floor coverings require a solid, strong subfloor or underlayment to rest on. Plywood and particle board are both used as subflooring materials because of their strength, smooth surface, and the ease and speed with which they are installed.

To allow for expansion, the panels must not be tightly butted together but be given a slight spacing (equivalent to the thickness of a match book cover). The joints of adjacent panels must be staggered to guarantee that all four corners never meet at the same point (Figure 18:1). Whenever two panels meet on the same joist, two beads of glue ($\frac{1}{4}''$ in diameter) must be applied to the joist to ensure that both panels ends are glued.

The two main reasons a concrete floor is covered with plywood are to insulate the floor with polystrene foam and to level and soften its hard, no-flex surface. The procedural steps involved are:

1. Wooden 2×2 strips, spaced $16''$ o.c. apart and running parallel to one of the walls, are glued to a sound, dry, clean concrete surface.

2. The styrofoam is cut to fit between the strips and is then glued to the concrete.

3. Each strip must be checked for level along its entire length and with the other strips.

4. Glue is then applied to the top surface of each 2×2.

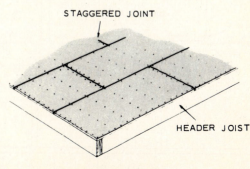

FIGURE 18:1

235

5. The plywood panels are positioned at right angles to the strips and nailed securely to the strips (Figure 18:2).

6. The finished floor covering can then either be nailed or glued to the plywood panels.

Wood Floors

Strip Flooring

Hardwoods, such as oak, maple, birch and beech, and softwoods, like pine and fir, are used for strip flooring. The individual pieces (ranging in thickness from $\frac{5}{16}''$ to $\frac{25}{32}''$ and available in widths of $1\frac{1}{2}''$, $2''$, $2\frac{1}{4}''$, and $3\frac{1}{4}''$, with $2\frac{1}{4}''$ being the most popular size) are normally laid perpendicular to the floor joists (Figure 18:3).

The various strips (ranging in length from 6″ to 6′ in a bundle (24 board feet) of number 1 common grade oak, for example) should be loosely arranged in the room to be covered five days before their actual installation. This allows them to become stable with their environment. The individual pieces comprising a run (one length across the room) can be cut and fitted together

prior to the installation to make certain that the lengths and the joints are staggered in successive courses. A $\frac{1}{2}''$ space is left next to the wall for expansion (Figure 18:4).

Each piece of strip flooring contains a tongue on a side and an end, and a groove on the other side and other end (Figure 18:5). During installation, the pieces can either be nailed or glued to the underlying surface.

The installation is started by positioning the first strip along a side wall with its groove edge facing the wall (Figure 18:6). It is imperative that this first run be perfectly straight to ensure that the floor will be straight. This first run must be face-nailed to the underlying material once it has been lined up perfectly straight. The nails are subsequently set beneath the surface and filled with wood filler. All successive runs are either blind nailed through the tongue (Figure 18:6) or glued to the underlying surface.

Occasionally, the groove of one strip will not slide easily over the tongue of the preceeding strip. To solve this problem, place a short piece of scrap flooring next to the difficult to end strip and tap it with a hammer until the two strips line up properly (Figure 18:7).

It is important to check the distance from the assembled floor to the wall from time to time to ascertain whether or not the runs are being kept in a straight line. This measurement is very important in the remaining foot of space so that the individual strips can be adjusted in tightness to correct any errors in alignment.

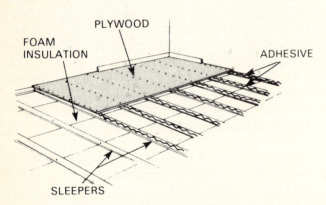

FIGURE 18:2

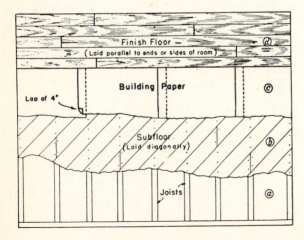

FIGURE 18:3

FIGURE 18:4

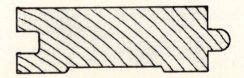

FIGURE 18:5

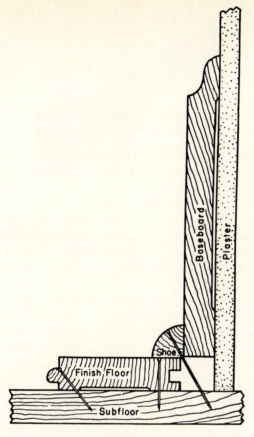

FIGURE 18:6

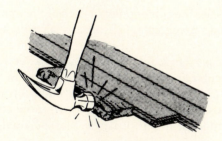

FIGURE 18:7

The final strip will probably have to be ripped to fit into the space remaining and then must be face nailed (This last piece should be rectangular rather than triangular in shape, if the floor was kept square). See Figure 18:8.

It is possible to install strip flooring over a concrete slab, provided the following steps and considerations are adhered to:

1. Attach water repellent 1 × 2 wooden strips, placed parallel to the longest wall at 16″ intervals, to a dry, sound concrete surface using the proper adhesive plus an occassional concrete nail.

2. Leave a short air space between the butt ends of each two strips.

FIGURE 18:8
Face nailing the last two pieces of strip flooring.

3. Cover these 1 × 2 strips with 4 mil polyethylene film, making certain to lap the edges of the plastic only over the edges of the wooden strips.

4. Nail a second layer of 1 × 2 wooden strips (sleepers), using 4d nails, directly on top of the first layer (Figure 18:9).

5. Install the strip flooring at right angles to the wooden strips (Figure 18:10), making certain that no two adjacent strips of flooring have their joints break within the same sleeper space. Also, each flooring strip must be supported by at least one sleeper.

Plank Flooring

The procedures for strip flooring are used to install this type of wood flooring. Each plank is $\frac{25}{32}$″ thick, with tongue and groove edges. The width may vary from 3″ to 9″ (1″ increments). The planks are placed on underlying surface in mixed or random widths.

To give the finished floor a rustic, aged appearance, holes are drilled near the butt edges of the two planks comprising the same run. The holes are then plugged with dowels to simulate wooden pegs (Figure 18:11).

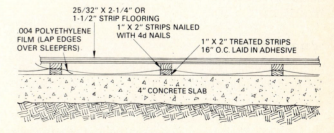

FIGURE 18:9

FIGURE 18:10
Courtesy of Bostich.

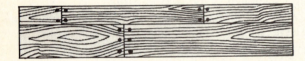

FIGURE 18:11

Every effort should be made to avoid several successive runs with similar widths, similar joints, or a predominance of peg holes in one area. This problem can often be solved by cutting a different length of plank from the end of the previous run and using the cut off portion to start the next run of the same width (Figure 18:12). Developing a plan of attack prior to the actual installation should help to avoid these common mistakes.

Parquet Blocks

Solid wooden blocks, constructed from short sections of hardwood flooring ($2\frac{1}{4}$" wide and $\frac{25}{32}$" thick), can be purchased as squares having dimensions of $6\frac{3}{4}$", $7\frac{1}{2}$", 9", or $11\frac{1}{4}$", with 9" being the most popular size. These blocks have their tongue edges at right angles to each other.

FIGURE 18:12

Laminated blocks, in the same sizes and $\frac{3}{8}$" thick, are available. They consist of three layers (plies) of hardwood bonded together. The tongue edges are situated opposite each other.

Parquet blocks have been formed by gluing together individual pieces of flooring to create a unit square. Either type is installed by nailing or gluing the squares to the underlying material. Solid blocks must be given a one inch expansion space wherever they adjoin a wall.

The installation pattern can be either square or diagonal, with the grain pattern of each block being placed at right angles to those surrounding it.

To successfully install a wood parquet floor in a square pattern, these steps must be carefully followed:

1. Locate the center point of each of the four walls in the room, disregarding all offsets, alcoves and stair areas.

2. Snap a chalk line on the floor between two opposite center lines. Repeat for the other walls.

3. Check the point of intersection of the two chalklines for square. A framing square, a square tile, or the method of a 3-4-5 triangle may be used (Figure 18:13).

4. Either lay a row of loose blocks end to end along each chalkline or measure the exact length and width of the room to determine the exact size the border tiles (blocks) should be. If possible, the tiles on opposite walls should be similar in size to give the finished floor a finely crafted appearance. The chalklines may need to be shifted somewhat to attain this important architectural feature.

5. Apply mastic to the underlying surface and cover it with builder's felt. The felt helps guarantee a smooth surface and provides the installer a clean surface on which to mark the location of each block.

6. Apply a second coat of mastic over the felt and then install the blocks. The first block is placed at the junction of the chalklines, with the balance of the blocks extending out in all directions from the first one.

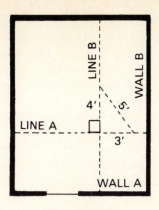

FIGURE 18:13

ESTIMATION OF WOOD FLOORING

Strip Flooring

To determine the number of board feet of strip flooring required to cover the floor surface of a room, the following equation must be used. It requires that you determine the area of the room in terms of total square feet of floor space.

_____Area (sq. ft.) + _____% of area =

_____ Board feet

Choose from Table 18:1 the number that goes into the blank for percent. These numbers relate to the flooring size used. To determine the number of bundles of flooring required, use this format:

$$\frac{\text{Total number of board feet}}{24 \text{ board feet per bundle}}$$

= Number of bundles required

For example, the calculations necessary to determine the number of board feet and the number of bundles of $\frac{3}{8}'' \times 2''$ flooring to cover a room $12' \times 15'$ (180 square feet) are the following:

180 square feet + 30% of 180 (54) = 234 board feet

$$\frac{234 \text{ board feet}}{24 \text{ bd. ft./bundle}} = 10 \text{ bundles}$$

Parquet Blocks

To determine the total number of $9'' \times 9''$ blocks required to adequately cover a floor area, the following calculations must be made:

TABLE 18:1
Relationship between Flooring Size and Percentage Used to Convert Square Feet to Board Feet

Percentage	Flooring size
55%	$\frac{25}{32}'' \times 1\frac{1}{2}''$
$42\frac{1}{2}\%$	$\frac{25}{32}'' \times 2''$
$38\frac{1}{3}\%$	$\frac{25}{32}'' \times 2\frac{1}{4}''$
	$\frac{3}{8}'' \times 1\frac{1}{2}''$
	$\frac{1}{2}'' \times 1\frac{1}{2}''$
30%	$\frac{3}{8}'' \times 2''$
	$\frac{1}{2}'' \times 2''$
29%	$\frac{25}{32}'' \times 3\frac{1}{4}''$

1. Multiply the square foot area by 1.8 to equal the number of blocks needed.
2. Multiply the number of blocks by the specific waste allowance in Table 18:2 for the floor area.
3. The total number of blocks required will equal the sum of the two calculations, (1) + (2). (Note: For any area up to 300 sq. ft., a rough approximation of the total number of blocks required is easily obtained by doubling the area in square feet).

RESILIENT FLOORS

Resilient flooring types are manufactured as sheet materials, up to 15′ wide, or as 9″ or 12″ square tiles. The materials of either type and the designs, colors, and textures available provide endless variations for the creation of an ideal floor scheme.

Whenever the subfloor is smooth, clean, and level, tiles or sheet materials can be installed directly to it. If

TABLE 18:2
Waste Allowances for Specific Area Ranges

Area range (sq. ft.)	Waste allowance
1–50	14%
50–100	10%
100–200	8%
200–300	7%
over 300	5%

the subfloor is questionable, it is advisable to install an underlayment designed specifically for resilient flooring. Remember, to leave a slight space between the underlayment panel butt edges for expansion, to stagger the panels' joints, and to check the entire surface for smoothness since slight irregularities or rough underlayment edges will eventually become visible on the surface.

Tiles

The procedure for installing floor tiles is very similar to that for parquet blocks. They must be installed with a floor tile adhesive that is water resistant, easy to clean up, and suitable for the material being installed. Most floor tile adhesives possess all these qualities, but in order to be effective, they must be applied to a dry surface.

Specific information concerning the four most popular tile materials is presented in Table 18:3.

Sheet Materials

Most sheet materials are cemented to the subfloor. They can be left loose, provided that some form of molding along the perimeter of the flooring holds it in place. The most important attribute of sheet flooring is its lack of seams. With some preliminary planning, it is possible to use sheet flooring to cover an average size room, from wall to wall, without the presence of a seam.

Sheet flooring can be installed above, below, and at grade level when equipped with the appropriate backing. It is advisable to consult manufacturer's specifications prior to the installation, with respect to location and underlying surface, so that the correct type of sheet flooring and adhesive can be used.

A comprehensive sketch of the entire floor area should be drawn prior to the actual installation. It should show all the pertinent measurements. These measurements are then transferred to the sheet flooring, making certain that its entire perimeter is increased by 3". The sheet flooring is then cut with a sharp utility knife along this outer line and carried into the room for custom fitting (Figure 18:14).

Align one of the material's factory edges along one wall to determine both proper fit and straightness. The material's flexibility will allow it to be positioned in the room with the excess material flared up against the walls and corners of the room (Figure 18:15).

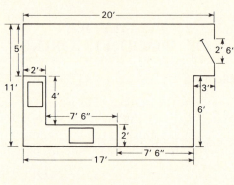

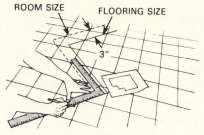

FIGURE 18:14

TABLE 18:3
A Comparison of the Four Most Popular Floor Tiles

Material	Thickness	Installation	Durability	Maintenance	How cuts are made
Asphalt	$\frac{1}{8}''$	Anywhere	Good	Difficult	Scored deeply with knife or awl; then snapped on line. Tile must be heated for irregular cuts (scissors used).
Vinyl Asbestos	$\frac{1}{16}''$ $\frac{1}{8}''$	Anywhere	Excellent	Easy to clean	Heated and then cut with knife or scissors
Vinyl	$\frac{1}{16}''$ $\frac{1}{8}''$	Anywhere	Excellent	Easy to clean	Tile must be warmed before cutting with scissors.
Cork		On or above grade	Good	Fair, good (when covered with vinyl)	Easily cut with a utility knife

FIGURE 18:15

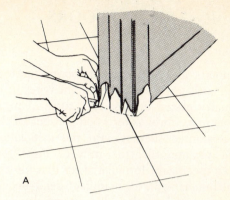

A

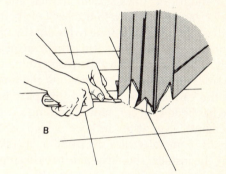

B

Start by cutting the excess material at the outside corners downward on a straight line to the junction of the floor. Once this has been accomplished, do the same for the inside corners. Then carefully trim the excess material resting on the wall until the flooring lies flat and fits snuggly against the wall (Figure 18:16).

Doorways are treated like outside corners. However, you must make a series of vertical cuts before the material can be pressed tightly into the door jamb area and the excess carefully trimmed off (Figure 18:17).

The floor covering must be completely fitted to the room before the adhesive can be applied. In large areas, it is advisable to roll up all or a portion of the covering and spread the adhesive with a trowel (Figure 18:18) before rolling the material back onto the adhesive. A push broom is used to smooth out any ripples in the floor covering and to expel any air pockets.

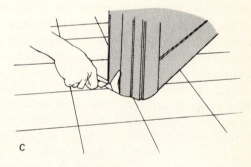

C

FIGURE 18:17

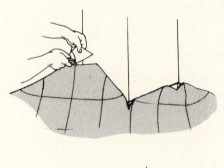

FIGURE 18:16

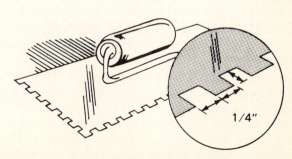

1/4"

FIGURE 18:18

RIGID FLOORS

Ceramic Floors

A smooth, structurally sound flooring surface can be covered with any of three different types of ceramic tiles: quarry, glazed, or mosaic. All three are installed by the same procedures as for wood parquet blocks.

To guarantee success with the installation of a specific ceramic tile, the installer must select an adhesive specifically formulated for that tile and must become familiar with the manufacturer's specifications with respect to how the adhesive should be applied; e.g., if it should be thicker than normal and if it is necessary to apply adhesive to the back of the tile.

It is advisable to take the time to lay out the tiles over the complete floor before setting them in place permanently. This procedure will help determine the size of the border tiles, as well as to pinpoint the exact locations of the various tiles, provided the space between each tile has been kept constant. By noting specific key measurements (e.g., the distance which a specific number of complete tiles is from a specific wall location), the installer should be able to easily reproduce the original layout.

Spread the adhesive over a 3 sq. ft. area of the floor that includes the junction of the two chalk lines. The first tile is set in alignment with the two lines and is then surrounded with additional tiles. Each tile must be firmly pressed downward into the adhesive to minimize its tendency to slide into another tile, causing adhesive build-up on the edges. Once one area has been tiled, another 3 sq. ft. area can be covered with adhesive.

It is advisable to check that the tile surface is level and the tile pattern is square periodically during the installation. Pieces of wood, or thick cardboard ($\frac{1}{16}$" thick) should be used as spacers to maintain a uniform spacing between the tiles.

Border tiles may require cutting. This can be accomplished by using a cutter (Figure 18:19) or by scoring the tile with a glass cutter and then snapping it in two on the scored line.

Ceramic mosaics are normally purchased as foot square sheets with individual 1×1 or 2×2 pieces embedded in a web backing. The squares are first set in place, without adhesive, to determine where the webbing should be cut to achieve the various shapes desired.

A minimum of 24 hours is necessary for the adhesive to thoroughly set. Any spacers used between the tiles must be removed before the grout can be pressed between the joints. Latex grout, in dark colors, is preferred over other kinds because it resists cracking when stepped on, and it remains clean for longer periods of time.

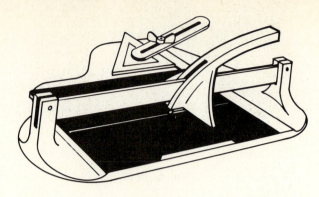

FIGURE 18:19

A damp sponge is used to clean the excess latex grout from the tile surface. The grout should be allowed to set up for at least 24 hours before walking on the tiles.

Slate

A box of gauged slate contains various rectangular-shaped pieces, $\frac{1}{4}$" to $\frac{1}{2}$" thick, in several subdued colors. It includes enough material to cover 10 sq. ft. of floor space.

The individual pieces can be set in a bed of adhesive, provided the underlying surface is sound. Small strips of $\frac{3}{8}$" lath, sheet rock, or plywood placed between the individual pieces provide uniform spacing that will simulate a mortar joint once the grout has been added (Figure 18:20).

Thicker, ungauged slate can only be set in a bed of mortar which rests on an underlying concrete slab.

Slate can be cut with a masonry blade; or it can be scored on a marked line with a cold chisel while resting on a flat bed of sand, and then be snapped along the scored line.

Brick and Flagstone

Any structurally sound flooring surface can be covered with real brick or stone veneers, or with plastic materials that simulate brick or stone.

FIGURE 18:20
Slate floor with $\frac{3}{8}$" mortar joints.

The $\frac{3}{4}''$ thick pieces are set in adhesive like ceramic tile. This material provides the installer with a variety of design possibilities. Figure 18:21 shows the three most popular design patterns for brick.

The "mortar joint" between each brick should be a uniform $\frac{3}{8}''$, and be filled with grout one day after the brick or stone is installed. After the grout has been applied and the surface thoroughly cleaned, a sealer, specially formulated for masonry floors, should be applied. It will improve the floor's appearance as well as protect it against spills and stains.

CARPETING

There are many reasons why floors are covered with carpeting. It looks good, is soft to walk on, absorbs noise, helps conserve energy, hides unsightly floors, and even helps to coordinate the appearance of the room with its contents.

The area to be covered with carpeting must be sketched out on graph paper showing all the pertinent measurements; e.g., the basic room doorways, alcoves, closets, and obstructions (Figure 18:22). The overall length is multiplied by the overall width to determine

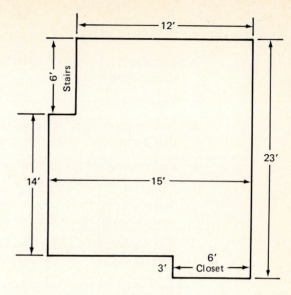

FIGURE 18:22

the room's area in square feet. Dividing this value by 9 yields the number of square yards of carpeting required to cover the floor.

Roll carpeting, available in widths up to 15' can be installed four different ways: using an adhesive, using pressure-sensitive double-faced tape, using carpet tacks, or by using metal tackless stripping (Figure 18:23).

No carpet should be installed, however, without first laying some form of padding on the floor. Its presence helps guarantee a full carpet life as well as a softer surface on which to walk.

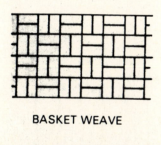

BASKET WEAVE

HERRINGBONE

RUNNING

FIGURE 18:21

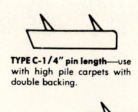

TYPE C-1/4" pin length—use with high pile carpets with double backing.

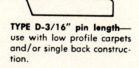

TYPE D-3/16" pin length— use with low profile carpets and/or single back construction.

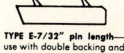

TYPE E-7/32" pin length— use with double backing and medium or high pile carpets.

FIGURE 18:23

QUESTIONS

1. List the four (4) major categories of finished floor coverings.

2. List three (3) benefits derived by gluing a plywood subfloor to its joists rather than nailing it in place.

3. Explain the procedures involved to cover a concrete floor with plywood.

4. List the procedures that must be performed on a concrete slab to prepare it so that it can be covered with strip flooring.

5. Determine the total number of board feet required to cover a 12′ × 16′ floor using the strip flooring size 25/32″ × 2 1/4″.

6. What are the three (3) critical considerations that should be made on any underlayment that will be eventually covered with resilient flooring?

7. List the four (4) popular floor tiles, explaining how each one is cut prior to its installation and how each is maintained as a floor covering.

8. How large an area should you cover with adhesive when installing ceramic tile?

9. List four (4) advantages that the installation of carpeting provides.

19

FINISH AND TRIM: EXTERIOR AND INTERIOR

Nothing detracts more from the appearance of a house than the improper installation of the wooden trim members on the inside or outside of the house that are supposed to give it a finished professional look.

EXTERIOR TRIM

The cornice and rake sections of a residential roof are important integral components of the exterior trim. How these two sections will be constructed must be determined by careful examination of the architectural plans long before the roof is begun. The size of the overhang, the pitch of the roof, and whether or not the rake section will be closed or extended must be learned before you cut the first roof rafter.

Cornice Section

The eave or cornice section is formed by the overhang of the roof. It connects the roof edge to the upright wall below. In most residential homes, the cornice area is completely boxed in to provide a finished appearance as well as to prevent the entry of pests into the house (Figure 19:1).

An open cornice, in which the overhanging portions of the rafters are exposed, requires the insertion of a nailer between the rafters so that the underlying sheathing and exterior covering can be extended to the roof sheathing. This design necessitates more work to properly cut and fit each individual component into these spaces (Figure 19:2).

It is important for the carpenter to learn both the name of each member of a cornice section and its structural purpose (Figure 19:3).

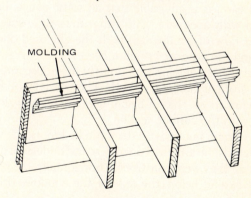

FIGURE 19:1
Sloped cornice soffit.

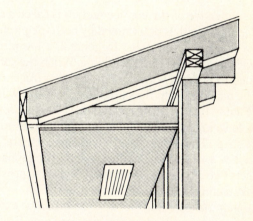

MOLDING

FIGURE 19:2

A *fascia board*, which serves as the face of the cornice section, is nailed across the vertical ends of the rafters. It should be installed flush with the top surface of each rafter, before the roof is sheathed, to guarantee that it will be covered by the sheathing.

245

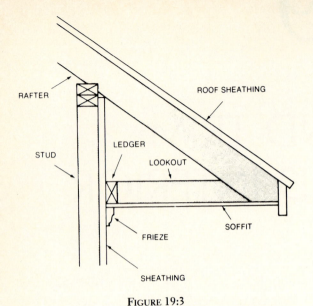

FIGURE 19:3

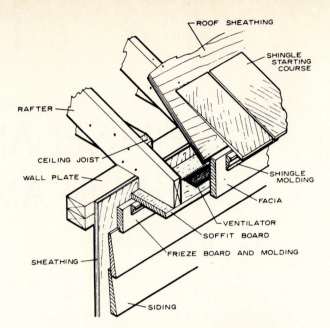

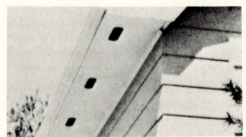

FIGURE 19:4

The underside of the cornice section is called a *soffit* (plancier). Its width is determined by both the pitch and the overhang of the rafters. In general, a low-pitched roof will have a substantial overhang. When the soffit is designed to be level with the horizontal, a *ledger board* must be nailed level to the upright studs of the wall section (Figure 19:3). This board provides support for the 2×4 *lookout studs* that connect the rafters to the ledger strip. The ledger board, the lookouts, and possibly a 2×4 nailer, located just behind the fascia that connects two adjacent rafters, are designed to provide a suitable nailing surface for the soffit section.

The material used for the soffit areas can be exterior plywood, custom fitted at the site, or some prefabricated material designed to fit a specific opening. In either case, it is advisable that all soffit panels be installed with adequate ventilating units to ensure the proper flow of air through both the enclosed cornice area and the attic area (Figure 19:4).

The soffit panels are supported at the house wall by a *frieze board*, a molding, or both. The frieze board has a second purpose; that of blending the exterior wall covering with the cornice section.

For certain homes with steep pitched roofs, the soffit panels must be attached directly to the underside of the rafters; if they were constructed horizontally, the panels might extend below the tops of the windows (Figure 19:5).

Rake Section

The rake is that portion of the roof that overhangs the gable end. A close rake is simple to construct. A nominal 2″ board, known as a *fascia block*, is nailed to the top edge of the rafter and is covered with the roof

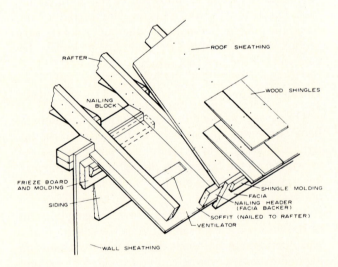

FIGURE 19:5

sheathing. The fascia board is then nailed to its fascia block. No soffit is required because there is only an extension from the side wall of $1\frac{1}{2}$″ (Figure 19:6).

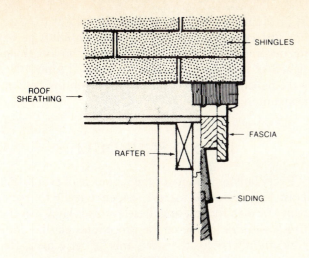

FIGURE 19:6

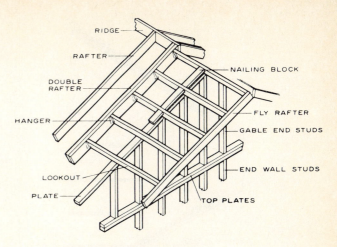

FIGURE 19:8

An extended boxed rake section must be constructed before the roof is sheathed. Short pieces of 2 × 4 or 2 × 6 lookouts are cut to the desired length and spaced at 16″ intervals along the top edge of the end rafter, before they are nailed to the rafters (Figure 19:7). If possible, the nails should be driven through the rafters into the lookouts, whose top edge must be flush with that of the rafter. A nominal 2 × 4 or 2 × 6 known as a fly rafter (lookout block) with its top surface cut the same as that of the rafters, is nailed to the lookouts and finally attached to the ridge board at its upper end (Figure 19:8).

A fascia board is nailed to this nominal 2″ lookout block, and the exterior rake section is completed by nailing a 1 × 2 trim member to the top of the fascia board. The soffit panels, with vents installed, are nailed to the lookouts and are supported along the wall by either a frieze board or molding (Figure 19:9).

The resulting nailing of the roof sheathing to the overhanging components of the rake section helps to give each of them additional strength.

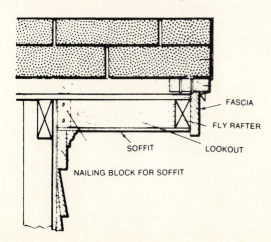

FIGURE 19:7

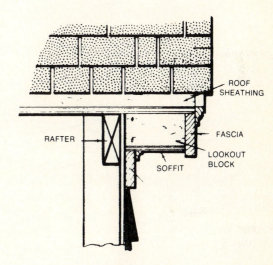

FIGURE 19:9

Finally, the rake fascia must be constructed to perfectly match the contours of the adjoining cornice section. This is normally accomplished by framing the area with 2 × 4s, before the trim members are added, to achieve the finished, boxed appearance (Figure 19:10).

Optional Exterior Trim Components

Wood Corners

Some carpenters prefer to cut horizontal wood siding or wood shingles and fit them against inside and outside corners that are constructed of either 1 × 2 or 1 × 4 wood trim rather than to use the more popular metal corners. Because of their exposure, these wooden corners must always be installed plumb and level to avoid the illusion of improper or faulty construction (Figure 19:11).

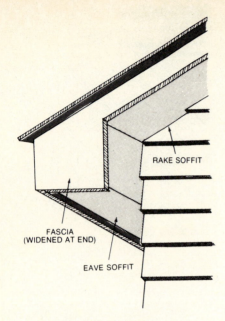

FIGURE 19:10
Horizontal cornice soffit—the cornice fascia and rake fascia meet at the corner of the roof.

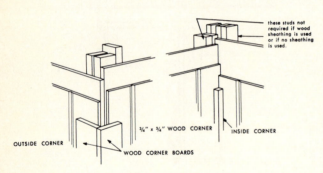

FIGURE 19:11

Entranceways

To accent the house entrance, pillars can be installed which extend from the eave section of the overhanging roof to the ground below (Figure 19:12).

Another method used to make an entranceway more appealing is to use extra thick wooden trim moldings, especially designed to outline the front door. The two upright members (pilasters) give the appearance of columns which are necessary to support the protruding cap above the door (Figure 19:13).

Shutters

Wooden or molded plastic shutters can be installed on each side of the house windows. They are anchored to the exterior covering with screws or fasteners, enabling them to be removed whenever the house must be painted. Shutters make the house more decorative by providing a contrasting color to the color of the exterior surface (Figure 19:14). Their installation is purely for decorative purposes, since they are no longer designed to perform a function (In colonial times, they were pulled closed as a means of preventing cold air from entering the windows.).

FINISH SCHEDULE

Every builder is aware of the sequence of events that must be followed to make a residence ready for occupancy. A plan for ordering these events is known as the *finish schedule*. Some of the procedures can be done concurrently while others can only be performed after some other phase of the finishing schedule has been completed.

The rough plumbing must be installed very early in the schedule. This includes running the hot and cold water lines (Figure 19:15), connecting the drainage system and sewer pipes, and setting the bathtub in position. The entire system must be checked for leaks and be inspected by the building inspector before any interior walls can be installed.

The electrical wires must be run through the wall and ceiling framework (Figure 19:16) to the various electrical boxes. These boxes will eventually house the switches, outlets, and lights for the various rooms within the house. The entire electrical installation must be inspected before the ceiling and walls can be covered.

The furnace and heating ductwork or pipes must be installed and its thermostat properly positioned before the walls and ceilings are covered.

All exterior walls can be insulated once all the lines for the various utilities have been run and connected (Figure 19:17 and 19:18). This practice guarantees that the entire exterior wall surface has been properly covered and sealed to the weather. Applying a polyethylene vapor barrier over the entire wall surface helps to ensure this condition. The plastic can be cut from the individual window areas once the walls have been covered and the window trim installed (Figure 19:19).

Ceiling areas and adjoining nonliving attic areas can be insulated before the finished ceiling is applied or after the ceiling has been covered.

Concrete should be poured for basement and crawl space floors at this time to allow sufficient time for the concrete to dry and cure properly.

The interior door frames should be installed before the walls are covered (Figure 19:20). Most builders prefer to hang the doors and then remove them until the walls and ceilings are completely finished.

Gypsum wallboard or plaster should be applied first to the ceilings and then to the walls, but not until all the

FIGURE 19:12

THE MONTICELLO

THE JAMESTOWN

THE EMPORIA

THE PRINCESS ANNE

THE BELMONT

THE AMESBURY

THE BRISTOL

FIGURE 19:13

FIGURE 19:14

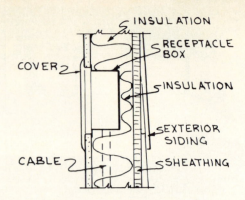

FIGURE 19:17
Fit insulation around electrical boxes and other openings.

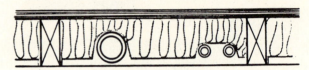

FIGURE 19:18
Place insulation between the sheathing and water and sewer pipes.

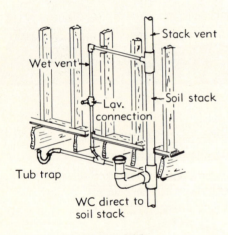

SINGLE FLOOR

FIGURE 19:15

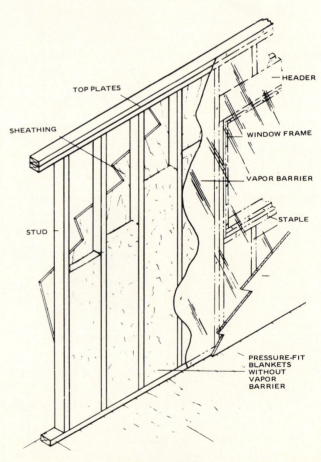

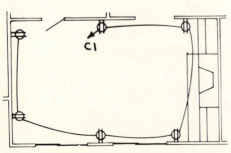

FIGURE 19:16

FIGURE 19:19
Install vapor barrier over pressure-fit blankets.

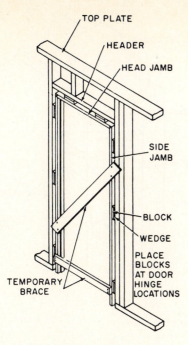

FIGURE 19:20
Installing a jamb for a new door.

FIGURE 19:21
Door framing detail.

utility lines have been inspected. A substantial block of time must be set aside to ensure it is sufficient to allow for the taping, drying, and sanding required between the finishing coats of joint compound or for the plaster to dry.

The appropriate moldings are applied to all window and door frames once the ceiling and wall areas have been completed (Figure 19:21). Note that the trim molding, applied to both window and door frames, should be positioned so that $\frac{1}{4}''$ of the jamb is visible.

Kitchen and bathroom cabinets can be installed after the walls have been covered. The plumbing fixtures are installed at the same time.

All the ceilings and walls can now be painted or covered with wallpaper. It is important to remember that sizing must be applied to the wallboard before the wallpaper is applied. This procedure ensures a good bond between the two materials.

The finished flooring is installed after the walls and ceilings have been painted or covered with wallpaper. Wood flooring has to be sanded thoroughly, have its crevices and holes filled with wood filler, and then be covered with a wood sealer. Baseboard molding is applied after the flooring has been installed, except when carpeting will be the finished floor. Then the baseboard molding is installed before the carpeting is laid.

The final procedure in the interior finishing schedule is to install all the light fixtures and the electrical switches and outlets, and to cover each electrical box with the correct cover plate.

INTERIOR TRIM

Moldings are used in the interior of all residential homes to cover various exposed seams and joints and give each room a finished appearance. These decorative trim strips, predominately made of wood, are available in a wide range of sizes and shapes that are designed to fulfill specific purposes. Every carpenter should know the various molding types that are available, and also know where each type should be used and why it is used in that particular location.

Window Trim

Figure 19:22 illustrates various types of window molding and where they are positioned. The *stool* is the first molding type to be applied to a window's interior surface. It is positioned directly over the window sill,

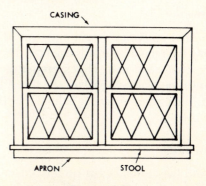

FIGURE 19:22

properly leveled, recessed $\frac{1}{16}''$ from the window sash, trimmed to provide the desired overhang, and is then nailed to the sill. The two *side casings,* one for the left side and the other for the right side, are installed next. The base of each is cut square to rest on the stool, while the top edge must receive a 45° miter cut. You must measure the length of the inside frame, from the stool to the base of the head jamb, to determine where to start the 45° miter cut (Figure 19:23).

After both side casings have been cut and nailed into place, the *head casing,* mitered at both ends, is placed between them and nailed in place (Figure 19:23).

To give the stool additional support, the *apron* is placed beneath it, trimmed if necessary to equal the dis-

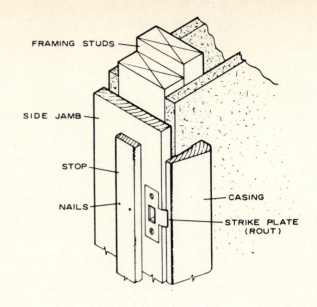

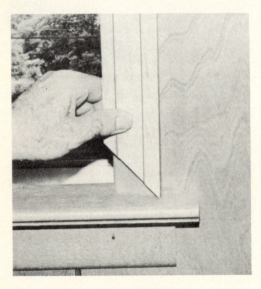

FIGURE 19:23

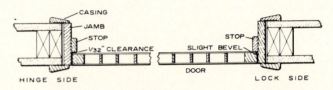

FIGURE 19:24

tance between the outside edges of the side casings, and then nailed into place.

For double or paired windows, a *mullion* (middle) *casing* is nailed to the vertical member separating the two windows (Figure 19:22).

Door Trim

Interior door openings should be trimmed with the same casing type molding that is used for the windows. To ensure that each door closes properly, *door stop* molding must be nailed to each of the three door jamb surfaces at the proper distances behind the strike plate (Figure 19:24).

Baseboard Trim

Baseboard (base) molding is used to cover the joint between the wall and flooring and, normally, is not in-

stalled until the floor is completely finished. Its hollow back is designed to accommodate slight irregularities in a wall's surface. It should be cut square to butt tightly against a door casing (Figure 19:25), coped at inside corners, and mitered at outside corners (Figure 19:26).

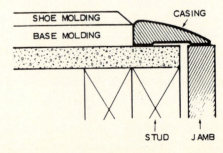

FIGURE 19:25

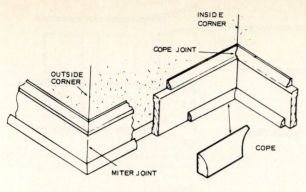

FIGURE 19:26

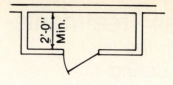

FIGURE 19:27

Closets

The interior of most closets need rod holders placed at the proper height to receive the closet rod for hanging various items in a wardrobe. These holders can be attached to the underlying wall studs, to horizontal blocking that connects two upright studs whenever a stud is missing, or to 1 × 2 or 1 × 4 wooden strips that are nailed to the studs around the perimeter of the closet (Figure 19:27). These strips also serve to support the closet shelves.

Optional Interior Trim Components

Chair Rail Molding

This specific type of molding (Figure 19:28) is applied horizontally to the walls to prevent the chair tops from harming the walls. In some homes, where the window sills are at the correct height, the chair rail molding replaces the apron beneath the window stool and runs continuously around the room (Figure 19:29).

Cornice Molding

One, two, or more pieces of trim molding can be employed to decorate the junction between the walls and ceilings (Figures 19:29 and 19:30). Since it is purely decorative, and not functional, this optional feature is not found in most residential homes.

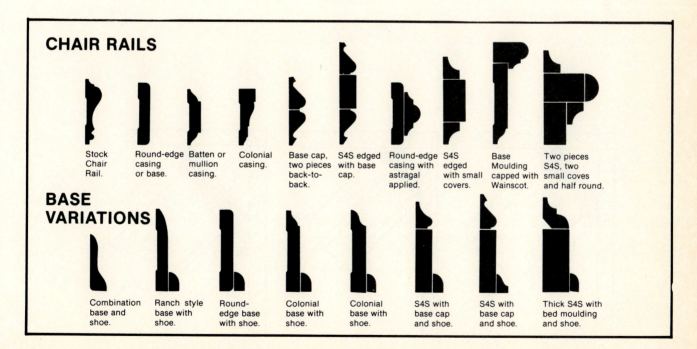

FIGURE 19:28

Wall Treatments–
from the floor to the ceiling– decorative and functional.

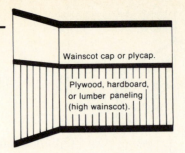

Wainscot cap or plycap.

Plywood, hardboard, or lumber paneling (high wainscot).

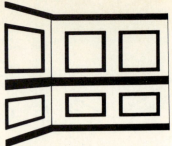

Panels over painted wall in same color, or painted or stained for contrast.

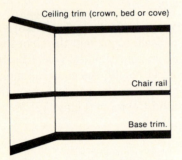

Ceiling trim (crown, bed or cove)

Chair rail

Base trim.

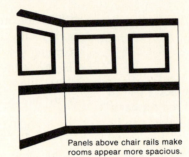

Panels above chair rails make rooms appear more spacious.

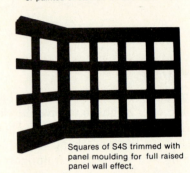

Squares of S4S trimmed with panel moulding for full raised panel wall effect.

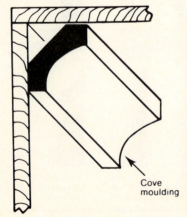

Apply moulding for panel look.

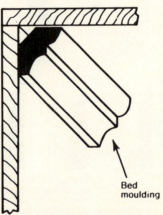

Mouldings configured around wallpaper strips.

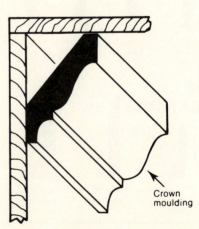

Mouldings framing wallcovering.

FIGURE 19:29

Cove moulding

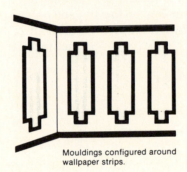

Bed moulding

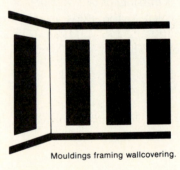

Crown moulding

FIGURE 19:30

254

Table 19:1 gives an alphabetical presentation of the various molding types, listing the common sizes and uses for each type. The thickness is stated first and the width second (Figure 19:31).

 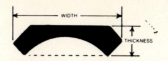

FIGURE 19:31

TABLE 19:1
Wood Moldings

Name	Shape	Size (width × height —in inches)	Use
Astragal		$\frac{3}{8} \times 1\frac{3}{8}$ $\frac{3}{8} \times 1\frac{5}{8}$	Center strip for double windows
Attic		$\frac{1}{2} \times 1\frac{1}{2}$	Fits the angle that the roof makes with the knee wall (attic storage converted to living space)
Back band		$\frac{11}{16} \times 1\frac{1}{8}$	Cap for baseboards and casings
Balusters		$\frac{1}{2} \times \frac{1}{2}$ $\frac{3}{4} \times \frac{3}{4}$ $1\frac{1}{8} \times 1\frac{1}{8}$ $1\frac{5}{16} \times 1\frac{5}{16}$ $1\frac{9}{16} \times 1\frac{9}{16}$	Extension jambs (windows, doors), hand railings
Band		$\frac{9}{16} \times 1\frac{5}{8}$ $\frac{9}{16} \times 2\frac{1}{2}$	Ceiling molding, garage trim
Bar rail		$1\frac{5}{16} \times 1\frac{7}{8}$ $1\frac{11}{16} \times 2\frac{1}{4}$	Decorative rail for home bars
Base Clam Colonial		$\frac{7}{16} \times 2\frac{1}{4}$, or 3 $\frac{1}{2} \times 2\frac{1}{4}$, or 3 $\frac{9}{16} \times 2\frac{1}{4}$, $3\frac{1}{4}$, $3\frac{1}{2}$, or $4\frac{1}{4}$ $\frac{11}{16} \times 2\frac{1}{2}$, or $3\frac{1}{2}$	Baseboards
Base molding		$\frac{5}{8} \times 1\frac{3}{8}$ $\frac{11}{16} \times 1\frac{5}{8}$	Placed on top of flat baseboard trim for decorative purposes
Bed		$\frac{9}{16} \times 1\frac{5}{8}$ $\frac{1}{2} \times 1\frac{1}{2}$	Wall and ceiling junction
Brick molding		$1\frac{5}{16} \times 2$	Brick and wood junction

Name	Shape	Size (width × height —in inches)	Use
Cap		$\frac{5}{8} \times 1\frac{1}{8}$	Edging to cover edge of paneling (either vertical where panel ends or as a chain rail for a partial panel)
Casing Clam Colonial		$\frac{9}{16} \times 2\frac{1}{4}$ $\frac{5}{8} \times 2\frac{1}{4}$ $\frac{11}{16} \times 2\frac{1}{4},\ 2\frac{1}{2},\ \text{or}\ 3\frac{1}{2}$	Window and door trim
Chair rail		$\frac{9}{16} \times 2\frac{1}{2}$	Placed horizontally around room to prevent chairs from hitting the wall surface
Corner (outside)		$\frac{3}{4} \times \frac{3}{4}$ $\frac{7}{8} \times \frac{7}{8}$ $1\frac{1}{8} \times 1\frac{1}{8}$ $1\frac{5}{16} \times 1\frac{5}{16}$	Covers and protects outside corners
Cove		$\frac{1}{2} \times 1\frac{1}{2}$ $\frac{11}{16} \times \frac{4}{16}$ $\frac{5}{8} \times \frac{5}{8}$ $\frac{11}{16} \times \frac{7}{8}$ $\frac{11}{16} \times 3\frac{1}{4}$	Inside corners, wall, and ceiling junction
Crown		$\frac{9}{16} \times 3\frac{1}{2},\ \text{or}\ 2\frac{1}{2}$ $\frac{11}{16} \times 4\frac{1}{2},\ \text{or}\ 5\frac{1}{4}$	Wall and ceiling junction, mantel trim, picture frames; used outside under eaves
Drip cap		$\frac{11}{16} \times 1\frac{5}{8}$ $1\frac{1}{16} \times 1\frac{5}{8}$	Protects top of window and door frames
Extension jamb		$\frac{23}{32} \times 2\frac{11}{16}$	Extends window jamb so it is flush with interior wall
Half round		$\frac{1}{2},\ \frac{5}{8},\ \frac{3}{4},\ \frac{7}{8},\ 1\frac{1}{8},\ \text{or}\ 1\frac{5}{8}$	Seam cover, where two paneling edges meet
Hand rail		$1\frac{9}{16} \times 1\frac{11}{16}$	Hall and stairway railings
Inside corner		$\frac{3}{8} \times 1$	Inside corners, wall and ceiling junction
Lattice		$\frac{1}{4} \times \frac{3}{4},\ \frac{7}{8},\ 1\frac{1}{8},\ 1\frac{3}{8},\ \text{or}\ 1\frac{5}{8}$ $\frac{1}{4} \times 2\frac{5}{8},\ 3\frac{1}{2}\ \text{or}\ 5\frac{1}{4}$	Lattice work; covers joints; build jambs out to finished wall
Mullion Clam Colonial		$\frac{1}{4} \times 1\frac{1}{2}$ $\frac{1}{4} \times 2$	Center trim for double windows
Nose and cove		$\frac{3}{8} \times \frac{1}{2}$ $\frac{5}{8} \times \frac{3}{4}$	Decorative trim for floors, doors, cabinets, ceilings
Picture frame		$\frac{11}{16} \times 1\frac{3}{8}$ $\frac{11}{16} \times 2\frac{1}{16}$	Picture frames

Name	Shape	Size (width × height —in inches)	Use
Picture molding		$\frac{11}{16} \times 1\frac{3}{8}$	Substitute for crown molding; positioned near ceilings to hang pictures from
Pilaster		$1\frac{1}{8} \times 5\frac{1}{2}$ $1\frac{1}{8} \times 7\frac{1}{2}$	Front door trim
Pole		$\frac{3}{4}$, $1\frac{1}{8}$, $1\frac{5}{16}$, $1\frac{3}{8}$, or $1\frac{9}{16}$	Clothes pole, curtain rods, banisters
Quarter round		$\frac{1}{4}$, $\frac{3}{8}$, $\frac{1}{2}$, $\frac{5}{8}$, $\frac{11}{16}$, $\frac{3}{4}$, $\frac{7}{8}$, or $1\frac{1}{8}$	Covers joints, inside corners, shelf cleats
Saddle		$\frac{5}{8} \times 3\frac{5}{8}$ $\frac{5}{8} \times 4\frac{5}{8}$ $\frac{5}{8} \times 5\frac{5}{8}$	Doorways
Screen molding		$\frac{1}{4} \times \frac{5}{8}$ $\frac{1}{4} \times \frac{3}{4}$	Hides screening seams, conceals or decorates exposed edges
Shelf edge		$\frac{1}{4} \times \frac{3}{4}$	Decorative shelf support; same as screen molding
Shoe		$\frac{7}{16} \times \frac{11}{16}$ $\frac{1}{2} \times \frac{3}{4}$	Floor base or together with base molding
Sill		$1\frac{5}{8} \times 7\frac{1}{4}$ $1\frac{3}{8} \times 6\frac{3}{4}$	Window sills
Stool		$\frac{11}{16} \times 2\frac{1}{4}$, $2\frac{1}{2}$, $3\frac{1}{4}$, or $5\frac{1}{4}$ $1\frac{1}{8} \times 2\frac{1}{2}$, $3\frac{1}{4}$, or $5\frac{1}{4}$	Interior window sill (double hung)
Stop Clam		$\frac{3}{8} \times \frac{5}{8}$, $\frac{3}{4}$, $\frac{7}{8}$, $1\frac{1}{8}$, $1\frac{3}{8}$, $1\frac{5}{8}$, $2\frac{1}{4}$	Door stop or base molding, window edging, and door trim
Colonial		$\frac{7}{16} \times 1\frac{1}{8}$, $1\frac{3}{8}$ or $1\frac{5}{8}$ $\frac{1}{2} \times 1\frac{3}{8}$, or $1\frac{5}{8}$	

QUESTIONS

1. Explain the basic difference between an open cornice and one that is closed (boxed).

2. List the various structural framing components of a reasonably wide soffit area.

3. List the various structural members used to construct an extended, boxed rake section.

4. Four (4) different types of molding are used to trim most windows. List each type, in order of its installation, stating quantities needed, and explain how each is cut, mitered, or trimmed before it is nailed in place.

5. Explain how and where to install door stop molding and its importance to the function and longevity of the door.

6. Compare the three (3) different cuts received by baseboard trim molding and tell where each is used.

7. List three (3) different molding types that can be used to decorate the junction between walls and ceilings.

8. Compare the three (3) types of cornices (open, closed, sloping) with respect to construction and materials.

9. Explain how the two (2) types of gable end projections are constructed with respect to conditions where (a) the rake's position is and (b) the rake is greater than 16″.

10. List three (3) ways to ventilate the soffits.

20

CHIMNEYS, FIREPLACES, WOOD BURNING STOVES

Any residential home, whose rooms will be warmed by the combustion of a fuel, must have a chimney installed to remove the smoke and exhaust gases from its interior.

Every chimney must have its own support and may not be at all dependent upon any other structural components. Its footings must extend below the frost line and extend at least 6″ beyond the dimensions of the actual chimney on all sides. A chimney located near the center of the house will prove to be more efficient and less costly to build than one positioned on an exterior wall (Figure 20:1).

Whenever a chimney is required to service more than one heating source, each with its own separate flue, its size must be large enough not only to house the flues but also to provide the proper amount of space between them (Figure 20:2).

A chimney is constructed during the framing procedure and before the roof area is completed. It must extend at least 2′ above the roof peak or 3′ above a flat roof (Figure 20:3).

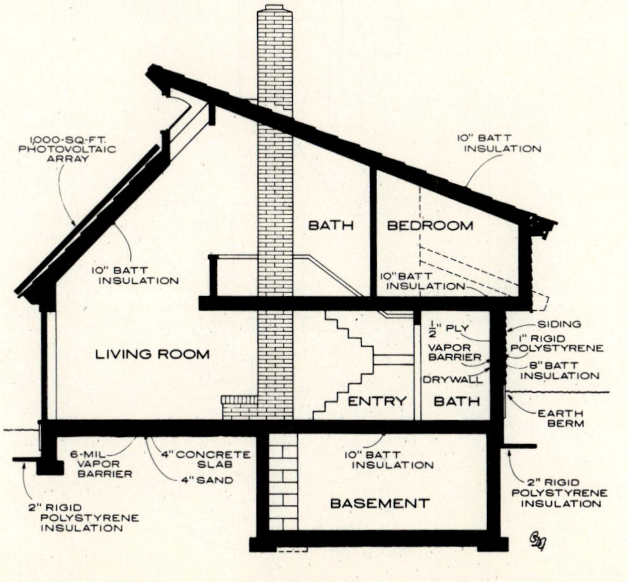

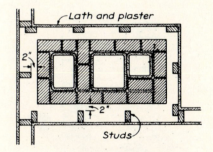

FIGURE 20:2
Double fireclay flues are separated by brick.

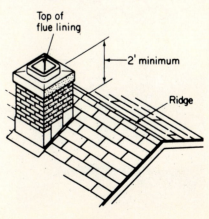

FIGURE 20:3

FIGURE 20:1

The area where the chimney passes through the roof must be properly and adequately flashed to guarantee that no rain water can enter the opening (Figure 20:4).

Flue linings, usually 2′ in length with a fireclay consistency, are installed plumb before they are covered with some form of masonry. When several linings are housed within the confines of the same chimney, they must be separated by at least one brick's thickness, the vertical joints must be staggered by at least 7″, and the height of each must be a minimum of 4″ above the top of the masonry chimney that encloses them (Figure 20:5).

Flue linings for furnaces must be kept straight, whereas those for fireplaces can change direction (Figure 20:6). All linings must be at least 2″ away from any wooden frame member and the segments of each flue lining must be bonded together with smooth mortar joints.

The size of a flue lining is critical to a chimney's effectiveness. For a specific heating system, it is advisable to follow the manufacturer's recommendations since its size is related to that of the firebox. The correct lining ensures the proper draft, as well as the absence of smoke within the rooms of the house.

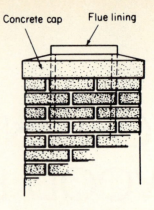

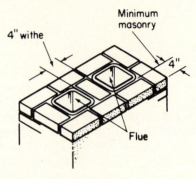

FIGURE 20:5

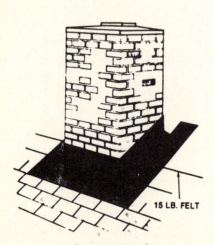

FIGURE 20:4

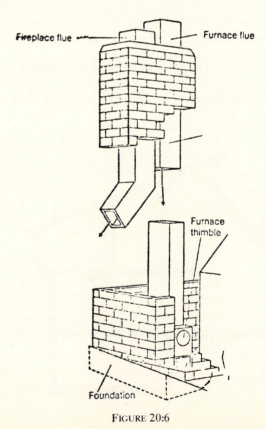

FIGURE 20:6

The proper size flue lining for a fireplace is determined by using one of the following three methods:

1. For a chimney 15′ or less in height, the flue area is 10 percent of open fireplace area.
2. For a taller chimney, the flue area is 12 percent of open fireplace area.
3. The alternative is to figure 1 sq. ft. of fireplace opening for every 13 sq. in. of flue area.

Table 20:1 relates flue size to fireplace width and may be helpful in selecting the correct flue lining.

FIREPLACES

It is important for anyone contemplating the installation of a fireplace to be cognizant of the following facts:

1. The average fireplace is extremely inefficient at heating the home since most of the heat goes up the chimney.
2. The performance of any fireplace can be improved by installing glass doors or fan type grates (Figure 20:7), both designed to return heated air to the room.
3. The installation of a steel "heatilator," which depends on an exterior air source, will also make the fireplace more efficient since it vents the heated air back into the room (Figure 20:8).
4. The installation of an air vent directly in front of the fireplace is the best means of eliminating any draft problem within the room (Figure 20:9).
5. The dimension of each fireplace component is critical and is related to the other components.

6. Extreme care must be exercised to make all interior mortar joints smooth to reduce the possibility of soot and creosote buildup.
7. Periodic cleaning of the entire chimney area is necessary to prevent a chimney fire.
8. A special mortar is needed for the firebrick. It consists of nine parts sand, three parts Portland cement, and one part fireclay. The mortar joint between the firebricks should never exceed $\frac{1}{4}$″.

FIGURE 20:7

TABLE 20:1
Recommended Flue Sizes

Fireplace width*	Standard liners		Modular liners		Round liners	
	Outside dimensions*	Area of passage*	Outside dimensions*	Area of passage*	Inside diameter*	Area of passage*
24	$8\frac{1}{2} \times 8\frac{1}{2}$	52.56	8×12	57	8	50.3
30–34	$8\frac{1}{2} \times 13$	80.50	12×12	87	10	78.54
36–44	$8\frac{1}{2} \times 13$	109.69	12×16	120	12	113.00
	13×13	126.56	16×16	162	15	176.70
46–56	13×18	182.84	16×20	208		
58–68	18×18	248.06	20×20	262	18	254.40

*All dimensions and diameters in inches; area, in square inches.

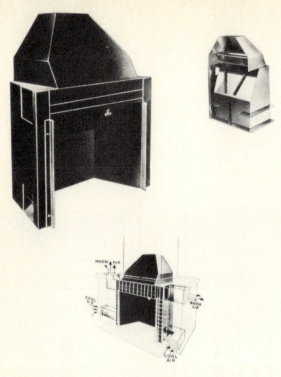

FIGURE 20:8

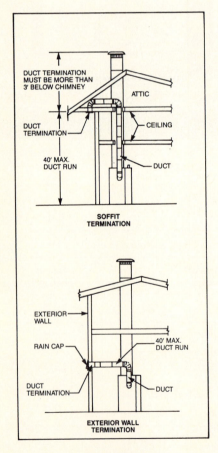

FIGURE 20:9
Fireplace air vent.

If you plan to personally tackle the complex task of building a fireplace, it will be necessary to become thoroughly familiar with each fireplace component, its purpose, and the reasons for its design. Figure 20:10 shows a cross section of a typical fireplace and its various components.

Ash Dump and Pit

Their installation allows the ashes to be passed through an opening (ash dump) into a storage compartment (ash pit) to eliminate the need for ash removal each time the fireplace is used. This convenience may be omitted whenever their installation is not structurally feasible.

Hearth

The *front* or outer hearth is constructed of some noncombustible material (e.g., brick, tile, flagstone) and is designed to provide the floor with protection against excessive heat and errant sparks. It must be 8″ to 12″ wider on each side than the actual fireplace opening.

The *back* or inner hearth is the actual floor of the firebox. It must consist of firebrick, at least 2″ thick, that are cemented together with the fireclay mortar. Both the firebrick and its special fireclay mortar are capable of withstanding high temperatures and open flames.

Firebox

This entire chamber is lined with firebrick and is bonded with fireclay. Both the side and back walls are sloped to radiate heat back into the room (Figure 20:11). The side walls slope inwards towards the back wall 3″ to 5″ per foot. The back wall is vertical from the floor to almost half the opening height of the fireplace and then it slopes forward.

The *width* of the fireplace opening can range from 24″ to 84″. The *height* should approximate $\frac{2}{3}$ to $\frac{3}{4}$ of the width, while the *depth* should be $\frac{1}{2}$ to $\frac{2}{3}$ of the height. To determine the most ideal fireplace opening for a specific room, use the above specifications plus the following rule of thumb: The fireplace opening should have an area of 5 sq. in. for every square foot of floor space.

Face

The material used to surround the fireplace opening constitutes the face. It can be stone, brick, tile, concrete, or wood. Whenever wood is used, it must be kept at least 8″ away from the firebox.

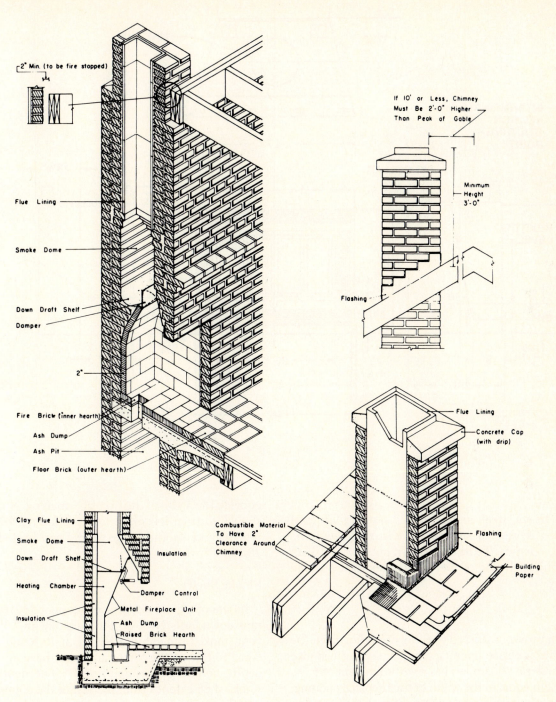

2" Min. (to be fire stopped)

Flue Lining

Smoke Dome

Down Draft Shelf
Damper

2"

Fire Brick (inner hearth)
Ash Dump
Ash Pit
Floor Brick (outer hearth)

If 10' or Less, Chimney
Must Be 2'-0" Higher
Than Peak of Gable

Minimum
Height
3'-0"

Flashing

Clay Flue Lining
Smoke Dome
Down Draft Shelf
Heating Chamber
Insulation

Insulation
Damper Control
Metal Fireplace Unit
Ash Dump
Raised Brick Hearth

Flue Lining
Concrete Cap
(with drip)

Combustible Material
To Have 2"
Clearance Around
Chimney

Flashing

Building
Paper

FIGURE 20:10

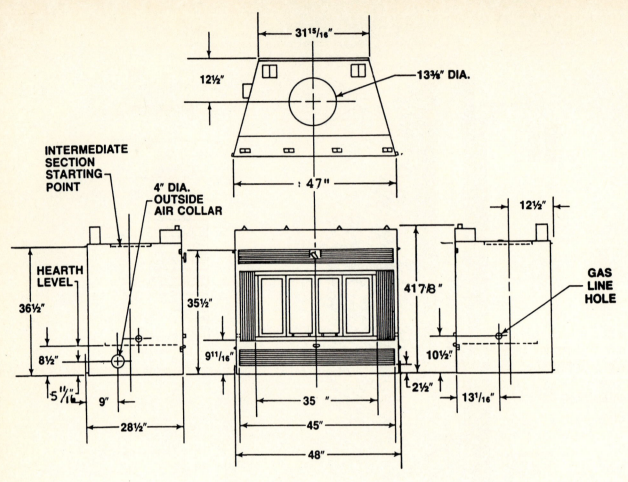

FIGURE 20:11

Lintel

This heavy steel angle iron, measuring at least $3\frac{1}{2}'' \times 3\frac{1}{2}''$, is used to support the masonry above the fireplace opening. To be effective, it must extend 3″ to 4″ beyond each side of the opening and rest on solid masonry.

Throat

This narrow constricted opening extending from the top of the fireplace opening (lintel area) to the smoke chamber (base of damper) has a vertical height of 6″ to 8″, a length equal to the width of the fireplace opening, and a cross-sectional area equivalent to that of the flue area.

Damper

This is a metal door assembly that controls the air flow and prevents the loss of heat up the chimney. It is positioned at the top of the throat area, with its base in the same plane as the smoke shelf. Its length must be equal to the width of the fireplace opening. No masonry should obstruct its complete operation nor hinder the tendency of its ends to expand when heated (Figure 20:12).

Smoke Shelf

Its smooth, concave surface is designed to redirect errant downdrafts back up the chimney and, thus, prevent smoke from entering the room. Its length must be equal to that of the throat and must have a minimum depth, behind the damper, of 4″. The depth of the smoke shelf extends from the rear of the damper to the rear flue wall.

Smoke Chamber

It is the area between the top of the throat and the base of the first flue lining. Both the front wall and the side walls are structurally tapered inward towards the flue lining to ensure that the smoke and exhaust gases are squeezed into the flue and subsequently emit-

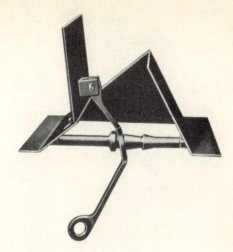

FIGURE 20:12

ted from the chimney. Smoke will definitely enter the room if this chamber is either missing or improperly constructed. All interior surfaces of this chamber must be covered with mortar and made as smooth as possible (Figure 20:13).

WOOD BURNING STOVES

Modern day "airtight," cast iron stoves (Figure 20:14) provide their owners with the following advantages:

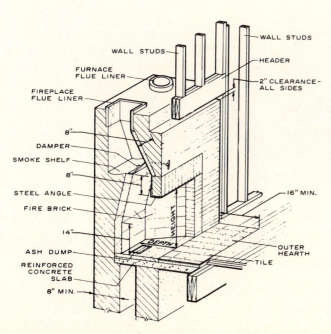

FIGURE 20:13
Note that the front and side wall taper towards the flue lining.

1. They are highly efficient and are capable of holding a fire for 12 hours or more.
2. They can sustain a fire for the entire winter, thus reducing the need to rekindle or start a new fire.
3. They burn wood so completely that ash removal is needed only once a week.
4. They are equipped with internal baffles, whose S-shaped pattern ensures excellent draft control by minimizing the air required for efficient combustion (Figure 20:15).

The accumulation of creosote in the chimney area, however, can become a serious problem with this type of stove, especially if the chimney is not sound. Regular creosote removal during the heating season is critical to the safe operation of any airtight stove.

Installation of a Wood Burning Stove

The position of a stove within a room is extremely important. Most local fire codes specify definite distances between the stove and the surrounding combustible surfaces. Figure 20:16 shows the suggested minimum distances between the stove and adjacent floors, ceilings, and walls.

These suggested minimum distances can be reduced provided noncombustible shields or wall and ceiling protectors are used. The most appropriate shields are either a $3\frac{1}{2}$" thick ventilated brick or masonry wall or a piece of sheet metal, 28 gauge or heavier. A sheet metal shield is attached to nonflammable, ceramic spacers whose purpose is to suspend the shield 1" in front of the combustible wall or ceiling. Both the top and the bot-

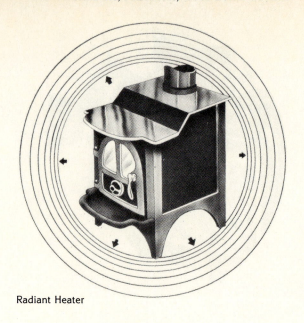

Radiant Heater

FIGURE 20:14
Radiant heater.

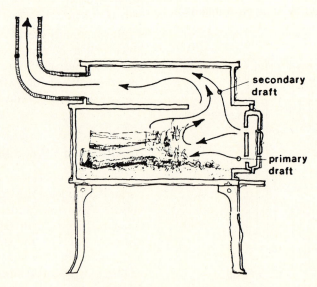

FIGURE 20:15

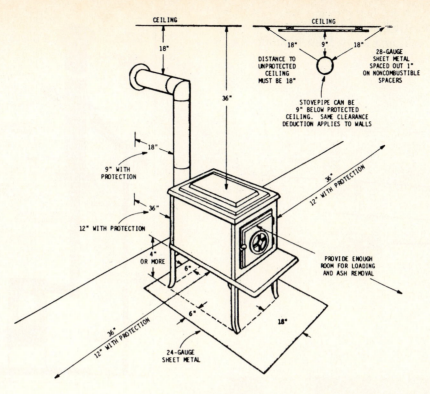

Figure 20:16

tom of the shield must be left open to ensure proper air circulation (Figure 20:17).

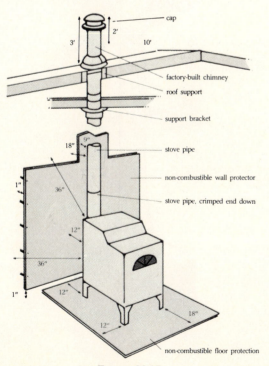

Figure 20:17
NFPA minimum clearances for safe installation.

Stovepipe Installation

The stovepipe's purpose is to connect the heater to the chimney. To be effective, as well as safe, installation requires close adherrence to the following specifics:

1. It should be at least 24 gauge steel (not galvanized).

2. Its diameter must not be different than that of the heater's pipe collar.

3. The pipes must be kept straight and as short as possible. The number of right-angle bends must be limited to two.

4. The pipes must overlap and have their crimped ends pointing towards the stove to reduce creosote buildup.

5. It must be 18″ away from any combustible materials, 9″ away from a noncombustible shield.

6. A fireclay or ventilated metal thimble must be used whenever the stovepipe passes through an interior wall or ceiling.

7. It must be positioned flush with the inside surface of the flue lining (Figure 20:18).

8. It must never be used as a chimney, but can be inserted into a metal chimney that has been tested to UL standard 103° high temperature requirements (Figure 20:19).

Every chimney must be capped or topped in some fashion to prevent rain and snow from entering the chimney (Figure 20:20).

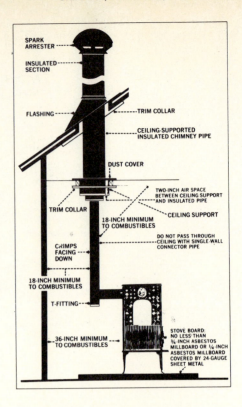

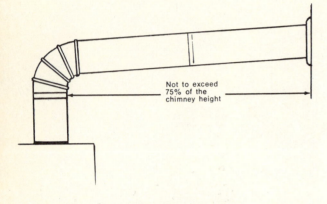

Not to exceed 75% of the chimney height

FIGURE 20:18

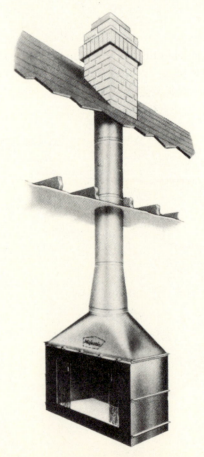

FIGURE 20:19

FIGURE 20:20

QUESTIONS

1. State the two (2) important criteria that dictate the dimensions and depth of the footings used to support a chimney in a northern climate area.

2. State two (2) reasons why a chimney should be centrally located within the house.

3. How far should a chimney rise above the highest point of the roof peak, or above a flat roof, to ensure the proper amount of draft?

4. Explain what may occur if a fireplace is serviced by an improper sized flue lining.

5. Calculate the most suitable fireplace dimensions to accommodate a 15′ × 20′ living room.

6. List three (3) desirable qualities that an airtight wood burning stove provides its owner.

7. State the minimum distance that a wood burning stove should be above the floors, and the distance that it should be away from the surrounding walls and ceilings.

21

HOME HEATING METHODS

The question of how the house will be heated must be thoroughly appraised and a decision made long before the construction begins. For most homes, some type of central heating system, designed to heat the entire building from one central location, is selected.

If solar power is one of the systems under consideration, you should be aware that the use of solar energy to heat residential hot water is a sound investment for most homes. For solar power to effectively heat an entire residence, however, the considerations and structural features become much more involved. The size of the house, its orientation to the sun, the latitude where the house is located, the size and location of its windows, the slope of its roof, its degree of weathertightness, as well as how thoroughly it is insulated, are all factors that must be considered before the house can be receptive to solar energy.

The purpose of every heating system is to maintain comfortable levels of temperature and humidity throughout every room in the house. All heating systems have been designed utilizing the knowledge that heat energy travels through the air in one of three ways: by *conduction* (from a warmer object to a colder one), by *convection* (warm air rises forcing cold air downward), and by *radiation* (heat rays radiate out in all directions from their source).

Gas, oil, or coal are used to fire most heating systems, while some homes are heated by electricity. Most heating systems operate automatically by the action of a thermostat. A thermostat switches the burner on when the temperature of the room falls below a predetermined setting, and just as efficiently shuts the burner off when the desired room temperature has been attained.

FORCED WARM AIR SYSTEM

Once the air in its chamber (expanded plenum) has been heated to a particular temperature, a sensing device activates the electrically powered blower which distributes the warm air throughout the house. The heat travels in galvanized steel ductwork to outlets located beneath windows on exterior walls. There are two basic types of duct systems used for warm air heating. The perimeter or radial system employs round ducts that radiate outward from the furnace (Figure 21:1), similar to the spokes on a wheel. The extended plenum system uses a large rectangular duct that extends in a straight line from the plenum. Individual ducts branch off this extended plenum to the registers in each room (Figure 21:2).

The ducts are rectangular-shaped when used in walls, and circular-shaped when carried through the joist areas. When the air cools, it returns to the furnace through cold-air return outlets, is filtered before it can be reheated, and then recirculated. To maintain the proper level of moisture within a home, a humidifier is usually installed on the furnace.

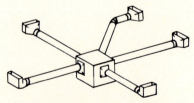

Individual supply system

FIGURE 21:1

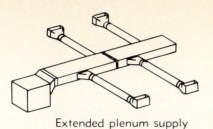

Extended plenum supply

FIGURE 21:2

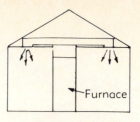

Highboy (upflow)

FIGURE 21:4

A schematic drawing of a typical forced warm air heating system is presented in Figure 21:3. Table 21:1 lists the four basic furnace designs for forced air heating and the desired location for each.

The highboy furnace (Figure 21:4), in conjunction with an extended plenum system (Figure 21:5), is frequently used when the basement area will be finished as a family room.

The horizontal furnace (Figure 21:6) is used in shallow crawl space or attic areas where all the ducts extend outward at the same level as the furnace (Figure 21:7).

The lowboy furnace (Figure 21:8) is used in partial or complete basements where all the ductwork lies above the furnace (Figure 21:9).

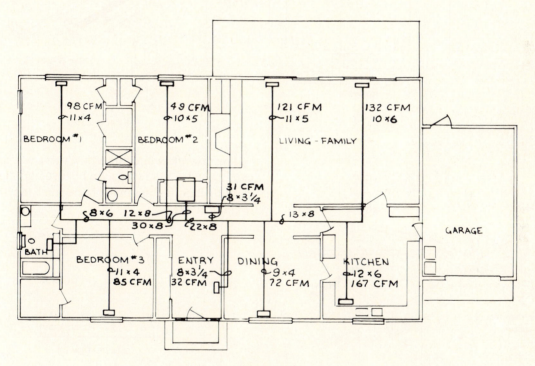

FIGURE 21:3
Warmed air enters through heating ducts and rises to ceiling; floor-level cold air goes out via return ducts for reheating.

TABLE 21:1
Forced Warm Air Furnace Designs and their Locations

Design type	Warm air	Cold air returns	Location
Upflow highboy	Out the top	In the back	Full basements
Horizontal	Out one end	In the other end	Crawl space, attics
Downflow	Out the bottom	Through the top	Main floor
Upflow lowboy	Out the top	Through the top	Partial basements

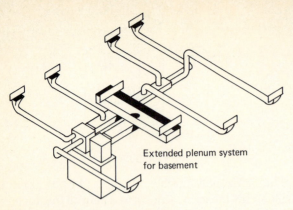

FIGURE 21:5
Extended plenum system (for basement).

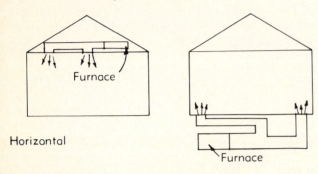

FIGURE 21:6

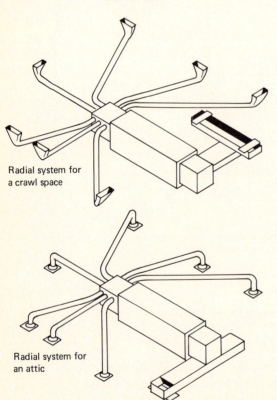

FIGURE 21:7
Radial system for (A) crawl space and (B) attic.

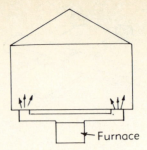

Lowboy (upflow)

FIGURE 21:8

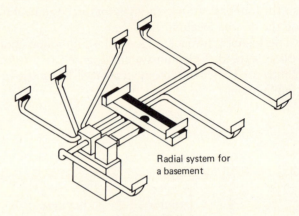

FIGURE 21:9
Radial system.

The downflow or counterflow furnace (Figure 21:10) can be used in a crawl space whenever its height is sufficient to allow the ductwork to be run beneath the floor (Figure 21:11).

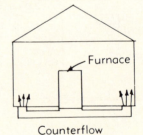

Counterflow

FIGURE 21:10

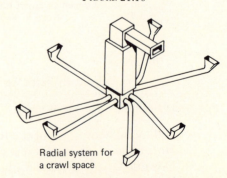

FIGURE 21:11
Radial system (for crawl space).

FORCED HOT WATER/ STEAM SYSTEMS

'For ease of comparison, the three most commonly used closed hot water piping systems and the steam system are presented in Table 21:2. Figure 21:12 presents a schematic drawing of each system to help you further understand the basic operation of each system.

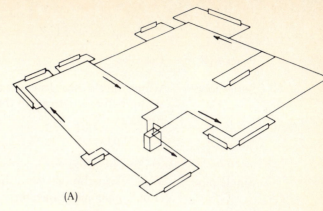

(A)

Single loop

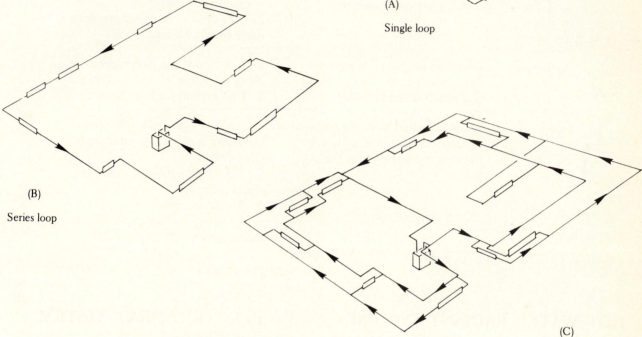

(B)

Series loop

Two-pipe direct return.

(C)

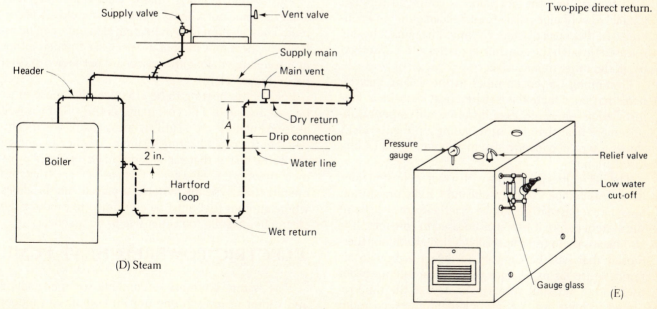

(D) Steam

(E)

FIGURE 21:12
(A) Single Loop. (B) Series Loop. (C) Double Pipe. (D) Steam.
(E) Steam Boiler.

TABLE 21:2
Comparison of Hot Water/Steam Heating Systems

Piping type	Operation	Comments
Series loop	Water travels through every convector before it returns to the furnace.	1. Not able to control the heat in a specific room 2. There is a decided drop in temperature along the route 3. Only practical in small homes
Single pipe	Water flows through both the supply pipe and convector.	1. Valves in system allow room temperature control 2. Less drop in temperature along route than with series loop
Double pipe	One pipe supplies hot water to the convector; the other returns cooled water to the boiler.	1. Very little drop in temperature along the route 2. Best system for large homes
Steam	Water is heated to steam and sent to radiators; it returns as water through another pipe to the boiler.	1. Radiators can be smaller than hot water type to give same quantity of heat. 2. They heat up faster than hot water radiators. 3. For protection, they require a safety valve, a water gauge, and a steam gauge.

HOT WATER RADIANT SYSTEM

In this system, the hot water is conveyed through parallel rows of pipes that are embedded either in a concrete slab floor or a plastered ceiling. The radiant heat is released through the concrete or plaster and is absorbed by all the other room surfaces.

Positioning the coiled pipes in the ceiling may be preferred since there tends to be a heat loss with a floor installation—large pieces of furniture and carpeting absorb more heat than the surrounding air. Figure 21:13 shows a typical concrete slab installation and a ceiling installation.

The difficulties associated with a radiant heated floor involve its complex initial installation. Extra precaution must be exercised to guarantee that there are no leaks in the system before it becomes encased in concrete. It is also necessary to use a specially designed exterior thermostat that has the capacity to detect a drop in air temperature far in advance of a typical interior room thermostat. Without this type of device, the time required to both heat up and maintain a desired room temperature would be too great.

ELECTRIC HEAT SYSTEM

This system employs unique electrical conductors, which become hot whenever they receive electrical current. These conductors, or heating elements, can be installed as heating cables or as specialized panels either in the ceiling or along the baseboard perimeter of each room (Figure 21:14).

For electric heating to be effective, the house must be well insulated. The cost to install this system as well as to maintain it are relatively inexpensive when compared to the other heating systems. Another advantage of electric heat is that each room can be regulated by its own thermostat. However, the cost of electricity to operate this type of system, especially in localities where another fuel must be consumed to produce the electricity, may be too expensive to be practical.

ELECTRIC POWERED HEAT PUMP

This system attempts to combine summer cooling and winter heating in one unit. It contains an indoor coil, an outdoor coil, and a compressor. It also has a

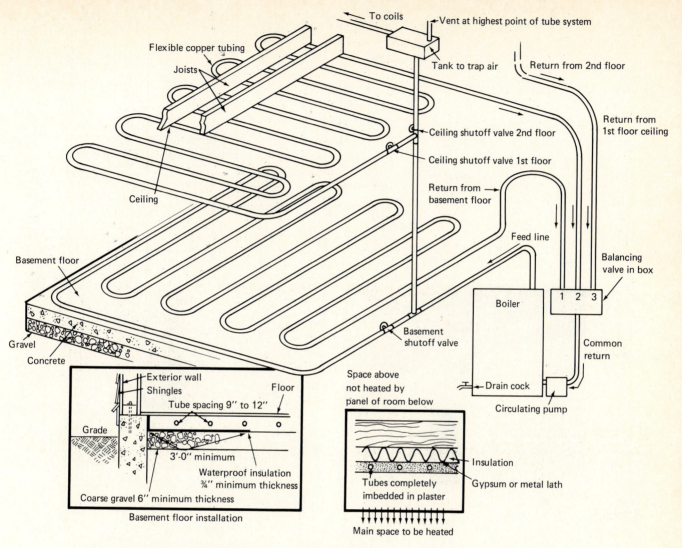

To coils

Vent at highest point of tube system

Flexible copper tubing

Joists

Tank to trap air

Return from 2nd floor

Ceiling

Ceiling shutoff valve 2nd floor

Ceiling shutoff valve 1st floor

Return from
1st floor ceiling

Return from
basement floor

Basement floor

Feed line

Balancing
valve in box

Gravel

1 2 3

Concrete

Boiler

Common
return

Basement
shutoff valve

Drain cock

Circulating pump

Exterior wall

Shingles

Floor

Space above
not heated by
panel of room below

Tube spacing 9″ to 12″

Grade

3′-0″ minimum

Waterproof insulation
¾″ minimum thickness

Coarse gravel 6″ minimum thickness

Basement floor installation

Insulation

Tubes completely
imbedded in plaster

Gypsum or metal lath

Main space to be heated

FIGURE 21:13

reversing control valve which dictates the direction the refrigerant will flow within the compressor (Figure 21:15).

This system loses its efficiency whenever the exterior temperature falls below 15°F., and thus necessitates the installation of electrical heating elements as a backup system. It relies on standard forced air system ductwork to circulate the air through the house, a power humidifier to maintain the proper humidity level, and an electronic air cleaner to purify the air. For a heat pump to function ideally, the house must be well insulated; e.g., 6″ in the attic ceiling, 4″ in the walls, and at least 2″ in the floors.

This system pumps warm interior air outside the house during warm weather, and extracts latent heat from exterior cold air and transfers this heat to the inside of the house during cold weather.

The initial cost of installation is more than other conventional heating systems, but because it uses less energy to perform its functions, the savings in fuel costs may offset the initial investment.

This thermostat-controlled system guarantees that a specific interior temperature will be maintained no matter what the outdoor weather is like, because the unit is able to switch from a heating phase to a cooling phase and back again automatically.

ALTERNATE ENERGY SOURCES

We are all aware that our dependence upon fossil fuels (coal, gas, and oil) to heat our homes will come to an abrupt halt when the earth's supply of these fuels is exhausted. Because of this concern, the scientific com-

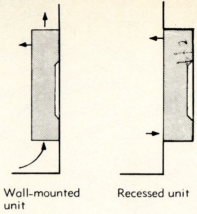

Wall-mounted unit   Recessed unit

B

FIGURE 21:14

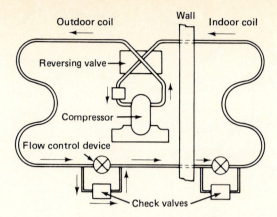

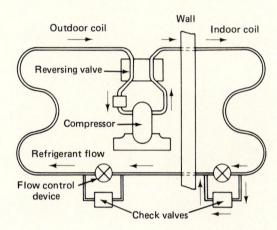

FIGURE 21:15

munity has endeavored to find alternate power sources to replace fossil fuels.

Many of these sources are "free" for the taking (e.g., sun, water, wind, geothermal), but the cost of the equipment required to harness any of these sources for home energy has been much more than the average household budget can handle.

In addition to cost, most of these alternate energy sources have other drawbacks which may curtail their installation, especially when it is understood that they all require a conventional energy system as a backup.

The balance of this presentation will discuss each alternate power source separately, listing the necessary components of each system and their function, the limitations involved, and the expected effectiveness with respect to energy output.

Energy from the Sun

New homes can be designed to receive the maximum benefits from solar energy without the installation of

any elaborate equipment, provided certain conditions are met. These include

1. The house must be energy efficient by being well insulated, caulked, and air tight.
2. The house must be properly positioned on the building site to achieve the following:
    a. The south facing side of the building remains unshaded during the heating season and consists of the living areas of the home.
    b. The bathrooms, garage, and storage areas be positioned on the north wall to buffer living areas from the cold.
    c. The bulk of the double or triple insulated windows be concentrated on the sunny side of the home. One way to do that is by constructing a greenhouse on the south side of the house (Figure 21:16).
3. The house should have a pronounced roof overhang on the south side of the house to prevent the sun's hot summer rays from entering the building (Figure 21:17), but allowing them in during winter.
4. Install a thermal mass to absorb the sun's rays during the day, which can then release this captured heat to the surrounding air during the cooler evening hours. This thermal mass can be achieved by installing thick masonry walls or floors or using water-filled containers.

Homes so designed have a passive solar system.

The four basic passive solar systems are distinguished from one another by the method used to gain the solar heat. The *direct gain* system (Figure 21:18) allows the sunlight to enter the house through large south facing windows and strike 4″ to 8″ thick masonry walls or floors that are dark in color. These masonry structures are good absorbers of heat during the day. At night, the heat stored in the masonry radiates back into the cooler room.

The *indirect gain* system uses either an 8″ to 16″ thick dark colored, masonry wall (Trombe) or a water wall. Either wall serves as a combination absorber and storage element that is located at least 4″ behind a large exterior expanse of glass or plastic. Most Trombe walls (Figure 21:19) have two sets of vents, one located at floor level and the other near the ceiling, that are used to distribute the heat evenly throughout the room. Drawing an insulating curtain between the wall and the glazed surface is very effective at minimizing any night time heat loss.

Another system, known as the *isolated gain* system, uses a separate space such as a greenhouse, atrium, or solarium (Figure 21:20) to capture the solar radiation. It is then absorbed and stored by masonry or by water-

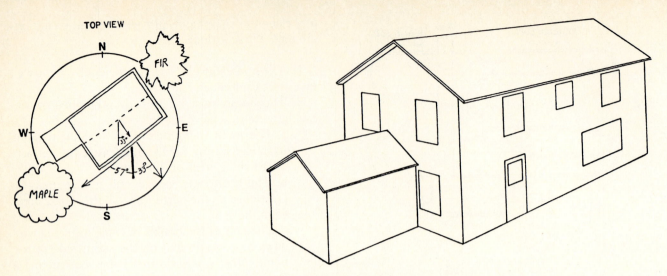

FIGURE 21:16
Sample home sketch. (Source: Northeast Solar Energy Center, Solar
Energy Education Department)

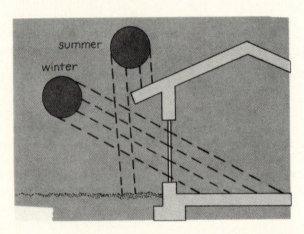

FIGURE 21:17

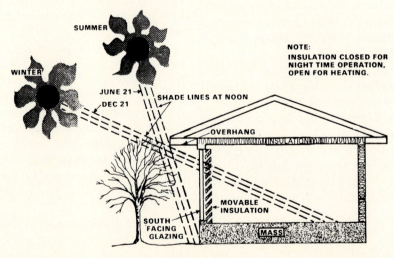

FIGURE 21:18
Direct gain.

Mass (Trombe) wall

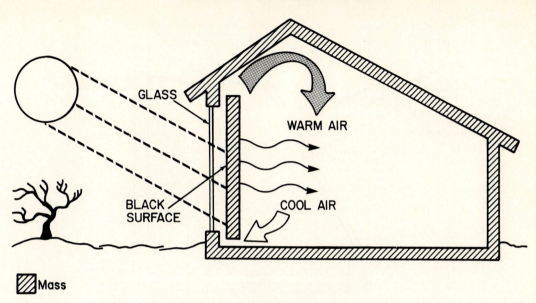

GLASS

BLACK
SURFACE

WARM AIR

COOL AIR

⧄ Mass

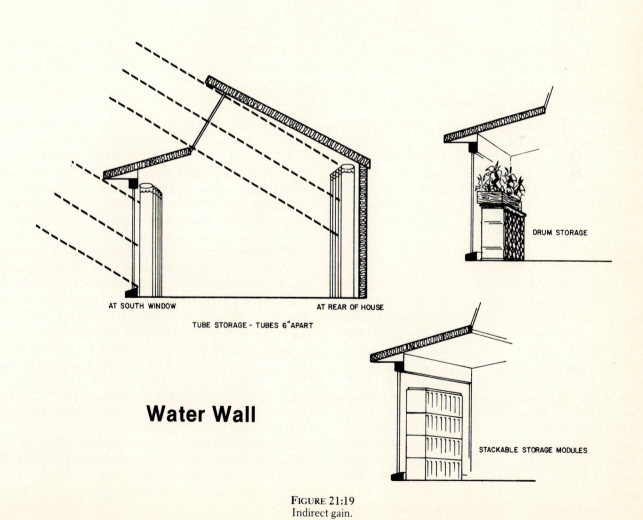

AT SOUTH WINDOW

AT REAR OF HOUSE

TUBE STORAGE - TUBES 6" APART

DRUM STORAGE

Water Wall

STACKABLE STORAGE MODULES

FIGURE 21:19
Indirect gain.

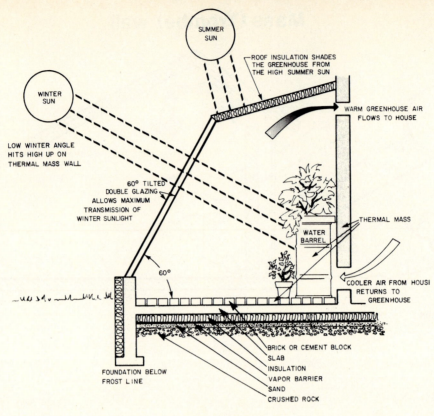

SUMMER SUN

ROOF INSULATION SHADES THE GREENHOUSE FROM THE HIGH SUMMER SUN

WINTER SUN

WARM GREENHOUSE AIR FLOWS TO HOUSE

LOW WINTER ANGLE HITS HIGH UP ON THERMAL MASS WALL

60° TILTED DOUBLE GLAZING ALLOWS MAXIMUM TRANSMISSION OF WINTER SUNLIGHT

THERMAL MASS

WATER BARREL

60°

COOLER AIR FROM HOUSE RETURNS TO GREENHOUSE

BRICK OR CEMENT BLOCK
SLAB
INSULATION
VAPOR BARRIER
SAND
CRUSHED ROCK

FOUNDATION BELOW FROST LINE

FIGURE 21:20
Isolated gain.

filled containers, each of which can be easily adapted to fit any type of residence.

The *sun-tempering* system differs from the direct gain system in that it lacks a definite storage element to absorb the radiant heat. It is designed to limit the dependency upon a conventional furnace during the daylight hours.

Energy from the Wind

Wind harnessed to produce electrical power for the home generally has not become a plausible replacement for the consumption of fossil fuels. Since most modern windmills are designed to develop maximum power in a 20 to 25 mph wind, it will be necessary, before an installation can even be considered, to collect data concerning the speed of the wind as well as its duration for at least one year. Purchasing an inexpensive anemometer and then positioning it on the exact spot where the windmill would be located should help you determine whether or not a windmill is feasible as an alternate energy source. Table 21:3 can also give you some qualitative help with respect to approximate wind speeds.

The components of a wind-driven power plant are the propeller, its tower, a generator, a governor, and a means of storing the electrical power produced.

The *propeller* has two or three blades and is designed to turn the generator whenever the wind blows. It should be positioned on a tower so that no item in the landscape obstructs the wind currents. As the tower height increases, the power output and the wind speed both increase.

The *generator* is able to supply special DC household appliances with electricity. Since most homes have AC power-driven appliances, a *static inverter* can be used to convert the DC power supplied by the generator to AC power. This deletes the storage requirement, since the utility company then acts as the storage facility.

Some systems require a *governor*. Its purpose is to control the output of electrical power by regulating the maximum blade speed. The objective is to maintain a steady voltage so that the storage batteries never become overcharged. Other systems require a voltage regulator as a further guarantee.

Surplus electrical energy is stored in a bank of stationary batteries whose combined total number of cells must equal the required voltage output of the system; e.g., a 110 to 120 volt system will require a minimum of 120 cells.

TABLE 21:3

Beaufort number	Windspeed (mph)	Mean windforce (lb./sq. ft.)	Qualitative description of wind
0	0–1	0	Calm. Smoke rises vertically.
1	1–3	0.01	Light air. Smoke drifts but wind vanes do not turn.
2	4–7	0.08	Light breeze. Wind vane turns. Leaves rustle.
3	8–12	0.28	Leaves and twigs in constant motion.
4	13–18	0.67	Moderate breeze. Raises dust and loose paper.
5	19–24	1.31	Small trees in leaf begin to sway. Small crested waves form.
6	25–31	2.3	Strong breeze. Large branches move. Telephone wires whistle.
7	32–38	3.6	Moderate gale. Large trees in motion. Walking is difficult.
8	39–46	5.4	Strong gale. Extreme difficulty in walking against wind.
9	47–54	7.7	Light roofs liable to blow off houses.
10	55–63	10.5	Hurricane. Even the strongest mills liable to be damaged.

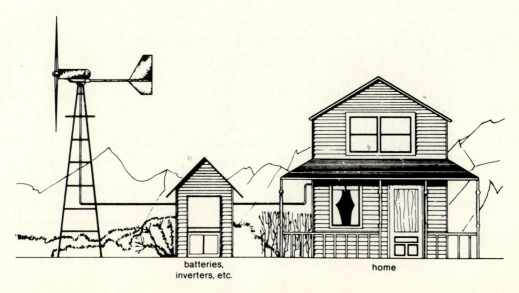

batteries, inverters, etc.

home

FIGURE 21:21

Energy from Water

If you are fortunate enough to have a source of water on your property, it may be possible for you to harness it to provide electrical power for your home. For this system to be effective, it will be necessary to have a continuous, year-round supply of water of sufficient magnitude flowing over a dam to turn a waterwheel.

Attempting to design an efficient system is extremely complex and should be delegated to a professional engineer, who must coordinate the power output of your water source with your home's electrical needs. To accomplish this feat, it will be necessary for the engineer to perfectly integrate each component of the system (e.g., the size and diameter of the wheel, the anticipated speed of the wheel, as well as the gearing ratio of the generator) to ensure maximum efficiency. Figure 21:22 shows a typical design for a simple overshot wheel.

Most water-powered systems require some personal monitoring to maintain a steady, sufficient flow rate throughout the year. During excessive rainy periods, it may be necessary to physically drop the dam's level to avoid damage to any of the system's components; during droughts, it may be necessary to perform the reverse.

Energy from Methane Gas

Bacterial action is capable of decomposing toilet wastes, animal manure, and other biodegradeable wastes whenever they are placed in a digester. During this decomposition, methane gas is produced, which can be stored and later used to heat the home, cook the meals, and even operate an electrical generator.

Even though it is possible for any homeowner to build a functioning methane gas plant, it is advisable to first check the zoning regulations and to assess specific energy needs. Most people may become discouraged with the idea of producing methane gas once they learn the restrictions placed on a system by the various authorities, and when they realize the magnitude of the plant required to generate a mere 50 percent of their energy needs.

Geothermal Energy

Water deep beneath the earth's surface is heated by magma and turned to steam. When captured, the steam has the capacity to turn turbines which produce electricity. Extensive knowledge and costly equipment are needed, placing this alternate energy source totally out of the reach of an individual homeowner. Also, any building lot where steam is actually emitted from the earth's surface (e.g., geyser) would be considered too dangerous for residential house construction.

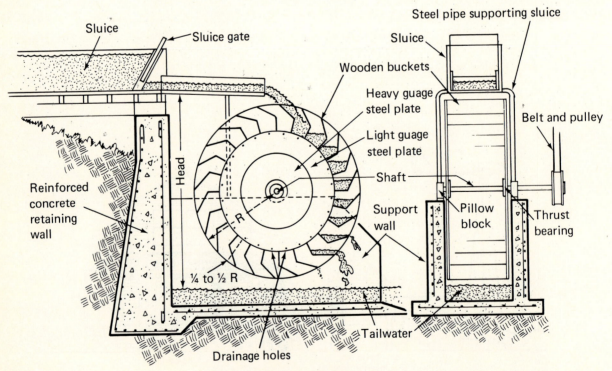

FIGURE 21:22
A simple overshot wheel.

QUESTIONS

1. List four (4) factors that must be considered to make a house receptive to solar power.

2. Compare the three (3) ways that heat energy travels through the air.

3. Explain why forced air registers should be located beneath windows.

4. State the basic advantages and possible disadvantage to using electric heat for a residential structure.

5. Explain the difference between a passive and an active solar system.

6. Explain the reasons why total dependence upon free energy (solar, wind, water) is not practical for most homes.

7. Explain the basic operational design of a heat pump.

22

PLUMBING BASICS

For many individuals, a residential plumbing installation appears to be too complicated to attempt. However, understanding the type and sizes of pipes to use, the method used to connect the various components, and where to locate each segment of the installation makes the job relatively simple.

Within every home there are two basic systems: one that supplies water to the various fixtures and one that removes the waste from these same fixtures. Even though each system is designed to perform its own separate function, every plumbing installation requires that there be a harmonious balance between the two systems.

PLANNING THE INSTALLATION

There are several types of pipe that can be used in the installation of each system. Before a proper selection can be made, learn what the local building code specifies as an adequate installation. Submission of plumbing diagrams is also required to learn whether or not they conform to the building code with respect to size of pipe, design of venting system, and distance between the components of each system (Figure 22:1).

In areas not serviced by municipal sewer lines, you must learn what the local building code requirements are for an adequate septic system.

In most localities, the cold water entrance line and the main hot and cold water lines running throughout the house employ either a $\frac{3}{4}''$ or a $1''$ pipe. Branch lines leading to fixtures and appliances are usually reduced in size to $\frac{1}{2}''$ pipes, while toilets and lavatories are serviced with $\frac{3}{8}''$ pipes.

In the waste system, $4''$ lines are used to drain toilets and $2''$ lines for showers, while washers, bathtubs, and

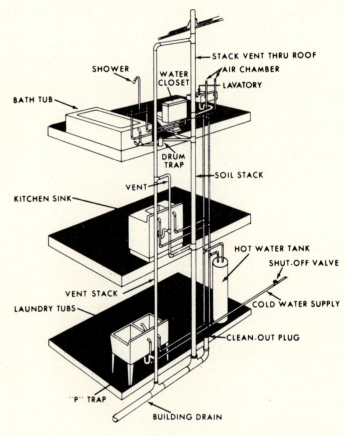

FIGURE 22:1

sinks drain through $1\frac{1}{2}''$ lines. Some lavatories drain through $1\frac{1}{4}''$ lines instead of the customary $1\frac{1}{2}''$ line.

The drainage vent system, which connects all the drainage components within the house and eventually passes through the roof, usually employs a $3''$ pipe positioned within the 2×4 stud walls. Where it enters the roof area, it is enlarged to a $4''$.

The decision of which kind of pipe to use (i.e., galvanized, copper pipe or tubing, cast iron, or plastic) in either or both of the two systems may be primarily dictated by the local building code. However, in areas where the code is not specific, pipe selection may be related to personal preference and the overall cost of the complete installation—with an eye to the expected length of service. Copper pipe or tubing is ideal for all above ground installations, but must be replaced with cast iron pipe for all below ground installations.

Complete installations with plastic pipe are acceptable in certain localities as long as CPVC (chlorinated polyvinyl chloride) is used for the hot water lines and special fixtures are employed to connect the hot water heater.

Each fixture drains into a U or S shaped trap whose purpose is to maintain a water seal within its confines so that gases, bacteria, and vermin are unable to enter the house through the drainage system. (Figure 22:2)

The drainage system, also known as the DWV system, is a three fold operation. It handles the removal of drainage water (D), the removal of waste (W), as well as the venting of sewer gases (V), so that the trap's water seal does not break. Within each horizontal run of this system, it is imperative that a plugged opening (cleanout) be installed for the purpose of removing any waste buildup or blockage.

PIPE JOINING TECHNIQUES

Once the type of pipe and materials, as well as the design and installation of the entire system, have been determined and accepted by the local plumbing code authority, the builder must be proficient in the techniques for joining the components of the system together, and in selecting the correct pipe fitting for each segment of the system.

Copper pipes are joined by sweat-soldering. The two surfaces to be joined must first be polished with either steel wool or fine emery paper. A stiff-bristle brush is used to liberally coat both surfaces with flux. This procedure is done to ensure a clean, water-tight bond between the pipe and its fitting.

A propane torch is then employed to heat the fitting until it becomes hot enough to melt the solid core wire solder. The solder is then held in the area between the pipe and its fitting until capillary action forces the solder to completely encircle the joint. (Figure 22:3)

Plastic pipes are joined together with a solvent specially designed for the type of pipe used. Because of the fast adhesive action of the solvent, a joint should be assembled dry first, before applying the solvent. The two components must be sanded smooth; burrs are removed; the joint components are tested for a snug fit, aligned properly and marked accordingly. If all the aspects of the union appear satisfactory, the two parts are disengaged and a thin coat of solvent is applied to

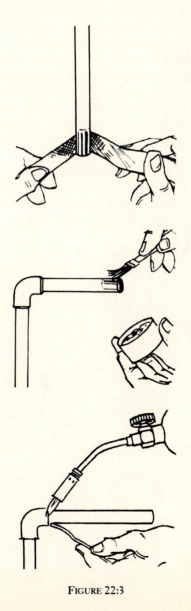

FIGURE 22:3

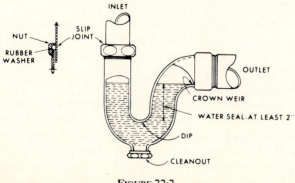

FIGURE 22:2
P trap.

the inside of the fitting while a thicker one is applied to the pipe. The pipe is inserted into the fitting, rotated to achieve the desired alignment, and then held firmly in the desired position for 30 seconds. (Figure 22:4).

ROUGH PLUMBING INSTALLATION

The rule of thumb for all plumbing layouts is to take the most direct and convenient route possible. Running most of the pipes parallel to the joists is advisable to avoid having to notch or cut holes in the joists, which reduces their strength.

The DWV system is laid out first because of the larger size of its pipes and components. The cost factor for this system is much greater than the supply system, and thus every attempt must be made to minimize runs of excessive length.

The closet bend, which connects the soil stack to the toilet flange, is the first item to be installed. Figure 22:5 shows the two most common methods of installation: the first, where the bend runs parallel to the joists and requires a brace beneath it to provide adequate support; the second, where a portion of a joist must be removed for its installation with double headers positioned at each cut end for extra support.

Whenever the first floor of the house will be a poured concrete slab, the drainage lines must be buried in soil beneath the concrete and must have a minimum slope of $\frac{1}{4}''$ per running foot. The closet bend, in this installation, is positioned so that its opening will be level with

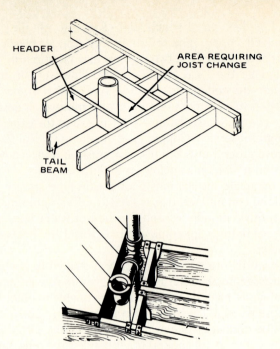

FIGURE 22:5

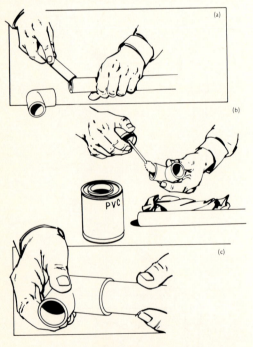

FIGURE 22:4
Cementing rigid plastic pipe.

the finished floor. Rags are stuffed into this opening before the concrete is poured to prevent any from entering the drainage lines.

Next, the soil stack is carried from the closet bend to the vicinity of the roof. There, it is enlarged before it passes through the roof. Flashing should be installed around the pipe and liberal amounts of roof cement used to seal the opening to insure against possible water leaks. (Figure 22:6)

Knowing in advance the number of branch waste and revent lines that will be needed for the complete system will enable you to assemble the correct number of T-connectors within the soil stack.

Next, the balance of the traps are installed in their respective locations. Make certain that the drain line for each fixture slopes downward towards the stack while each vent line slopes upward towards the roof. (Figure 22:7)

All rough plumbing installations will require the services of a saw and drill to make the necessary notches in the joists and studs. Limit the number of notches and the depth of each notch to avoid weakening any of these structural members.

If possible, a joist should be notched along its top edge with the cut no deeper than one quarter of its overall width (a 2″ notch would be maximum for a 2 × 8 joist). A piece of 2 × 2 or a steel brace should be positioned across each notch to prevent the pipe's weight from weakening the joist. (Figure 22:8) Holes drilled in a

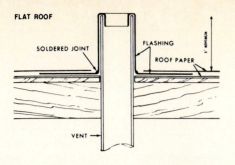

FLAT ROOF

SOLDERED JOINT
FLASHING
ROOF PAPER
VENT
MINIMUM 6"

PITCHED ROOF

FLASHING (IN)
FLASHING (OUT)
VENT INCREASER (IF USED)
12" MINIMUM
VENT

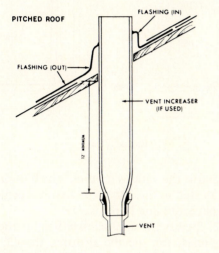

2x4 STUD
2x6 OR 2x8 PLATE
SOIL STACK

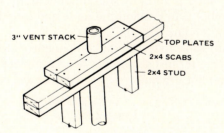

3" VENT STACK
TOP PLATES
2x4 SCABS
2x4 STUD

FIGURE 22:6

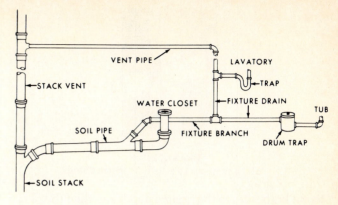

VENT PIPE
STACK VENT
LAVATORY
TRAP
WATER CLOSET
FIXTURE DRAIN
SOIL PIPE
FIXTURE BRANCH
TUB
DRUM TRAP
SOIL STACK

FIGURE 22:7

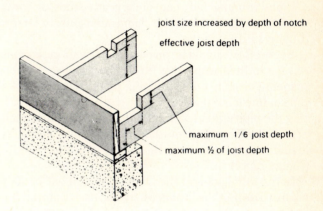

joist size increased by depth of notch
effective joist depth
maximum 1/6 joist depth
maximum ½ of joist depth

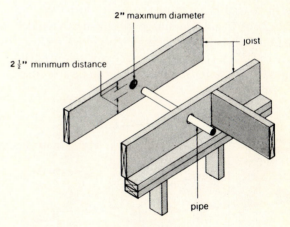

2" maximum diameter
joist
2½" minimum distance
pipe

FIGURE 22:8

joist should be located in its center and should be no larger in diameter than one quarter of the joist's width.

Studs on bearing walls should not be notched deeper than one quarter of their depth; holes should be drilled no larger in diameter than $1\frac{1}{4}''$. Installing a steel or wooden brace across the notch reinforces the stud and protects the pipe from a nail being driven into it when the wallboard is installed. (Figure 22:9)

To make a notch in either a stud or joist, a saw is used

FIGURE 22:9
Metal strap placed across the notched studs.

to cut both sides to the desired depth and then the plug is removed with a wood chisel.

Pipes running between or beneath studs or joists must be supported by pipe clamps, hangers, or straps. (Figure 22:10)

Every effort should be made to join the short sections of pipe to their respective fittings before they are permanently installed. Determining the actual length of a pipe before it is cut will avoid the embarrassment of cutting the pipe either too short or too long. Because each end of a length of pipe must be inserted into a fitting, its actual length must be greater than its visible length. (Figure 22:11)

In the installation of the DWV system, it will be necessary to restrict the distance between each trap and the soil stack to a maximum length. Table 22:1 suggests the maximum distances for specific vent pipe diameters.

The drainage fittings must also be the flush type (Figure 22:12). A regular drainage fitting has shoulders which could obstruct the passage of solid wastes.

Once the DWV system has been completely installed, the lines for the water supply system can be assembled

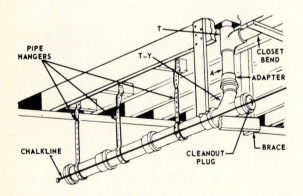

FIGURE 22:10

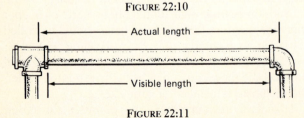

FIGURE 22:11

TABLE 22:1

Diameter of vent pipe	Maximum distance
2″	5′
$1\frac{1}{2}$″	$3\frac{1}{2}$′
$1\frac{1}{4}$″	$2\frac{1}{2}$′

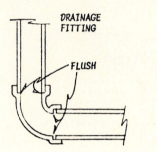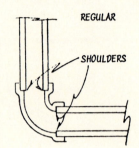

FIGURE 22:12

and installed at the various fixtures. The hot water line is always located to the *left* of the fixture's center line, and the cold water line to the *right*.

Table 22:2 gives the approximate heights the rough plumbing lines, for specific features should extend above the finished floor.

A shut off valve should be installed in each water line just before the fixture so that the water to each fixture can be turned off instead of having to completely shut down the entire system for one fixture. (Figure 22:13) Installing a union in a water line that services an appliance (water heater, dishwasher), between the shut off valve and the appliance will be beneficial whenever repairs to the appliance are needed. (Figure 22:14)

The installation of an air chamber, at least 18″ long, in every pipe that leads to a fixture prevents "water hammer," a noise which can occur whenever rapidly flowing water is abruptly shut off. The purpose of the air chamber is to cushion the noise and to reduce the strain on the line (Figure 22:15).

It is advisable to test all the components of the rough plumbing after installation has been completed. Leaks in a water line are much easier to detect and correct at

TABLE 22:2

Fixture	Height above finished floor
Lavatory faucets	20 inches
Toilet line	8 inches
Tub spout	22 inches
Tub and shower faucets	32 inches
Shower head	60 inches

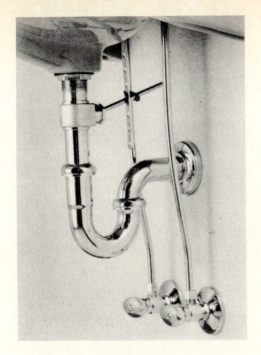

FIGURE 22:13

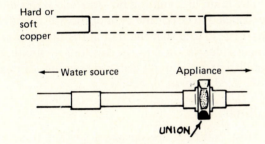

FIGURE 22:14

this stage than after the finished walls are installed. The installation of the various fixtures, however, must be performed after the walls are up and completed.

The house plans will tell the plumber exactly where to locate the pipes for the various plumbing fixtures. Figure 22:16 is an example showing the roughing-in dimensions used for a typical bathroom installation. Figure 22:17 shows the actual rough plumbing installation before the interior walls are installed.

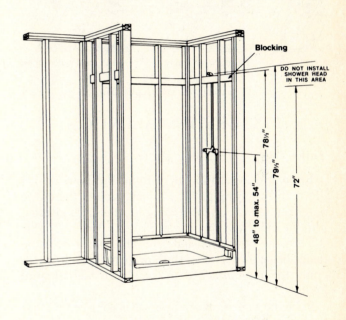

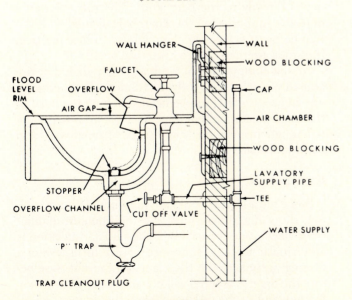

FIGURE 22:15
Note the air chamber located behind the wall.

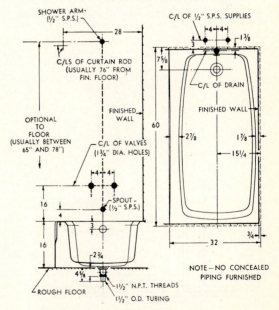

FIGURE 22:16
Typical bathroom roughing-in dimensions.

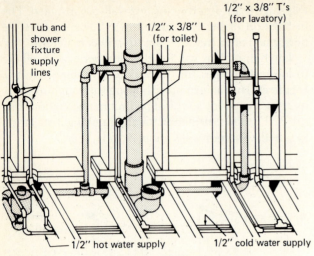

FIGURE 22:17
Typical bathroom rough plumbing.

FIXTURE INSTALLATION

Toilet

Begin the installation by turning the bowl upside down and applying putty, 1″ thick, around the entire outer rim. A wax or rubber ring must also be inserted into the discharge opening at this time (Figure 22:18).

Turn the bowl over and carefully align it over the flange bolts; then, anchor the bowl to the floor by tightening the nuts attached to the flange bolts (Figure 22:19).

Place the water tank on the bowl, making certain that the gasket and the two bolts align properly with their respective holes in the bowl. Hand tightening the bolts should guarantee a water tight seal (Figure 22:20).

FIGURE 22:18
Courtesy Sears Roebuck and Co.

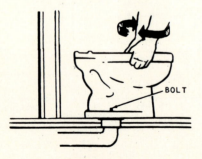

FIGURE 22:19

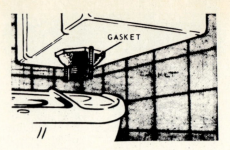

FIGURE 22:20
Courtesy Sears Roebuck and Co.

A toilet can receive its cold water from either a wall or a floor installation (Figure 22:21). A wall installation is usually preferred, especially if it is known in advance that the bathroom floor will be covered with either tile or carpeting.

Lavatory

A lavatory can be inserted into a cabinet unit or it can be supported by hangers anchored to two wall studs. It is advisable to install a 2 × 4 or 2 × 6 support block between the wall studs if the lavatory is to be hung on the wall.

The drain line and the hot and cold water lines may be installed in either the floor or the wall (Figures 22:22 and 22:23). Note that the trap in the installation has its longer side attached to the lavatory. It should also be noted that a wall type water supply line can easily be assembled with an air chamber.

Bathtub

Most bathtubs are designed so that their exterior flanges rest on 1 × 4 wooden supports which have been nailed directly to the bare studs. The tub is lowered onto the supports and the drain hole aligned with the trap before it can be anchored in place. Once in place, the studs are then covered with wallboard.

All the components of a tub assembly should be anchored securely to a 2 × 4 stud framework. Whether or not an access panel will be needed behind the head of the tub is a decision that must be made prior to the initial rough installation. Figure 22:24 shows a typical bathtub installation.

Kitchen Sink

Since most kitchen sinks are mounted within a cabinet, the water supply lines and the drain are hidden from view inside the cabinet. The actual installation is very similar to that mentioned earlier for a lavatory. Placing putty between the drain and the sink base before the drain is secured will prevent water from slowly draining out of the sink even when the drain is closed. Figure 22:25 shows a kitchen sink installation.

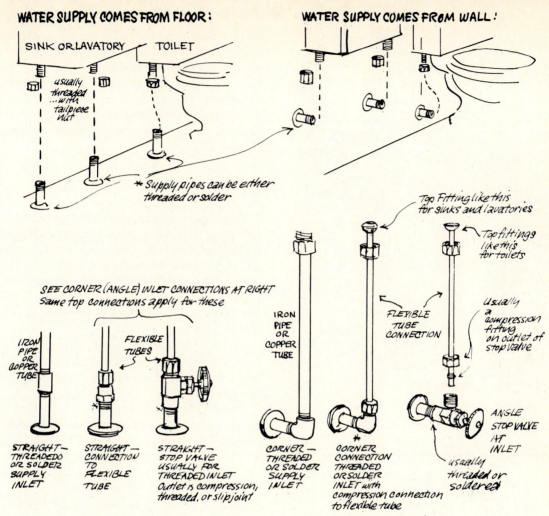

WATER SUPPLY COMES FROM FLOOR:

SINK OR LAVATORY   TOILET

usually threaded ...with tailpiece nut

Supply pipes can be either threaded or solder

WATER SUPPLY COMES FROM WALL:

Top fitting like this for sinks and lavatories

Top fittings like this for toilets

SEE CORNER (ANGLE) INLET CONNECTIONS AT RIGHT
Same top connections apply for these

IRON PIPE OR COPPER TUBE

FLEXIBLE TUBES

IRON PIPE OR COPPER TUBE

FLEXIBLE TUBE CONNECTION

Usually a compression fitting on outlet of stop valve

STRAIGHT — THREADED OR SOLDER SUPPLY INLET

STRAIGHT — CONNECTION TO FLEXIBLE TUBE

STRAIGHT — STOP VALVE USUALLY FOR THREADED INLET Outlet is compression, threaded, or slip joint

CORNER — THREADED OR SOLDER SUPPLY INLET

CORNER CONNECTION THREADED OR SOLDER INLET with compression connection to flexible tube

usually threaded or soldered

ANGLE STOP VALVE AT INLET

Most of these fittings are available with either threaded or solder inlets

FIGURE 22:21

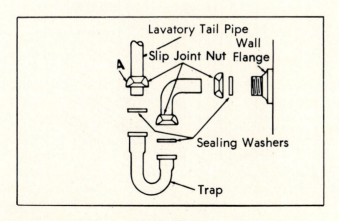

FIGURE 22:22
Courtesy Sears Roebuck and Co.

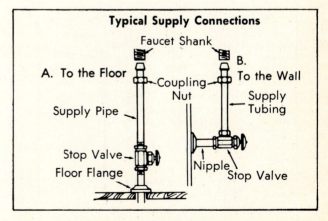

FIGURE 22:23

291

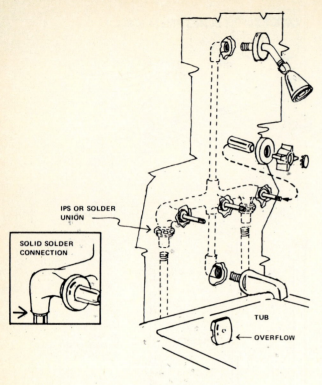

IPS OR SOLDER UNION

SOLID SOLDER CONNECTION

TUB
← OVERFLOW

FIGURE 22:24

Washing Machine

The plumbing installation for a washing machine requires a drain pipe to receive the drain hose from the machine, as well as separate hot and cold water supply lines equipped with threaded shut off faucets and air chambers. The washer's hot and cold water hoses are connected at one end to their respective threaded faucets, and to the washing machine at their other end (Figure 22:26).

SEPTIC SYSTEMS

Building a house in an area not serviced by a municipal sewer line will require the installation of an adequate septic system. The sewer line is carried usually 25' to 50' from the house to a steel septic tank (Figure 22:27). The capacity of the tank depends on the proposed number of bedrooms in the home. Table 22:3 relates the capacity of a tank to the number of bedrooms within a house.

From the tank, a watertight line must be run to a concrete distribution box (Figure 22:28), whose purpose is to distribute the tank's effluent into the system's drainage lines or "fields."

These drainage lines usually consist of either 10' lengths of perforated plastic pipe or foot long clay tiles.

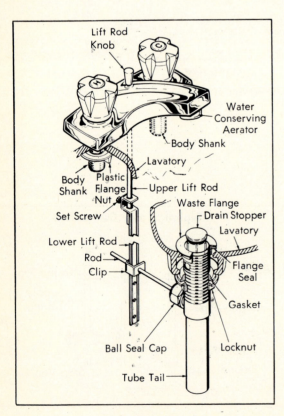

Lift Rod Knob

Water Conserving Aerator

Body Shank

Lavatory

Body Shank

Plastic Flange Nut

Upper Lift Rod

Set Screw

Waste Flange

Drain Stopper

Lavatory

Flange Seal

Lower Lift Rod

Rod Clip

Gasket

Ball Seal Cap

Locknut

Tube Tail

FIGURE 22:25

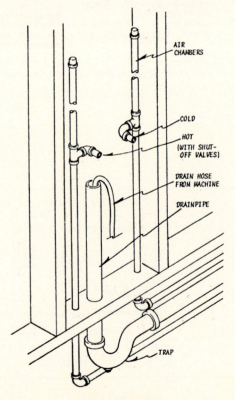

AIR CHAMBERS

COLD

HOT (WITH SHUT-OFF VALVES)

DRAIN HOSE FROM MACHINE

DRAINPIPE

TRAP

FIGURE 22:26

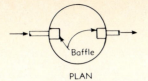

PLAN

SECTION

FIGURE 22:27

Crushed stone or gravel

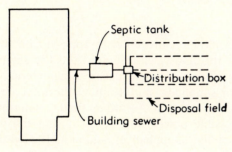

Septic tank

Distribution box

Disposal field

Building sewer

FIGURE 22:28
Distribution box (3 outlets).

TABLE 22:3
Relation of Tank capacity to Number of Bedrooms

Tank capacity	Number of bedrooms
750 gallons	One or two
900 gallons	Three
1000 gallons	Four
1250 gallons	Five

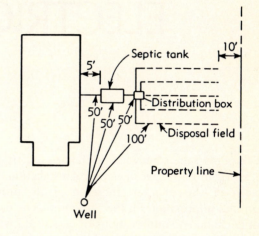

Septic tank

Distribution box

Disposal field

Property line

Well

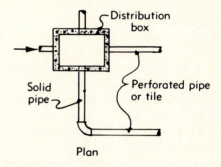

Distribution box

Solid pipe

Perforated pipe or tile

Plan

FIGURE 22:29
Septic system (top view).

They are laid in gravel lined trenches that can be perfectly level or slightly sloped (6″ in 100′) away from the distribution box (Figure 22:29).

After the lines have been installed in compliance with the local code, they must be covered with gravel before they are covered with soil, which will eventually be planted with grass seed.

QUESTIONS

1. Name and describe the two basic plumbing systems found in every home.

2. Explain what the letters D W V stand for in a home's plumbing systems.

3. Describe, in detail, the procedures and materials required to join two plastic pipes together.

4. Explain why a pipe's actual length must be greater than the apparent distance it is expected to span.

5. Explain the reason why both a shutoff valve and an air chamber should be installed in every pipe leading to a fixture.

6. List the steps involved in the installation of a toilet.

7. Explain the relationship between septic tank size and the number of bedrooms within the house.

23

ELECTRICAL WIRING

Most carpenters are somewhat hesitant to undertake a project involving any type of electrical installation. The intention of this chapter, therefore, is to provide background information for the carpenter so he or she can understand the ramifications of electrical installations and thus proceed with confidence.

An electrical installation for a house is not as difficult as it might seem because

1. The materials are standardized and readily available.
2. No special tools or skills are required.
3. Each of the various wiring tasks follows a simple, straight-forward series of steps suggested by the National Electrical Code.
4. The work is safe provided the installer understands and adheres to the necessary safety precautions for both initial installations and modifications to existing systems.

ELECTRICAL SYSTEM PLANNING

Before adequate electrical service can be provided for a new home, the total wattage must be determined and a plan must be devised for each wiring circuit within the house. This plan should include a wiring diagram for each room which shows the exact location of all the electrical components.

There are several methods used to calculate total wattage. One way is to add the total amperage provided by the two fuses or circuit breakers on each side of the main input line and then multiply this sum by 240, provided the house has three-wire service. This figure is the total watts available to the house at any specific mo-

ment. Another approach is to approximate the total wattage the house will require. Add the square foot areas of all the occupied spaces, using exterior dimensions for simplicity, and multiply the total area by 3 watts per square foot. A third method to determine the total possible power wattage is to sum the wattage ratings of all the appliances, and of all the lighting fixtures, within the house. Table 23:1 lists these ratings. For reasons of safety, this value should be 20 percent less than the total wattage available. If it isn't, the utility company will have to increase the incoming wattage.

The purpose of a home's service panel (Figure 23:1) is to separate the incoming main electrical power into individual circuits. Each wiring circuit is independently protected by its own fuse or circuit breaker. A circuit breaker can be either a push type or a toggle switch. It will remain in the "on" position unless the load within that circuit exceeds its rated amperage and causes it to trip automatically to the "off" position.

In most residential electrical systems, there are four different types of circuits. They are presented in Table 23:2.

Wires associated with a specific general purpose circuit may ramble throughout the residence and connect lights and outlets in various locations. Their path is usually determined by assessing the estimated combined total amperage of the components on the circuit so that it does not exceed a prescribed amount. Small appliance circuits, however, service only receptacle outlets in specific locations; i.e., kitchen, dining room, laundry, workshop, and pantry.

The initial wiring plan, drawn to scale on graph paper, should show the location of the various components of each circuit. Electrical wiring symbols (Figure 23:2) can be used to give additional clarity to a

Wattage Ratings for Appliances Found in the Home

Room	Appliance	Wattage rating
Kitchen	Blender	250–450
	Broiler	1,500
	Can opener	100
	Coffee percolator	600–750
	Crock pot	75–150
	Deep fryer	1,350
	Dishwasher	1,800
	Fan (exhaust)	70
	Floor polisher	300
	Frying pan	1,000
	Garbage disposal	900
	Iron	1,000
	Microwave oven	650
	Mixer	150
	Range	8,000–16,000
	Refrigerator	300
	Rotisserie	1,400
	Toaster	1,200
	Waffle iron	1,100
Bedroom	Air conditioner (window type)	800–1,500
	Blanket (single)	150
	Blanket (dual)	450
	Clock	3–10
	Dehumidifier	400–600
	Fan (table type)	80
	Hair dryer	400
	Radio	10
	Stereo	300–500
	Sun lamp	250–400
Living room	Heater, portable	1,300
	Television (color)	250
	Television (black and white)	50
	Vacuum cleaner	600
Laundry and utility	Air conditioner (central, 240v)	5,000
	Clothes dryer (gas)	1,400
	Clothes dryer (240v)	5,500
	Freezer	600
	Furnace (gas)	800
	Furnace (oil)	600–1,200
	Hot water heater (240v)	2,500
	Washing machine	900
Workshop	Drill (portable)	200–400
	Drill press	500
	Saw (radial)	1,500
	Saw (table)	600
	Soldering iron	150

Note: If an appliance's name plate states amperage only, multiply this number by the line voltage (120 or 240). A one horse power motor uses approximately 1,000 watts.

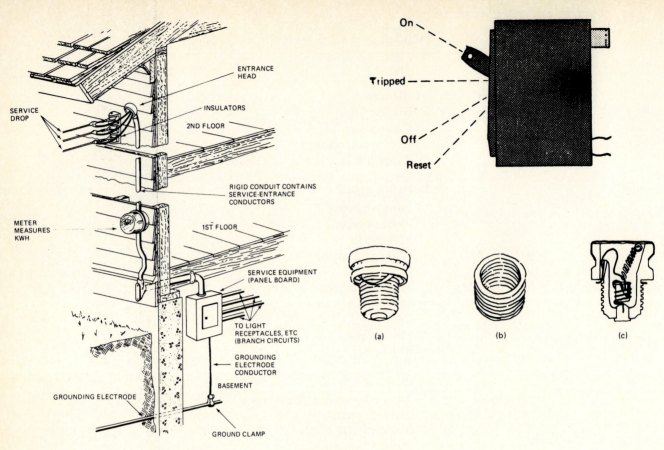

FIGURE 23:1

plan. A representative plan is presented in Figure 23:3. The finalized plan for the house can be used to number and identify each circuit on the service panel box. It can help eliminate the confusion that new homeowners often experience attempting to untangle the mysteries of a blank panel box.

The electrical load of a circuit will dictate the amperage capacity of the wire (conductor) and the rating of the fuse or circuit breaker protecting that circuit. Circuits for general lighting usually employ a #14 wire with a 15-amp breaker, while a #12 wire with a 20-amp breaker is used in the small appliance circuits. Table 23:3 gives the relationships between wire size, amperage capacity, and their applications. Note that as the number of the wire or conductor decreases, the diameter of the wire or the number of strands increases.

TABLE 23:2

Circuit type	Purpose/location	Amperage
General purpose	Lights, wall outlets	15 ampere
Small appliance	Outlets in kitchen, laundry, workshop, dining room	20 ampere
Large (special) appliance	Furnace, washer	20–25 amperes
240 volt line	Range, air conditioner, Clothes dryer	3 wire circuit with paired fuses or breakers

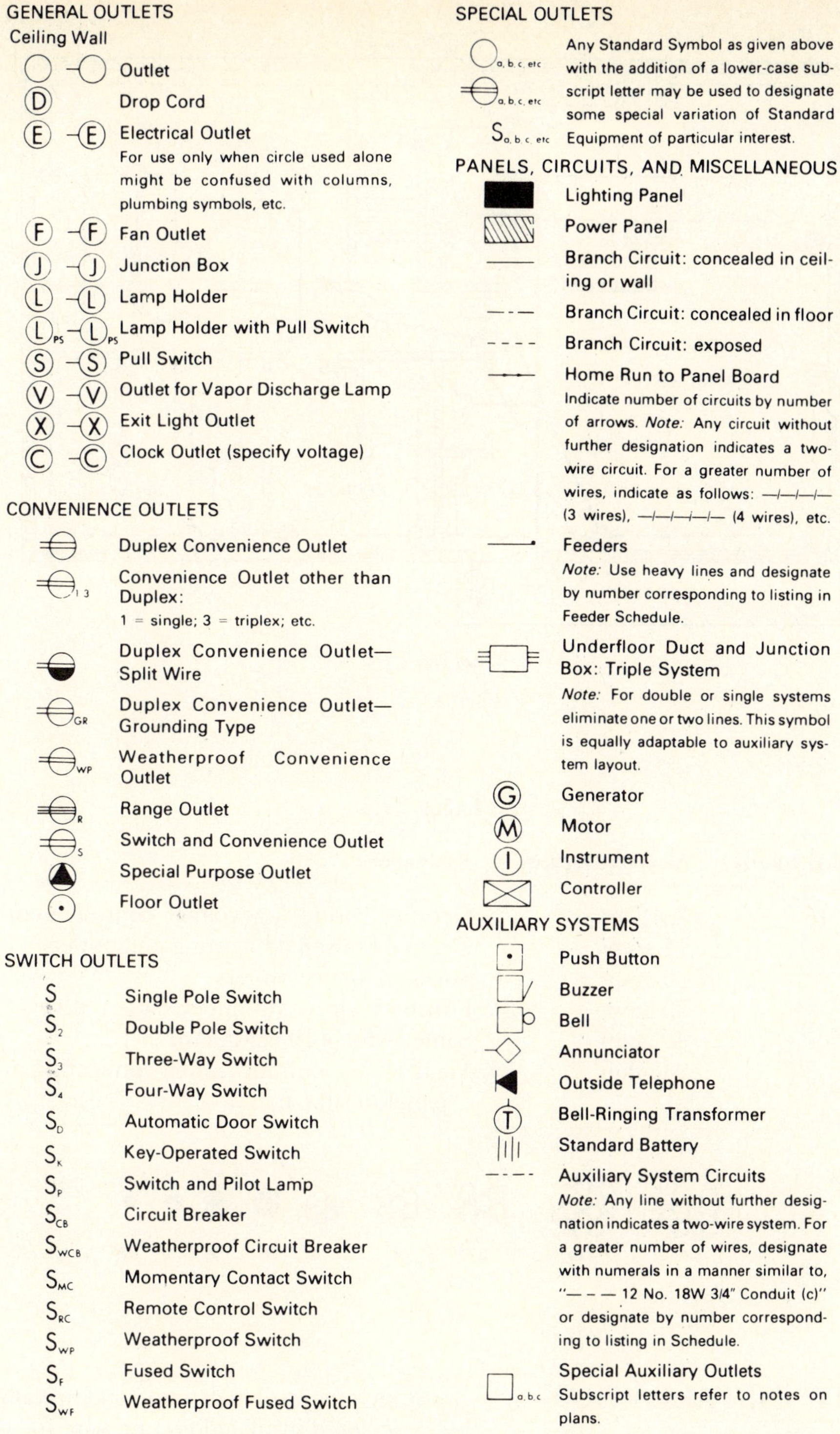

GENERAL OUTLETS
Ceiling Wall

Outlet

Drop Cord

Electrical Outlet
For use only when circle used alone might be confused with columns, plumbing symbols, etc.

Fan Outlet

Junction Box

Lamp Holder

Lamp Holder with Pull Switch

Pull Switch

Outlet for Vapor Discharge Lamp

Exit Light Outlet

Clock Outlet (specify voltage)

CONVENIENCE OUTLETS

Duplex Convenience Outlet

Convenience Outlet other than Duplex:
1 = single; 3 = triplex; etc.

Duplex Convenience Outlet—Split Wire

Duplex Convenience Outlet—Grounding Type

Weatherproof Convenience Outlet

Range Outlet

Switch and Convenience Outlet

Special Purpose Outlet

Floor Outlet

SWITCH OUTLETS

S — Single Pole Switch

S_2 — Double Pole Switch

S_3 — Three-Way Switch

S_4 — Four-Way Switch

S_D — Automatic Door Switch

S_K — Key-Operated Switch

S_P — Switch and Pilot Lamp

S_{CB} — Circuit Breaker

S_{WCB} — Weatherproof Circuit Breaker

S_{MC} — Momentary Contact Switch

S_{RC} — Remote Control Switch

S_{WP} — Weatherproof Switch

S_F — Fused Switch

S_{WF} — Weatherproof Fused Switch

SPECIAL OUTLETS

Any Standard Symbol as given above with the addition of a lower-case subscript letter may be used to designate some special variation of Standard Equipment of particular interest.

PANELS, CIRCUITS, AND MISCELLANEOUS

Lighting Panel

Power Panel

Branch Circuit: concealed in ceiling or wall

Branch Circuit: concealed in floor

Branch Circuit: exposed

Home Run to Panel Board
Indicate number of circuits by number of arrows. *Note:* Any circuit without further designation indicates a two-wire circuit. For a greater number of wires, indicate as follows: —/—/—/— (3 wires), —/—/—/—/— (4 wires), etc.

Feeders
Note: Use heavy lines and designate by number corresponding to listing in Feeder Schedule.

Underfloor Duct and Junction Box: Triple System
Note: For double or single systems eliminate one or two lines. This symbol is equally adaptable to auxiliary system layout.

Generator

Motor

Instrument

Controller

AUXILIARY SYSTEMS

Push Button

Buzzer

Bell

Annunciator

Outside Telephone

Bell-Ringing Transformer

Standard Battery

Auxiliary System Circuits
Note: Any line without further designation indicates a two-wire system. For a greater number of wires, designate with numerals in a manner similar to, "— — — 12 No. 18W 3/4" Conduit (c)" or designate by number corresponding to listing in Schedule.

Special Auxiliary Outlets
Subscript letters refer to notes on plans.

FIGURE 23:2
Wiring symbols.

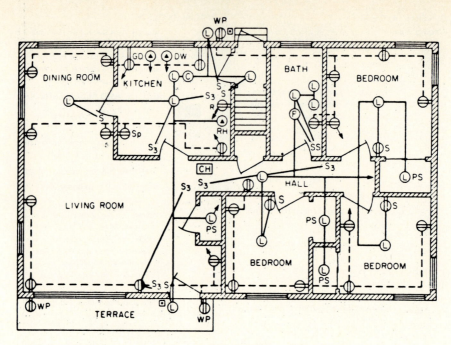

FIGURE 23:3

TABLE 23:3

Conductor size	Amperage capacity	Applications
18, 16	7, 10 amps	Doorbell wiring, low voltage control circuits
14	15 amps	General household lighting and outlets
12	20 amps	Small appliance outlets
10	30 amps	Furnaces, air conditioners, ranges, dryers
8	40 amps	Same as for #10 conductor
2	95 amps	Main service entrances, line connecting main panel to subpanel

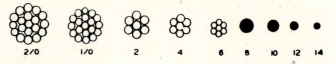

CIRCUIT INSTALLATION

Each indoor electrical circuit is installed by running cable inside or along ceilings, walls, and floors between the various switches and outlets. The forms of cable available, their specific uses, and code designations are presented in Table 23:4.

The cable most often used indoors is Romex, a flat, white plastic sheathed cable which encloses the hot conductor (black insulated wire), the neutral conductor (white insulated wire) and usually a bare or green-colored insulated ground wire for safety. The notation "12-2G NM" on the wire provides the following information: #12 wire size, two wires (black, white) plus ground wire, to be used in a dry indoor location. (Note: the cable actually contains three wires.)

Armored cable, commonly known as BX, has a flexible, galvanized steel armor casing surrounding the insulated wires, each of which is individually wrapped in kraft paper for protection, and a bare ground wire.

TABLE 23:4
Available Electrical Cable Types

Where used	Exterior coating	Code designation	Common name
Indoors, dry	Nonmetallic sheathed cable	NM	Romex
Indoors, dry	Flexible, steel armor	AC	BX
Indoors, damp	Waterproof plastic	NMC	
Underground	Heavy plastic	UF	

Note: All four types of cable are available in the following four conductor sizes: 12–3G, 12–2G, 14–3G, 14–2G.

The use of thin-walled conduit may be required by the existing electrical codes for certain specific installations; e.g., a furnace hookup. It must be properly positioned and completely installed before any wires are inserted and pulled through it. The size of conduit selected will depend on the number of wires that it will encase. Table 23:5 lists the available sizes of conduit and the number of conductors each can handle.

Table 23:6 describes the features and installation procedures for Romex, BX, and thin wall conduit.

Wire Connections

There are several means of connecting wires while running residential circuits. The most common method of connection involves attaching a wire to a screw type terminal, which can be either an end terminal or continuous one (Figure 23:4). Long-nosed pliers are most suited for the job of forming the wire's straight end into a hook which can then be held securely under the screw head. It is important to position the hook so that when the screw is tightened, the hook will also tighten. Also, remember that only the black or hot wires should be connected to *brass* screw terminals.

Solderless connectors are most commonly used when the ends of two or more wires must be joined together. The wires to be connected are first positioned parallel to each other, and then the plastic wire tie (solderless connector) is slid over the stripped ends of the wires and screwed on tightly. It is important to tug on each connection made with solderless connectors to make certain that all the wires are fastened securely (Figure 23:5).

Holes in the back plate of some receptacles are designed to receive the stripped end of a wire as the means of connection instead of a screw connection (Figure 23:6).

TABLE 23:5
Conduit Diameters and Wire Capacities

Diameter	Conductor size	Quantity
$\frac{1}{2}''$	# 12	3
	# 14	4
$\frac{3}{4}''$	# 12	5
	# 10	4
$1\frac{1}{4}''$	# 6	4
	# 4	2
$1\frac{1}{2}''$	# 1	3

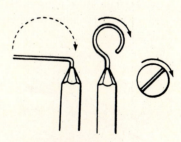

INSULATION CLOSE TO
TERMINAL SCREW

FIGURE 23:4

Romex	BX	Conduit

Appearance

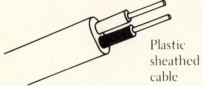

Plastic sheathed cable

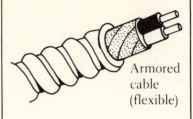

Armored cable (flexible)

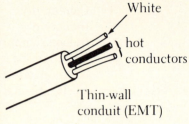

White

hot conductors

Thin-wall conduit (EMT)

Anchorage between connections

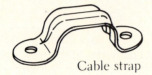

Cable strap

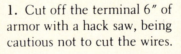

Staple

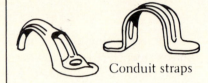

Conduit straps

Preparation prior to connection (in steps)

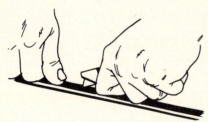

1. Two different techniques, each involving a razor knife, can be employed to remove the terminal 6″ of plastic covering, thus exposing the actual wires. The knife is used to make either a lengthwise cut or a circular one.

2. Once the covering has been cut and removed, the paper wrapping surrounding each wire is removed.

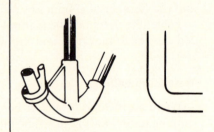

1. Cut off the terminal 6″ of armor with a hack saw, being cautious not to cut the wires.

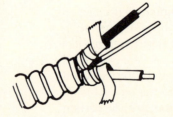

2. Remove the protective paper surrounding each wire.

1. An electrician's hickey is used to make any necessary bends in the conduit.

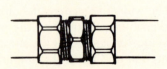

2. A coupling is used to join two pieces of unthreaded conduit. It consists of a nut on each end and a threaded coupling joint between the two nuts.

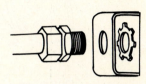

3. The wire strippers are used to remove one inch of insulation from the tip of each wire.

3. Slide the insulating fiber bushing into the end of the armor.

3. A connector with a single nut and coupling joint is used to attach conduit to a box. It is anchored to the box with a locknut.

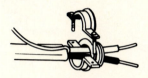

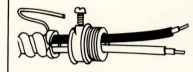

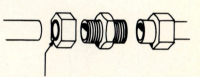

4. A connector, designed especially for nonmetallic cable, is installed over the wires and is held in place by tightening its two locking bolts.

4. Install the specially designed BX connector and tighten with its set screw once the ground wire has been pushed back against the armor.

4. A fiber ring inside each nut holds the conduit in place once the coupling joint is tightened with a wrench.

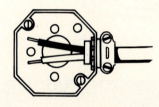

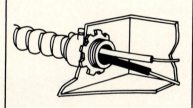

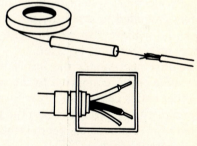

5. The threaded end of the connector is inserted into the electrical box and anchored in place by tightening the lock nut with a screwdriver.

5. Push the threaded portion of the connector into the electrical box and tighten lock nut with a screwdriver. Attach ground wire to connector's set screw.

5. A fish tape is used to pull the wires through the conduit. The wires are stripped after they enter the metal electrical box.

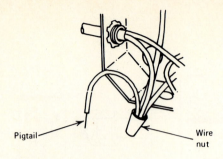

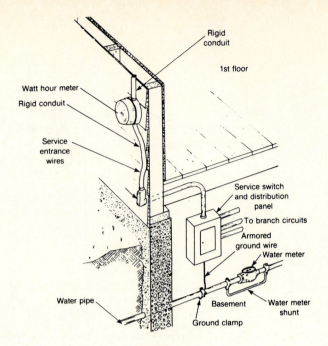

Pigtail

Wire nut

FIGURE 23:5

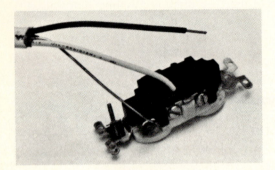

FIGURE 23:6

FIGURE 23:7

FIGURE 23:8

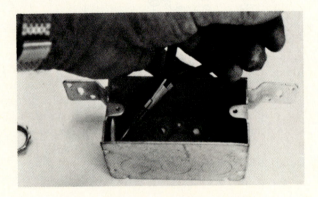

FIGURE 23:9

Grounding

Every electrical system must be properly and adequately grounded to avoid the possibility of electrical shock. To achieve this requirement, certain precautionary measures must be taken.

The first necessitates the installation of a ground cable. One end is connected to the service panel and the other end is connected either to an incoming cold water pipe that is buried beneath the ground or is connected to a $\frac{1}{2}''$ diameter copper rod embedded at least 8" into the ground (Figure 23:7).

The balance of the system requires that all lights, switches, outlet boxes, and receptacles be properly grounded.

Thin wall conduit and BX cable, provided they are securely grounded to their components, may not require the additional grounding wire that is required of nonmetallic Romex.

In all residential installations where Romex is used, each metal electrical box must be equipped with some means of attaching a grounding conductor. Some boxes have a tapped hole made to accept a screw for anchoring the ground wire to the box (Figure 23:8).

All electrical boxes have knockout holes through which the cables enter the box. Some can be twisted out by using a screwdriver or pliers while others must be punched out with a hammer and punch (Figure 23:9).

A specially designed grounding clip is another means of grounding an electrical box (Figure 23:10).

Figure 23:11 displays how a solderless connector is used to connect several ground wires intercepting at an electrical box, and also how the box itself is grounded.

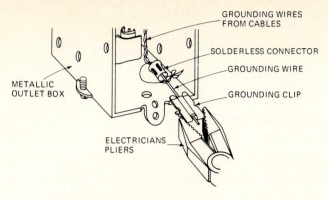

FIGURE 23:10

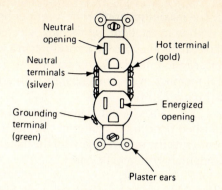

FIGURE 23:12

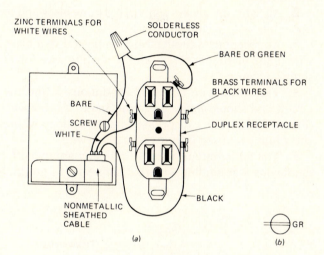

FIGURE 23:11

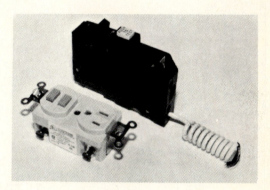

FIGURE 23:13

The installation of three-prong electrical outlets throughout the house increases the safety of a well-grounded system, provided the bare ground wire is attached to the green terminal screw on the receptacle (Figure 23:12).

The installation of ground fault circuit interrupters (GFCIs, Figure 23:13) in all bathroom, outdoor, and garage circuits is required by the National Electrical Code (NEC) in all new residential wiring systems. They act as supersensitive monitors of the electrical current flowing within a circuit, and will immediately break the circuit whenever there is a current difference between the black and white wires. Their installation in the areas already mentioned, where the presence of water coupled with an uneven electrical flow could be hazardous or fatal, provides the ultimate in protection against electrical shock.

Electrical Boxes

Table 23:7 presents the basic types of electrical wiring boxes, their sizes and characteristic shapes, and the maximum number of wires each can house. (The range

TABLE 23:7

Box type	Size	Depth range	Shape	Maximum number of wires	
				# 12	# 14
Wall	2 × 3	$2\frac{1}{5}''-3\frac{1}{5}$	Rectangular	5–8	6–9
Junction	4 × 4	$1\frac{1}{5}''-2\frac{1}{8}''$	Square	8–13	9–15
Ceiling	4 × 4	$1\frac{1}{4}''-2\frac{1}{8}''$	Octagonal	5–9	6–10
Weatherproof exterior box (same as a wall box)					

given for the maximum number of wires per box is due to the variation in the depth of the box selected.) Each type of box can be anchored in its position in a variety of ways; surface mounted, nailed or screwed to a wooden structural member, with box clips, or with adjustable hangers (Figure 23:14).

Table 23:8 gives the height in inches from the surface of the finished floor to the center of the electrical box.

There are specific rules that govern the installation of every residential system. They are:

1. All switches are to be wired only with black wires (Figure 23:15). Exceptions would occur with *switch*

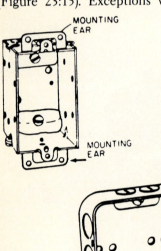

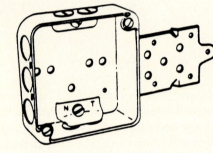

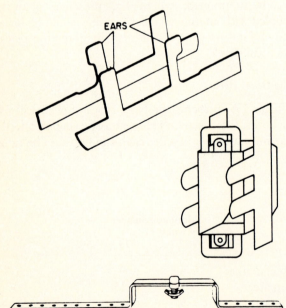

FIGURE 23:14

TABLE 23:8

	Kitchen area*	Other areas
Electrical switches	44"–46"	48"–50"
Electrical outlets	44"–46"	12" (garage 48")

*Dependent upon the height of the back splash above the counter tops

loops and with multiple wire switching. These occur whenever a white wire must be used for reasons of economics instead of the customary black wire. It must either be painted or taped black at both ends for proper identification.

2. All lights are always connected to one black and to one white wire (Figure 23:16).

3. All receptacles (outlets) are connected to a black wire (brass or copper screw) and to a white wire (silver screw)—Figure 23:17.

4. The white wire is always continuous within the circuit and should never be connected to a black wire (Figure 23:15). The exception is the switch loop where the white wire used is in reality a black wire.

5. The entire circuit must be properly grounded by employing a grounding wire that connects the ground-

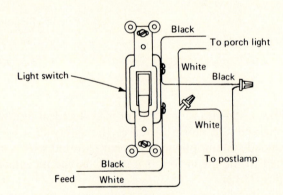

FIGURE 23:15

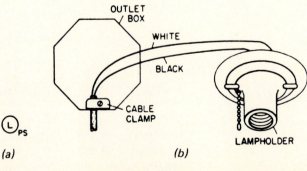

FIGURE 23:16

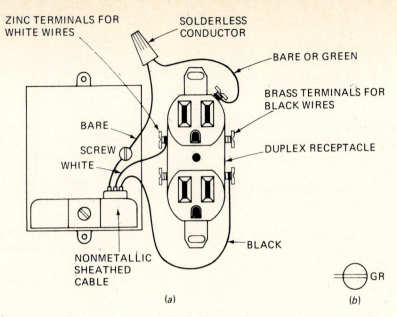

ZINC TERMINALS FOR WHITE WIRES

SOLDERLESS CONDUCTOR

BARE OR GREEN

BRASS TERMINALS FOR BLACK WIRES

BARE

SCREW

WHITE

DUPLEX RECEPTACLE

NONMETALLIC SHEATHED CABLE

BLACK

GR

(a)

(b)

FIGURE 23:17

ing terminal of each switch, light, and receptacle within the circuit to its accompanying grounded electrical box (Figures 23:17 and 23:18).

6. To properly install three-way switches, at least four wires (black, white, red, and ground) are required. The common terminal on each switch must be connected to a black wire or a white one painted black (switch loop).

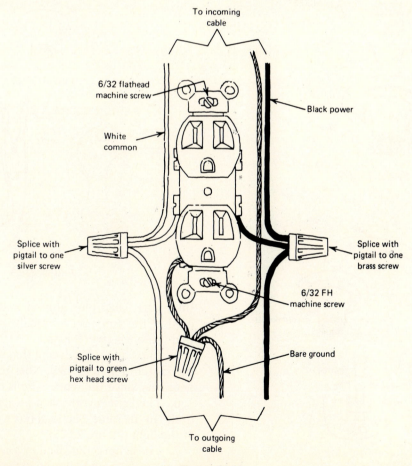

To incoming cable

6/32 flathead machine screw

Black power

White common

Splice with pigtail to one silver screw

Splice with pigtail to one brass screw

6/32 FH machine screw

Splice with pigtail to green hex head screw

Bare ground

To outgoing cable

FIGURE 23:18

Lights

The following line drawings and electrical diagrams illustrate the various electrical installations for lighting.

Wiring a light fixture that is *not controlled by a wall switch* (light with its own pull chain) is shown in (Figure 23:19).

Most light fixtures, however, are *controlled by wall switches*. The various wiring situations are described as follows:

1. The power supply is at the switch (Figure 23:20).
2. The power supply is at the light (Figure 23:21). Note the switch loop.
3. One switch controls two lights (Figure 23:22). Note the switch loop.
4. One light is controlled by two separate switches, with the source at the light (Figure 23:23). Note that a

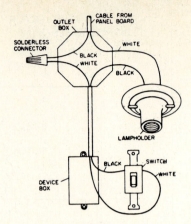

FIGURE 23:21

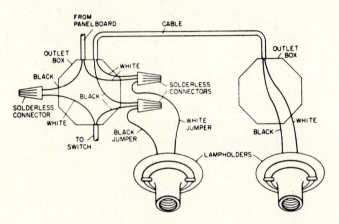

FIGURE 23:22

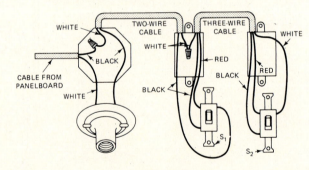

FIGURE 23:23

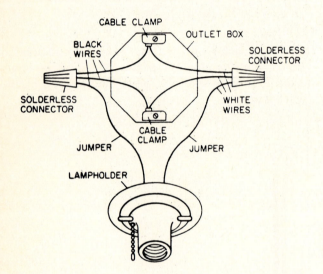

FIGURE 23:19

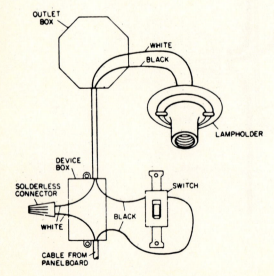

FIGURE 23:20

three-wire cable plus ground is used, black wires are used at the common terminals of both switches, and that there is a switch loop.

5. The same as (4), except that the power supply is at the switch (Figure 23:24). Note that no switch loop is necessary since the white wire is continuous.

6. The light is controlled by a switch, with an extra outlet in the same electrical box with the power supply at the light (Figure 23:25).

7. The light is controlled by a switch, with the power supply at the switch plus an extra outlet (Figure 23:26).

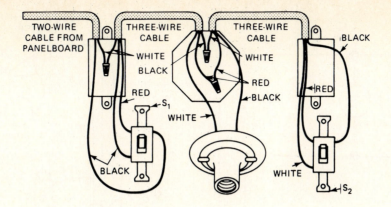

FIGURE 23:24

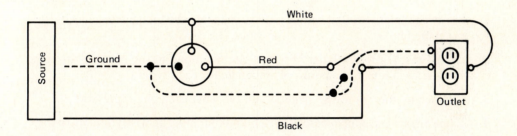

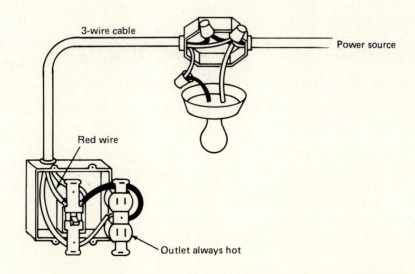

FIGURE 23:25

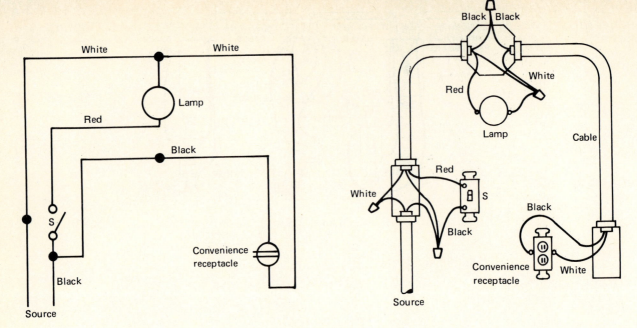

FIGURE 23:26

QUESTIONS

1. List two (2) methods that can be used to calculate the total wattage needs for a specific home.

2. List three (3) different types of wiring circuits found in the home. State the purpose, location, and amount of amperage for each type.

3. Choose any four (4) wiring symbols and tell what each one represents.

4. There are several appliances within a home that must be wired directly to the panel box. Name three (3) that must be on their own circuit.

5. State the size wire and amperage capacity for most electrical outlets and lighting fixtures within the home.

6. Compare how Romex and BX cable are prepared prior to their actual connections.

7. List two (2) methods used to connect or join several wires together.

8. What are the locations within the house that must be installed with ground fault circuit interrupters?

24

ALTERNATE BUILDING METHODS

In many geographical locations, conventional framing methods are being altered and replaced by other construction techniques and practices in an effort to make modern residences much more energy efficient. There are presently a variety of new building techniques that are being tested and constantly upgraded to lessen the consumption of fossil fuels.

ENERGY CONSERVATION DESIGN

A *sun-tempered solar home* (Figure 24:1) is the simplest, although not the most effective way, to design a home to reduce fuel consumption. To be effective, the following features must be incorporated during its construction:

FIGURE 24:1
Sample sun tempered home. (Source: Northeast Solar Energy Center)

1. Proper alignment of the house on the site to achieve optimum solar efficiency

2. Bulk of the windows situated on the south side of the house to capture the sun's rays

3. A sufficient roof overhang on the south side

4. The garage positioned to serve as a buffer against the cold northern winds.

An owner of a *super-insulated home* (Figure 24:2) can expect to spend very little to heat and cool the house. A home, so designed, will cost approximately 5 to 10 percent more than one not built with the following characteristics:

1. A high R-value in its ceilings, walls, and floors

2. A thick, continuous vapor barrier that is thoroughly sealed around all plumbing and electrical openings

3. A total window area that is 2 to 10 percent of the total exterior wall surface area, with the windows double- or triple-glazed and located on the south side of the home

4. A double 2 × 4 or a single 2 × 6 stud wall, spaced 24″ o.c., to accommodate more insulation in the walls (Figure 24:3)

5. Moveable insulation designed to be drawn at night to reduce heat loss through the windows

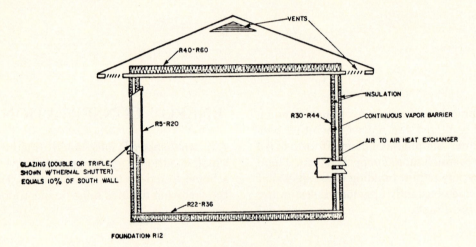

FIGURE 24:2
Sample superinsulated home. (Source: Northeast Solar
Energy Center)

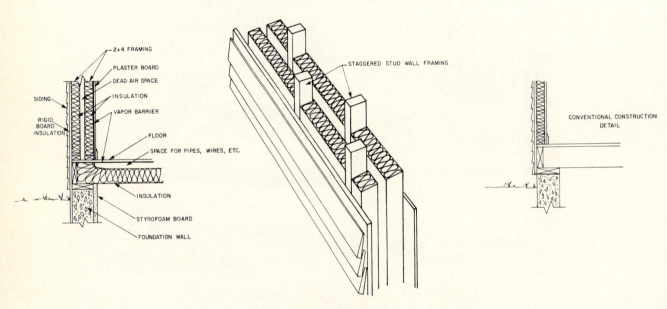

FIGURE 24:3
Superinsulated homes framing details. (Source: Northeast Solar
Energy Center)

6. An air-to-air heat exchanger (Figure 24:4) whose function is to maintain both a flow of fresh air and comfortable humidity levels throughout the interior (needed because of the high degree of airtightness)

7. A structure having small floor and exterior surface areas

8. The knowledge that a conventional furnace with its related distribution system can be replaced with a much less expensive system

The *thermal envelope home* (Figure 24:5) is also called the double envelope house or "house within a house" because of its structural design. This house is constructed with the following features:

1. 6″ to 12″ wide continuous airspace that surrounds the house on four of its six sides (the north wall, the south wall, the roof, and the floor)

2. The construction of two distinct exterior walls, a crawl space or sub-basement beneath the floor, and an attic air space below the roof

3. Conventional single walls on the east and west

4. A greenhouse or other solar receptive feature for the south facing airspace

The circulation of warm air through the airspace is very effective at keeping the house warm. The sun's rays, entering the house through the glazed greenhouse area, heat the air, which then rises and enters the roof airspace. It proceeds down the north wall airspace, into and along the crawl space beneath the floor, finally returning to the greenhouse area to complete this "convective loop."

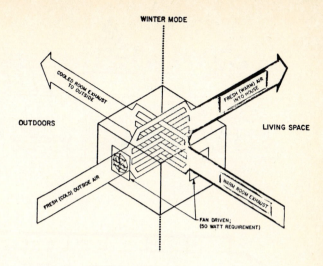

FIGURE 24:4
Air-to-air heat exchanger. (Source: Northeast Solar Energy Center)

Excess heat can be stored during daylight hours in the crawl space area, to be given up and circulated through the airspaces during evening hours when there is no direct solar heating.

Additional benefits from this house design are excellent noise control, a low rate of air infiltration, and the ease with which the structure can be cooled.

A truly *earth-sheltered home* (Figure 24:6) is one that has 60 percent of its exterior wall surface and 80 percent of its roof covered by earth. There are other homes that do not exactly fit the above description but still are labeled as earth-sheltered, underground, or bermed homes. Some are carved into hillsides, leaving their

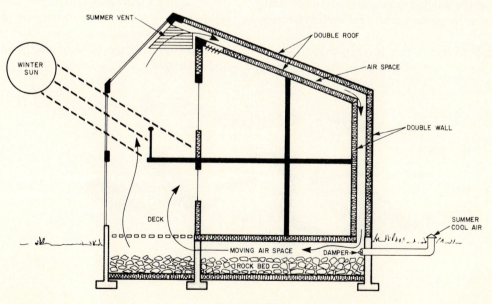

FIGURE 24:5
Double envelope house. (Source: Northeast Solar Energy Center)

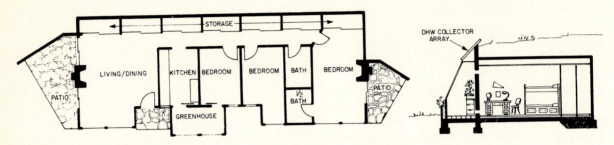

FIGURE 24:6
Earth sheltered home. (Source: Northeast Solar Energy Center)

south-facing wall exposed to the rays of the sun, while others are built as conventional structures with earth bermed up against one or more of their sides (Figure 24:7).

An earth-sheltered home's most advantageous feature is that it is much easier to heat and cool than any above ground structure. It is also resistant to rot, vermin attack, and is reasonably fire resistant. It provides excellent shelter from the wind, is intensely quiet inside, and, when designed with passive solar features, extremely easy to heat.

It is important to realize that it is not easy to locate the ideal site for an earth-sheltered home. The most conducive site would be one having a gentle, south-facing slope with excellent drainage (Figure 24:8).

The structural design for an earth sheltered home requires thick, massive reinforced concrete walls. The exterior surfaces of the walls must be covered with butyl sheeting before the extruded polystyrene insulation is applied. The sheeting helps to waterproof the walls, while the insulation helps prevent loss of heat through the walls and is sturdy enough to withstand the pressure of the back-filling during construction (Figure 24:9).

A dehumidifier is usually a necessary appliance in an earth sheltered home to keep any humidity problems under control.

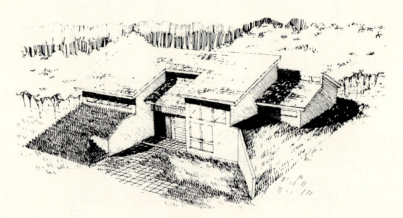

FIGURE 24:7
Buildings on flat ground can be "bermed" on one or more sides.

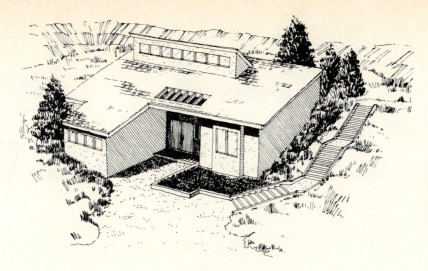

FIGURE 24:8
A residence built into a south-facing slope.

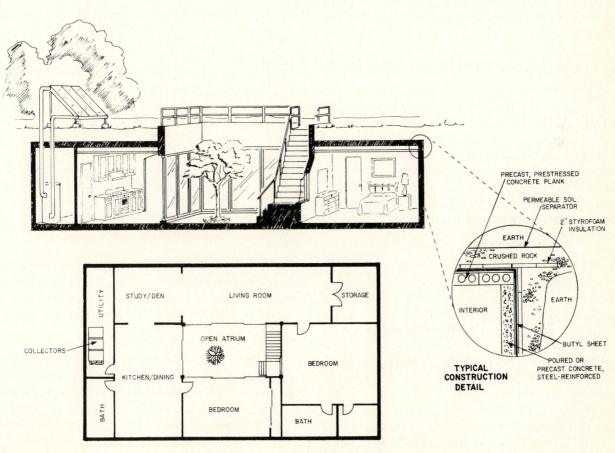

PRECAST, PRESTRESSED
CONCRETE PLANK

PERMEABLE SOIL
SEPARATOR

2" STYROFOAM
INSULATION

EARTH

CRUSHED ROCK

EARTH

INTERIOR

BUTYL SHEET

POURED OR
PRECAST CONCRETE,
STEEL-REINFORCED

**TYPICAL
CONSTRUCTION
DETAIL**

UTILITY

STUDY/DEN   LIVING ROOM   STORAGE

COLLECTORS

OPEN ATRIUM

BEDROOM

KITCHEN/DINING

BATH

BEDROOM

BATH

FIGURE 24:9
Underground home. (Source: Northeast Solar Energy Center)

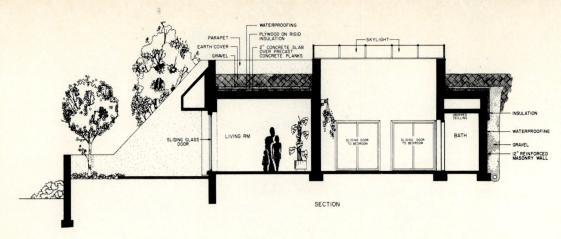

FIGURE 24:10

Roof areas can be covered, first with gravel and then with soil 18″ to 48″ deep. The soil can then be seeded with grass or planted as an integral part of the landscape (Figure 24:10).

It is important for any builder planning an earth-sheltered home to become familiar with the local building requirements in order to avoid problems that might result and to offer solutions to any objections that the building inspector may have. The most typical obstacles encountered in this style of home are the lack of windows in bedroom areas and the absence of a quick exit in case of fire.

POST AND BEAM CONSTRUCTION

In many contemporary homes, exposed beams and enlarged glass areas are an integral part of the design. Post and beam framing (using posts, beams, and planks as shown in Figure 24:11) is more conducive to this style home than conventional framing methods. It provides the following advantages:

1. Additional ceiling height is realized
2. Construction time is reduced because there are fewer pieces to assemble
3. Wider roof overhangs are possible
4. Headers are not required above the exterior doors and windows
5. The distance between the exposed beams can be greater because of the thickness of the solid wood decking used as the roof material
6. The cost of the materials is less whenever the base of the roof planks serves as the interior ceiling
7. The wooden components used have a much higher fire resistance factor than that of metal components.

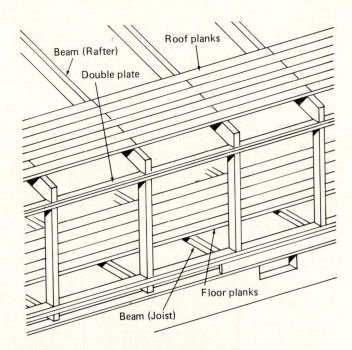

FIGURE 24:11
Post-and-beam framing. Note there are no headers beneath the plate. (National Forest Products Association)

Even though the difficulties associated with post and beam construction are usually slight, as a carpenter, you must be aware of them so that you can both plan and carry out the construction accordingly. The items requiring special attention are

1. Floor areas expected to support heavy, concentrated loads (e.g., bearing partitions, a refrigerator, a freezer, a bath tub) must be reinforced with additional support beneath the weighted area (Figure 24:12).
2. To eliminate the sway and lack of rigidity in the external walls, it will be necessary either to brace the corners or to use conventional stud framing in certain portions of the wall (Figure 24:13).

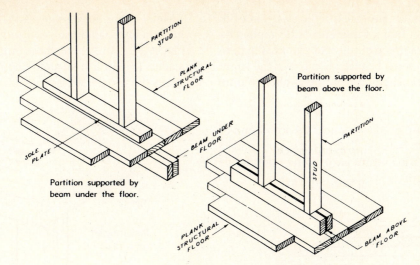

FIGURE 24:12

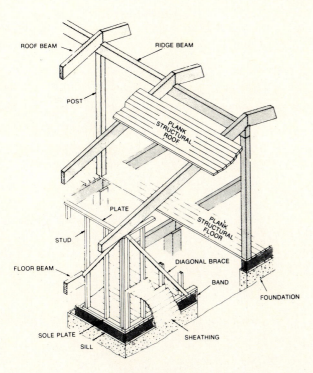

FIGURE 24:13
Note corner has both diagonal bracing and conventional stud framing.

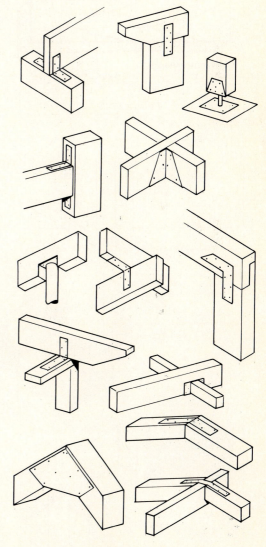

FIGURE 24:14
Metal fasteners and connectors for post-and-beam construction.
(Western Wood Products Association)

3. The installation of electrical wiring, plumbing lines, and components of the heating system may present a problem, since there is little space to conceal them in the outside walls and in the ceilings.

4. Normal nailing procedures to connect a beam to a post does not provide the strength necessary to resist the various stresses exerted upon the joint formed between post and beam. Instead, metal fasteners are needed to reinforce this connection (Figure 24:14).

5. Using post and beam construction for homes in colder areas requires that a vapor barrier be applied directly to the upper surface of the roof planks. Then, fiberboard insulation must be placed on top of the vapor barrier before the finished roof can be applied (Figure 24:15).

Metal fasteners are used to connect the lower end of each support post to the sole plate and its upper end to the top plate, the roof beam, and/or the continuous header (Figure 24:16).

Metal straps are employed at the exterior corners of the building, whenever continuous headers are used (Figure 24:17), to connect the butt ends of two beams or headers (Figure 24:18) and to connect each pair of roof beams (Figure 24:19).

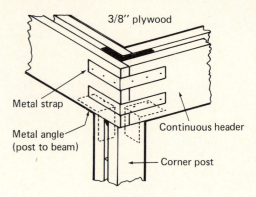

FIGURE 24:17

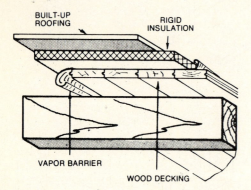

FIGURE 24:15

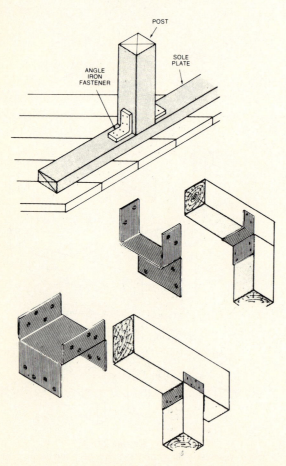

FIGURE 24:16

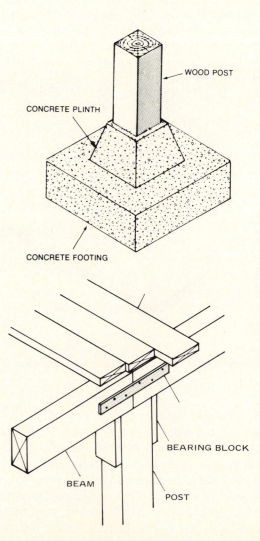

FIGURE 24:18

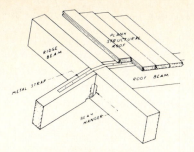

FIGURE 24:19

The beams used to support the roof can run from the side walls to the ridge beam (transverse beams), or they can extend from one end of the structure to the other end (longitudinal beams), as in Figure 24:20. Note that in each design, the roof planks are assembled at right angles to the roof beams. It is also important to position every roof beam directly over a supporting post (Figure 24:21).

The solid tongue and groove wood roof planks must be toe-nailed and face-nailed to the roof beams, as well as edge-nailed to each other, by driving long nails into predrilled holes (Figure 24:22). If possible, the roof planks should be long enough to extend over more than one span (the distance between two beams) to ensure a stronger, less flexible roof surface.

PREFABRICATED HOMES

The term prefabricate means to construct house components in standardized sections at the manufacturer's for shipment and quick assembly at the building site. Each one of the different methods of fabrication used for house construction relies on the use of mass production techniques to prepare the components for their eventual assembly. The advantages which the various types of manufactured housing offer are

1. The erection is fast and easy, with less on-site hours required.
2. The purchase price is reasonable.
3. They all provide flexibility in design.
4. Construction delays are eliminated, with very little pilferage of materials.
5. The total cost is less because the manufacturer is able to purchase the materials in large quantities and able to construct them with machines, with much less waste than is customary with a frame built house.

Precut Homes

All the components used to assemble a *precut home* are first cut in the factory to specified lengths. The

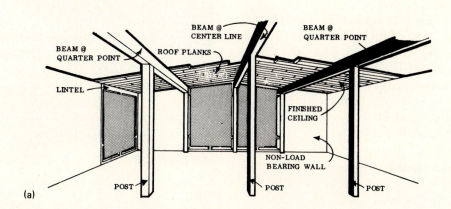

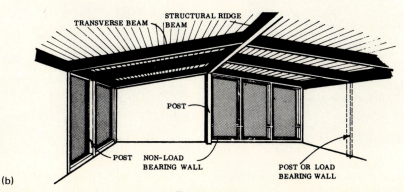

FIGURE 24:20
(A) Longitudinal Beam. (B) Transverse Beam.

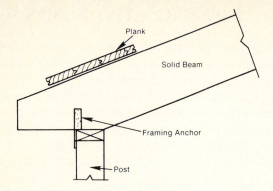

FIGURE 24:21

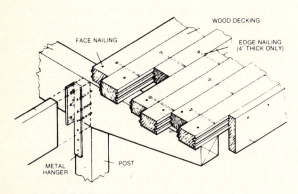

FIGURE 24:22

Panelized Homes

Commonly referred to as "prefab homes" by most carpenters, they are more correctly called *panelized prefabricated homes*. These homes are constructed of flat panels which are first fabricated and then assembled as they move along the factory's production lines (Figure 24:24).

pieces are then stamped with symbols (Figure 24:23) so that the carpenter can ascertain, with the aid of a thorough set of step-by-step instructions, when and where each component should be positioned. The advantages of this method of prefabricated house construction are the elimination of the time required to measure and cut each component and the ability to assemble the entire house without the need for electricity.

FIGURE 24:23
On-site assembly of house shell from prefabricated components; all parts are identified with markings.

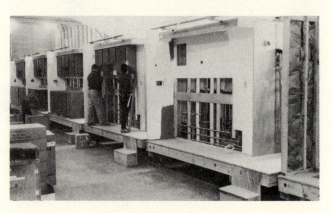

FIGURE 24:24
Panelized prefabricated house.

After each panel is completed and inspected, it receives a number, which is used to designate its loading position on the delivery trailer and its sequence in the actual assembly procedure (Figure 24:25).

Most manufacturers require that the foundation and subfloor be completed before the panels are delivered. Some manufacturers, however, supply fabricated floor panels that are assembled directly on top of the foundation, thus eliminating the need to construct a subfloor (Figure 24:26).

FIGURE 24:25
Finished wall loads onto large open-top trucks; completed panel includes door, window, and heating unit; electric cable at left will connect to adjacent wall.

FIGURE 24:26
Stressed-skin floor panels are lowered onto foundation; their finished underside (basement ceiling) is a feature not found in conventionally built homes.

The biggest advantage associated with panelized homes is the tremendous savings in time and labor personnel. Another plus for this method of construction is that the panels can be designed to fit both the size and style of home desired (Figure 24:27).

Modular (Sectionalized) Homes

The plan of most modular homes is based on a standardized grid system which is divided into 4″ squares or segments. Groups of squares form standard units of measurement called *modules*. This method is advantageous because its incorporates the common building dimensions of 16″, 24″ (minor module), and 48″ (major module) into the overall plan (Figure 24:28).

Using the MOD 24 framing system, the overall house dimensions are based on 24″ modular units. This 24″ module unit is used as the spacing to frame the floors, walls, and roof. The savings in materials realized with this system is the elimination of one framing member (stud, joist, truss, or rafter) for every 4′ of length (Figure 24:29).

A window framed with this method requires 23 percent less framing than a conventionally framed window (Figure 24:30).

Some modular designs are able to achieve maximum flexibility in the positioning of doors and windows by superimposing a 16″ module to the MOD 24 framing system (Figure 24:31).

It is possible, with the aid of a crane, to position entire house sections on a foundation when modular prefabrication methods are used. One floor homes usually involve only two completed sections, the front and rear halves, while two floor homes are assembled with four sections. One possible drawback to modular homes is that individual sections are limited to a 14′ width because of transportion restrictions (Figure 24:32).

FIGURE 24:27
Gable end wall is moved into position with windows whole and intact —evidence of efficient and careful loading and transport.

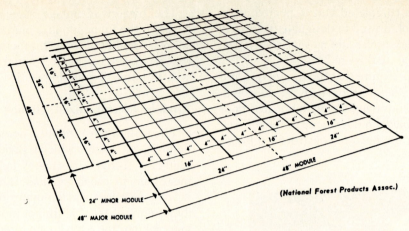

(National Forest Products Assoc.)

24" MINOR MODULE

48" MAJOR MODULE

FIGURE 24:28
Courtesy of National Forest Products Association.

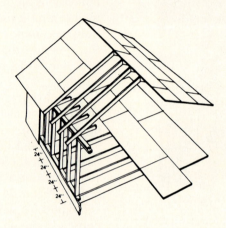

FIGURE 24:29
All framing members align in the 24" module system.

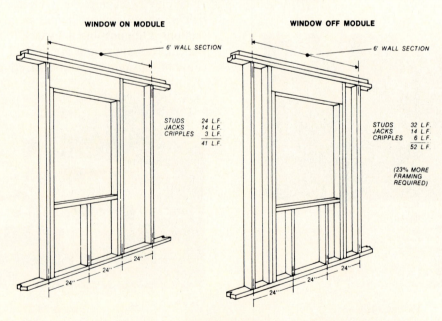

WINDOW ON MODULE   WINDOW OFF MODULE

6' WALL SECTION   6' WALL SECTION

STUDS   24 L.F.   STUDS   32 L.F.
JACKS   14 L.F.   JACKS   14 L.F.
CRIPPLES   3 L.F.   CRIPPLES   6 L.F.
             41 L.F.   52 L.F.

                          (23% MORE
                          FRAMING
                          REQUIRED)

FIGURE 24:30
Proper design can cut cost and labor. On-module construction at the
left requires 23 percent less framing than unit on the right.

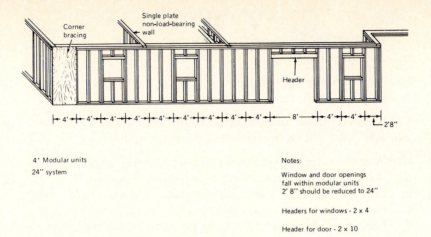

Corner bracing

Single plate non-load-bearing wall

Header

|← 4' →|← 4' →|← 4' →|← 4' →|← 4' →|← 4' →|← 4' →|← 4' →|← 4' →|← 8' →|← 4' →|← 4' →|
2'8"

4' Modular units
24" system

Notes:

Window and door openings
fall within modular units
2' 8" should be reduced to 24"

Headers for windows - 2 x 4

Header for door - 2 x 10

FIGURE 24:31
National Lumber Manufacturers' Association.

FIGURE 24:32
Sectionalized house unit about to be loaded on a ferry for shipment to building site.

The predominate advantage that a modular home has over the other types of prefabricated homes is that almost all of the construction has been completed at the factory, leaving very little to be done after the sections have been connected. The bathrooms are finished, the kitchen cabinets have been installed, and the totally assembled utility lines only need to be connected to their sources (Figure 24:33).

Roof shingles must be installed where the individual roof sections join to complete the roof. Where the wall sections meet, the connecting plywood pieces must be removed before the trim molding and carpeting can be completed.

FIGURE 24:33
Clean, simple and attractive, completed house features 32" wide overhang for extra weather protection.

QUESTIONS

1. List three (3) advantages that an all-weather wood foundation has over a masonry foundation.

2. List four (4) advantages that post and beam construction has over conventional framing methods.

3. List three (3) items of concern that the contractor should be aware of whenever post and beam construction is employed.

4. Compare the three (3) types of prefabricated homes with respect to:
    a. The number of pieces needed to assemble each home,
    b. The time required to assemble the various pieces,
    c. The man power and equipment needed for the assembly, and
    d. The list of operations that must be done before the home can be inhabited.

25

CABINET MAKING

A carpenter needs special skill and familiarity with the procedures in order to properly assemble the components of kitchen and bathroom cabinets. This expertise is also required for constructing other pieces of furniture or storage facilities throughout the house.

Cabinets may be constructed at the building site, custom built in a woodworking shop, or be purchased as fully assembled, mass produced factory units.

For new home construction, the cabinet maker must closely adhere to the architectural plans. The plans usually include a top view of the entire layout (Figure 25:1), individual cabinet elevation drawings (Figure 25:2), and the specifications defining the type and quality of the materials, as well as the type of joints, to be used in the construction.

When contracted to construct new cabinets for an older home, the cabinet maker can either copy the original master plan or make slight to radical changes in it. Whenever the location of the appliances will be moved to enhance the overall design, making the area more pleasant and efficient to work in, a change in the original cabinet layout can be made.

CABINET SIZES

Most cabinets are constructed to conform to standard sizes. Kitchen base cabinets are normally 36″ high, 24″ deep, with widths ranging from 12″ to 48″ in 3″ increments, and a toe space 4″ in height and $3\frac{1}{2}$″ in depth

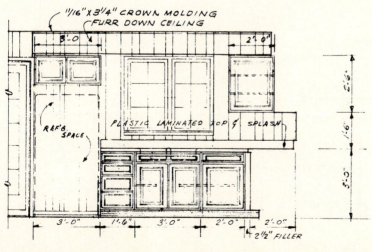

FIGURE 25:1

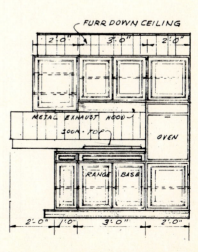

FIGURE 25:2

(Figure 25:3). Kitchen wall cabinets may be 12", 15", 18", or 30" high, with widths similar to the base cabinets (Figure 25:4). The distance between the base and wall units ranges from 15" to 18", except over a sink or range where it must be 24".

Bathroom lavatories should be 31" high, 18" to 22" wide, with a 3" recess for knee space and an additional 3" recess at the base of the cabinet for toe space (Figure 25:5).

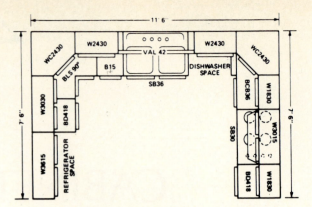

FIGURE 25:4
Typical use of wall and base cabinets for a "U" shaped kitchen.

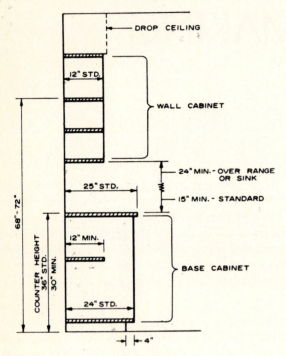

FIGURE 25:3
Kitchen cabinets.

CABINET TERMINOLOGY

Any carpenter interested in building cabinets must have a basic understanding of the terms used to describe: (1) the cabinet's components, (2) the various methods of joining the components together, and (3) the various types of cabinet hardware.

Base cabinets usually are constructed with a *base* that is designed to rest on the floor, support the *bottom shelf*, define the *toe panel*, and provide an anchoring surface for the two *side panels*. The side panels must be *dadoed* on their inner surfaces where permanent *shelves* will be positioned to ensure support (Figure 25:6), and *rabbeted* on their rear edges to receive the *back panel* (Figure 25:7).

The vertical facing strips, *stiles*, and the horizontal facing strips, *rails*, are both attached to the front edge of

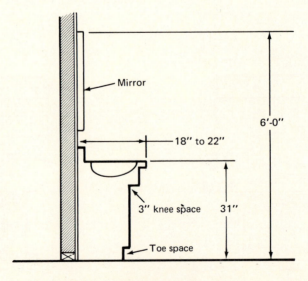

FIGURE 25:5
Bathroom lavatory.

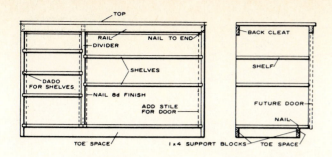

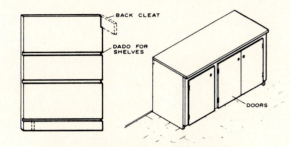

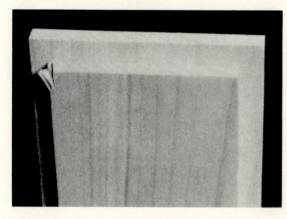

FIGURE 25:6

FIGURE 25:7

the side panels. The *top rear rail* or wall strip is used to connect the top rear surfaces of the side panels, as well as to anchor the cabinet to the wall. Additional front and rear rails are used to define the *drawer openings* and support the *drawer guides*.

A *drawer front* can be joined to its sides by one of several types of joints. The bottom of the drawer is inserted into dado grooves cut in all four drawer pieces—front, back, and both sides (Figure 25:8).

WOOD JOINTS

There are eight basic types of joints that can be used to assemble cabinets. Most cabinetmakers have a preference for some of these joints over the others and will construct their cabinets using these joints exclusively. The joints, presented in Figure 25:9, are (1) butt, (2) dado, (3) rabbet, (4) lap, (5) dovetail, (6) mortise and tenon, (7) miter, and (8) tongue and groove.

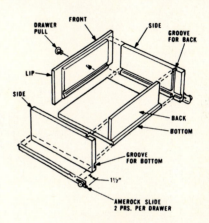

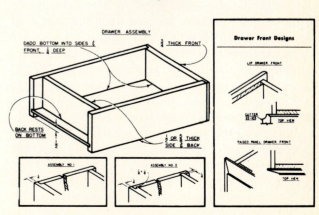

FIGURE 25:8

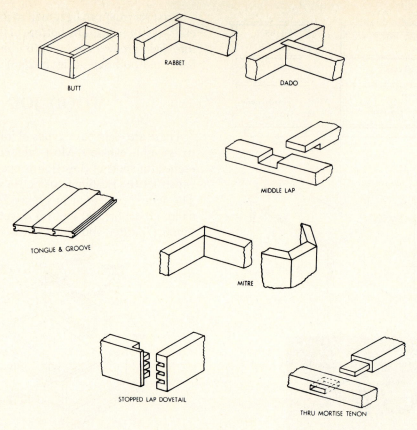

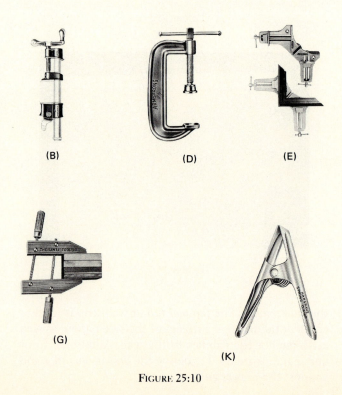

FIGURE 25:9

The cabinetmaker must decide during the planning stages which joints to use, giving special consideration to the cabinet's eventual appearance and structural soundness, and the time required for construction.

Clamps are needed to connect some of the joints, especially those that are joined together with glue. Some of the types used are presented in Figure 25:10.

Butt joints are only used for simple construction. They may be strengthened by the use of dowels. Both the *dado* and *rabbet* joints, involving groove cuts in one of the joining members, are used when the joint must be stronger than the simple butt joint.

Lap joints are used whenever the two pieces join at a right angle in places where appearance is not a factor. *Dovetail* joints are commonly used on corners, where both strength and stress are involved. Most drawer components are joined with this type of joint.

Whenever extra strength and durability are desired, especially when connecting legs and rails together, *mortise-and-tenon* joints are used.

Miter joints connect pieces cut at 45° angles. They are used to join two pieces of molding or trim as well as picture frame corners.

Corrugated fasteners are used to join two flat pieces

FIGURE 25:10

of wood together, especially in miter joints and furniture. Made of steel, they may be purchased with either plain or saw edges, and they may have either parallel or divergent ridges. Figure 25:11 shows the four different types and some applications.

To join the underlying structural edges for table tops and other large surfaces, a *tongue-and-groove* joint is commonly used.

CABINET HARDWARE

Swinging cabinet doors that are installed flush with their opening can be hung with simple butt hinges. These are cut into the door and opening, leaving only the center pin exposed. Decorative hinges may be mounted on the door's surface, or a semiconcealed hinge type may be used (Figure 25:12). These doors must be fitted so that they provide at least $\frac{1}{16}$" clearance on each edge.

A door that overlaps its opening can be hung with butt hinges, or decorative ones designed to accommodate this type of opening (Figure 25:13).

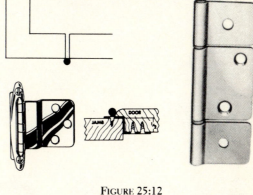

FIGURE 25:12

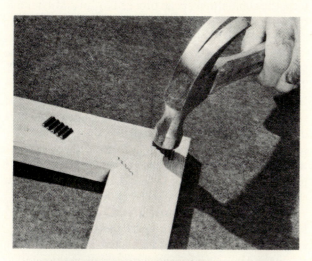

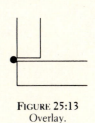

FIGURE 25:13
Overlay.

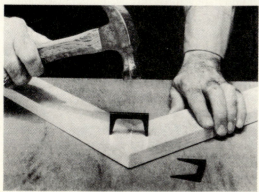

FIGURE 25:11
Corrugated fasteners.

The lip type of swinging cabinet door has a $\frac{3}{8}$" lip cut in its back surface, making it easier to fit the opening than a flush door. They can be hung with either a semiconcealed or a fully concealed hinge (Figure 25:14).

Hinges should be prefitted and then removed before applying the final surface finish to the doors. Decorative, yet functional, door and drawer hardware (Figure 25:15) can be installed after they have been finished. Drawer pulls or knobs must be centered on the door's

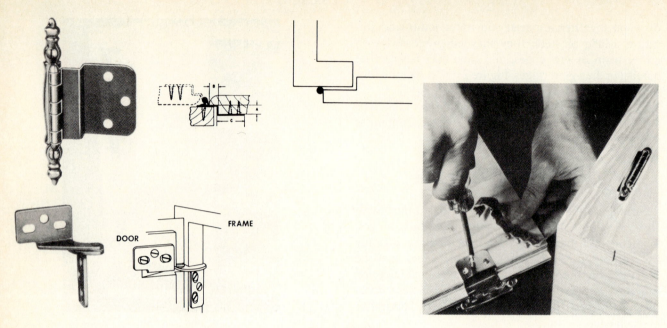

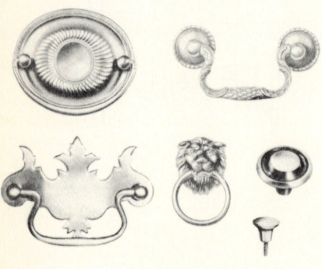

FIGURE 25:14

FIGURE 25:15

surface and will have a better appearance if positioned slightly above the middle of the drawer. Door pulls should be placed on the bottom third of wall cabinets for easy access, and on the top third of base cabinets.

Figure 25:16 shows the common types of catches used to keep swinging cabinet doors closed.

CABINET ASSEMBLY

After determining whether the exterior surface of the cabinets will consist of wood, veneer, or a laminate sur-

face, the cabinetmaker can proceed to cut and assemble the components. In most instances, the assembly procedure follows a definite sequence of steps.

The base of the cabinet is assembled first, using the appropriate measurements for the size cabinet desired. The two end panels are then attached to the base with finishing nails or a contact adhesive, depending on the finish surface selected. These end panels must be dadoed at the exact height that the shelves will be inserted before they are assembled; or the shelves can be made adjustable by installing shelf brackets to both side panels. Extreme care must be exercised to ensure that these adjustable brackets are level (Figure 25:17).

Some cabinetmakers use the wall surface as the back for cabinets, while others would never construct one lacking a back panel. When a back panel is used, both side or end panels must receive a rabbet cut along their entire rear vertical edge. Prior to insertion of the back panel, both the top rear rail (wall strip or ledger) and the shelves (including the bottom panel or shelf) must be assembled. The bottom shelf must be flush with the front of the sides before it is secured (Figure 25:18).

Next, the partitions are assembled using nails driven through the back panel and an adhesive. Note that the partition is notched to fit around the ledger strip. It is advisable to attach a temporary anchor strip, nailed to the top edges of the end panels, to ensure that the partition remains both level and plumb (Figure 25:19).

The facing strips are then applied to the front of the cabinet. The vertical strips, *stiles*, are assembled first, and then the horizontal strips, *rails*, are added. They are

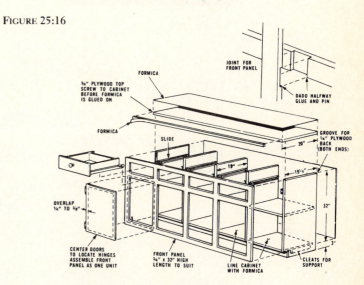

FIGURE 25:16

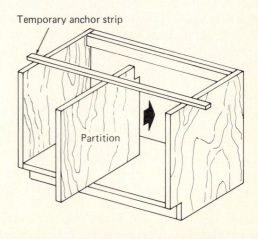

FIGURE 25:17
Adjustable shelving.

FIGURE 25:18

FIGURE 25:19

best installed by, first, placing the cabinet on its back and, then, using both glue and recessed finishing nails to secure them (Figure 25:20). These facing strips are also used to define the drawer openings.

There are three basic types of drawer guides. The corner guide uses wooden strips attached to the end panels and partitions as the means of keeping the drawer in its opening. An additional strip, located above the top of the drawer's side panel, may be needed to prevent the drawer from tilting in its opening (Figure 25:21).

The center guide, located in the middle of the drawer's base, can be constructed with a wooden runner built into the cabinet's frame. The drawer base has a

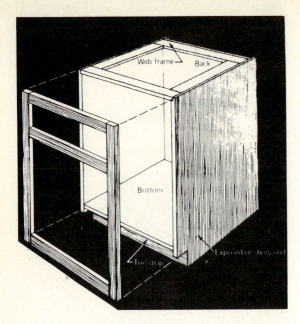

FIGURE 25:20

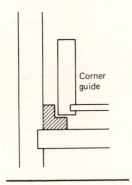

FIGURE 25:21
Corner guide.

slot, extending from front to rear, that slides along the wooden center guide (Figure 25:22).

Most cabinetmakers prefer to use side-mounted metal drawer slides for the cabinet drawers. This type offers more support, prevents the drawer from dropping out of its opening, and is long-lasting (Figure 25:23).

CABINET INSTALLATION

Prior to the actual installation, it is advisable to locate and mark the exact locations of the vertical wall studs, mark out each cabinet's outline on the wall, and check with a level to determine how plumb the wall surface is and how level the floor is. This last procedure will help determine how much and where the units must be shimed to attain an attractive cabinet installation (Figure 25:24).

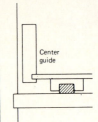

FIGURE 25:22
Center guide.

The wall cabinets should be hung first. Several units should be connected together before they are raised into position. (Two or more people may be needed.) Installing a temporary ledger strip to the wall, whose top edge is level with the base of the cabinets, will make the installation easier. Once the joined units are raised into position and are resting on the ledger strip, they must be carefully checked for plumb and level; shims are placed where necessary (Figure 25:25).

Since wall corners are often not perfectly square, wood shims are needed to make the cabinets square (Figure 25:26).

Once the cabinets are plumb and level, pilot holes are drilled through the ledger strips so that long wood screws can be inserted to secure each unit to the wall (Figure 25:27).

The base cabinets are installed after the wall units have been thoroughly anchored to the wall. They must be installed plumb and level and positioned at the desired height below the wall cabinets. Shims are used to attain these requirements (Figure 25:28).

SHELVES, BUILT-INS, WALL UNITS

All of these structures are designed to provide organized storage in an attractive setting. They can be constructed as permanent additions to an existing room or as temporary, collapsible structures that can be reassembled to match a mood or new location.

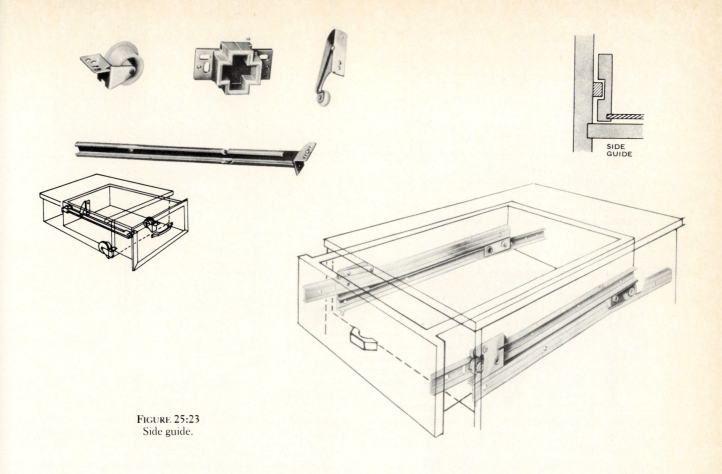

FIGURE 25:23
Side guide.

SIDE
GUIDE

FIGURE 25:24

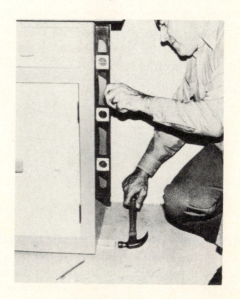

FIGURE 25:25
Cabinets plumb.

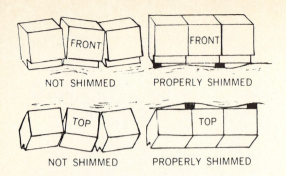

FIGURE 25:26
Cabinets must be shimmed to ensure proper alignment.

FIGURE 25:27

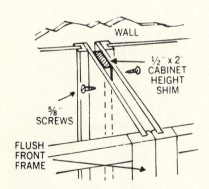

FIGURE 25:28

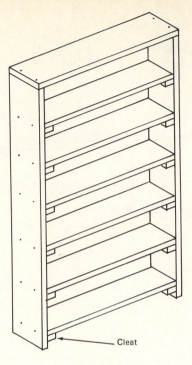

FIGURE 25:29

positioned against a wall, the bookcase does not require a rear panel.

Wall units can be designed with bookshelves, a shelf to house a television set or stereo, and storage compartments located below the shelves (Figure 25:30).

FIGURE 25:30

Shelves can be constructed to fit within the thickness of a wall, be attached to the wall, rest directly on the floor, or extend in height from the floor to the ceiling. Bookcase type shelves can be easily assembled by attaching shelving to wooden 1 × 2 cleats that have been screwed to the two side panels (Figure 25:29). When

QUESTIONS

1. List all the items that are usually included in the architectural plans which are related to the installation of the kitchen and bathroom cabinets.

2. State the standard sizes for kitchen base cabinets, for kitchen wall cabinets, and for bathroom lavatories.

3. Explain how a dado cut differs from a rabbet cut.

4. Describe five (5) of the eight basic wood joints.

5. Compare the installation and eventual appearance of butt, semi-concealed, and fully concealed hinges.

6. Tell the proper position where door pulls should be located on wall and base cabinets.

7. Give the names of the two (2) facing strips used on cabinets and two reasons for their use.

8. Name the three (3) types of drawer guides and give reasons why you prefer one over the other two.

26

ALTERATIONS AND ADDITIONS

There are many carpenters that specialize in making alterations and additions to existing dwellings. Even though the basic procedures involved are the same as those performed during the home's actual construction, the carpenter must be familiar with the types of problems that are associated with alterations or additions and must perform the necessary tasks to make each project structurally sound and cosmetically acceptable.

To simplify the information in this chapter, the most common projects requested by homeowners were divided into five headings.

EXTERIOR WALLS

Window and Door Installation

One way to make a home more energy efficient is to replace all metal, loose-fitting, drafty windows and doors with fully insulated units that are designed to resist heat loss and air infiltration.

To the novice, the procedure of removing one unit and replacing it with a more energy efficient one seems simple, but this rarely is the case.

The carpenter can obtain a rough approximation of the size of the unit to be replaced in its trimmed frame but is unable to know exactly what the rough opening is until the old unit is removed.

Since most lumber dealers do not stock large quantities of windows and doors, it is necessary to approximate the size needed and place the order long before the actual installation takes place. To be reasonably safe, the carpenter should order a unit whose height and width dimensions are slightly *smaller* than those an-

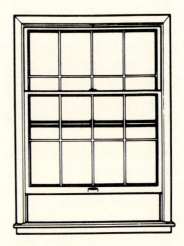

FIGURE 26:1

ticipated for the existing unit, since it is much easier to fill in than it is to cut out.

If the interior dimensions of the window, as measured from sash to sash (Figure 26:1) are 36" × 48", the replacement window selected should be slightly smaller (e.g. 2' 8" × 3' 10" or 32" × 46").

To remove the old window or door, first remove all the interior trim surrounding the unit and a sufficient amount of the exterior wall to expose the entire unit. Once this has been accomplished, place a short length of 2 × 4 against the unit's interior jamb and hit it repeatedly with a hammer until the unit is loosened enough to lift it from its opening (Figure 26:2). Removal and subsequent window replacement from a second floor is not one of the simpler tasks expected of a carpenter.

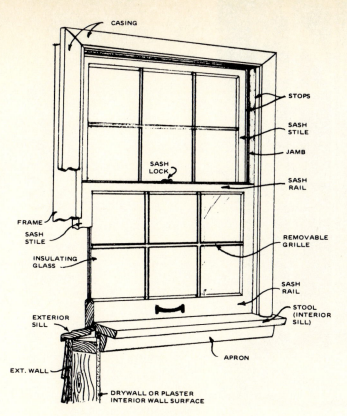

FIGURE 26:2
Parts of a typical double-hung window.

butt edges cannot be installed properly unless the siding is removed from the entire wall.

Installing a new window, door, or sliding glass door in an area where one did not exist before is best started on the inside wall. A sufficient area of wall surface must be removed to expose the upright studs. Extreme care must be exercised while removing the wall surface so that no utility lines are cut or disturbed. If uncovered carefully, they can usually be rerouted out of the way (Figure 26:4).

The rough opening is then framed out. In most cases, one or more studs must be cut and removed to frame the opening (Figure 26:5). Before this is done, especially for large openings on a ground floor with rooms above, it is advisable to make a temporary support frame for the floor above (Figure 26:6). Make certain to frame the opening as previously suggested in Chapter 14 (Windows and Doors). In short, use a double header supported by trimmer studs nailed to full length studs.

After the frame has been completed, small pilot holes are drilled from the inside through the exterior wall surface to pinpoint the exact location of the opening and to help the carpenter realize how much of the exterior wall surface must be removed (Figure 26:7).

Once sufficient wall surface has been removed, the exterior sheathing can be cut out and removed, exposing the unit's rough opening. The unit can then be installed, and nailed into place when plumb and level. The exterior wall is replaced; followed by a restructuring of the interior wall surface.

It is necessary to place the new window into the existing opening to check how well it fits before it can be nailed into place. A $\frac{1}{2}''$ clearance on each of the four sides is acceptable and may be needed to properly square the unit. Any gap greater than $\frac{1}{2}''$, however, should be shimmed out before installing the unit (Figure 26:3).

Additional problems may result whenever the exterior wall finish has aluminum siding. It is difficult to cut in a straight line, and the j-bar used to conceal the

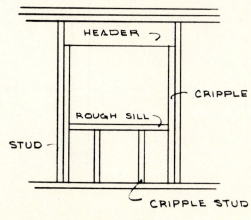

FIGURE 26:3

FIGURE 26:4

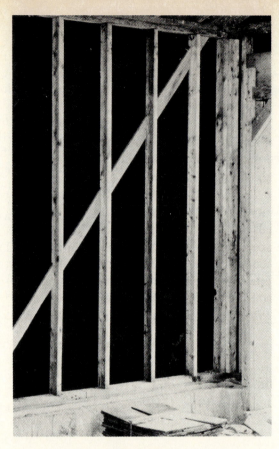

FIGURE 26:5
The two center studs must be either cut, top and bottom,
or removed completely

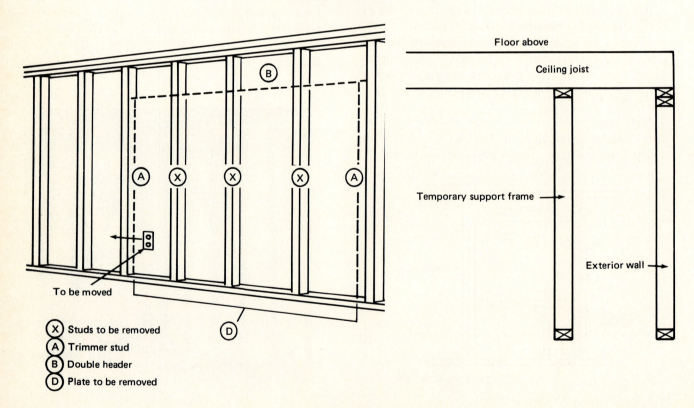

To be moved

- (X) Studs to be removed
- (A) Trimmer stud
- (B) Double header
- (D) Plate to be removed

Floor above

Ceiling joist

Temporary support frame →

Exterior wall →

FIGURE 26:6

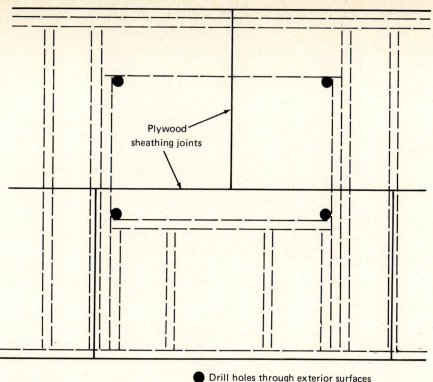

Plywood
sheathing joints

● Drill holes through exterior surfaces

FIGURE 26:7

Wall Air Conditioner Installation

The window styles in some homes make it necessary for an air conditioner to be installed in the wall. The procedure to follow is practically the same as that mentioned earlier for windows and doors (Figure 26:8).

The unit can be installed in proximity to any convenient wall outlet, but perhaps would function more effectively if wired directly to the electrical panel.

It should be supported on the outside wall to keep it level and operating efficiently (Figure 26:9). Since most units come equipped with some type of insulating housing, it may be possible to carefully cut directly through the exterior sheathing and wall surface and avoid a major removal of the exterior surface.

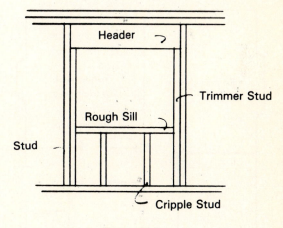

FIGURE 26:8

INTERIOR WALLS AND CEILINGS

Removal of a Bearing Wall for Room Enlargement

When a home has been constructed with a kitchen and a dining room that are both too small for one family or another, the only solution is to remove the wall between the two rooms and create a much larger, more serviceable single room.

If the wall is a bearing partition, upon which most of the rooms upstairs are resting, simply removing the wall is out of the question. A suitable solution, however, is to install a double or triple wooden header, a steel I-beam or a channel beam just beneath the doubled top plate. Prior to the removal of the existing wall, a temporary 2 × 4 shoring wall is required to support the ceiling while the header is being installed. Both ends of the header must be adequately supported to prevent it from sagging (Figure 26:10).

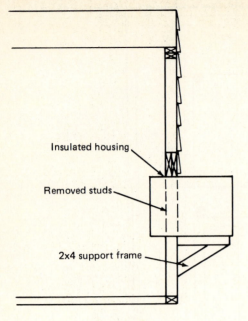

FIGURE 26:9

Insulated housing

Removed studs

2x4 support frame

Once installed, the header and its supports can be covered with some type of wall covering to make their presence less noticeable.

Build a New Closet

The closet space in many homes is definitely inadequate. The solution to this problem is usually to construct a closet inside an existing room.

The new closet should be at least 24″ wide (deep) for clothes to hang properly. The length of the closet is not critical and usually is determined by the space available and the location of the ceiling joists above.

When positioned in a corner, only two walls must be constructed since the existing walls are used as the back and one side of the closet.

The biggest problem with the installation of a closet involves the positioning of one of the two walls to be constructed. One of the two will be easy to build because it can be positioned perpendicular to the ceiling joists. The other wall, however, must be installed directly over (parallel to) a joist in order to be structurally sound. This requirement may, therefore, alter either the desired length or the width, depending upon the closet's location and its relationship to the ceiling joists (Figure 26:11).

Lower a Room Ceiling

To reduce the expense of heating and cooling an older home with high ceilings, a simple solution is to drop the height of its ceilings. To accomplish this, a 2 × 4 gridwork must be constructed at the lower height.

Begin by nailing 2 × 4s around the perimeter of the room. They must be nailed to the upright studs hidden in the existing walls. Then, 2 × 4 ceiling joists are cut to size so that they extend across the room's width from

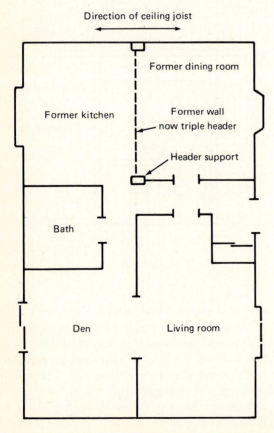

Direction of ceiling joist

Former dining room

Former kitchen

Former wall now triple header

Header support

Bath

Den   Living room

FIGURE 26:10

FIGURE 26:11

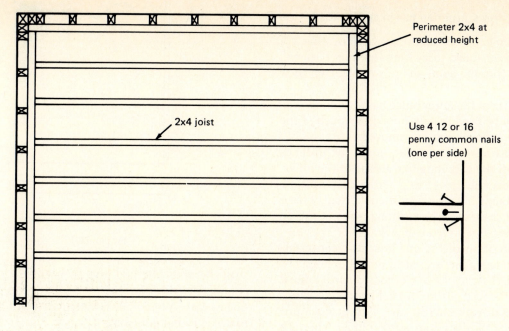

Perimeter 2x4 at
reduced height

2x4 joist

Use 4 12 or 16
penny common nails
(one per side)

FIGURE 26:12

one wall to the opposite one. They are then toe-nailed to the 2 × 4s previously nailed to the wall (Figure 26:12).

To prevent the lowered ceiling from sagging, it is necessary to connect the structural members of the new ceiling to those of the old ceiling. This is best accomplished by inserting eye hooks in both the old and new ceiling joists and connecting them with either heavy wire or cable (similar to installing a suspended ceiling)—Figure 26:13.

The 2 × 4 gridwork can then be covered with wallboard, paneling, or ceiling tile.

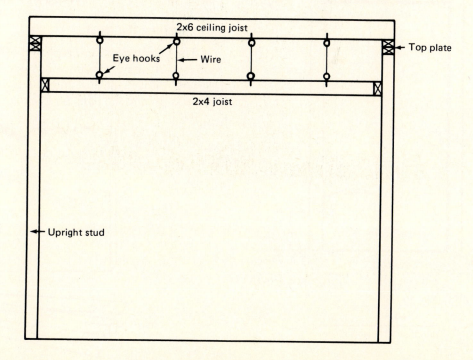

2x6 ceiling joist

Top plate

Eye hooks   Wire

2x4 joist

Upright stud

FIGURE 26:13

Trim a Door after Carpeting has been Installed

Most carpet mechanics remove all the doors that will conflict with their ability to lay the carpeting. They will not, however, trim the doors and rehang them on their hinges.

Since the base of the lower hinge has not changed its location, it can be used to determine how much of the door's bottom surface must be removed.

A tape measure is used to determine the distance from the top of the carpeting to the base of the hinge. Most carpenters subtract $\frac{1}{8}''$ to $\frac{1}{16}''$ from this measurement to ascertain where the door should be cut. This is to ensure that the door, when rehung, will not rub on the knap of the carpet (Figure 26:14).

Once the location of the cut line has been established, a framing square is used to draw the line. The door can be cut by one of two methods. A razor knife is used to score the face of the door before it is cut with a plywood blade in your circular saw. This practice prevents the face of the door from being chipped when it is cut (Figure 26:15).

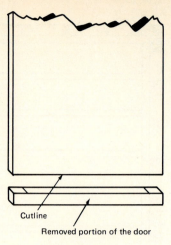

Cutline

Removed portion of the door

FIGURE 26:15

In the other method, the razor knife is used to cut completely through the veneer to the underlying hollow core. A hand saw is then used to cut through the two side framing members.

First, remove the veneer from the solid wood piece that has just been cut off (Figure 26:16). Then, pry off the two side framing pieces that are attached to the solid piece with corrugated fasteners. The hollow-core cardboard sandwich design must be cut and pushed out of the way so that the bottom block can be reinserted into the base of the door. Check its fit before you glue it in place (Figure 26:17).

ROOF

Reshingling the Roof

The longevity of most asphalt roof shingles is approximately 20 years. Evidence of ripped or torn shingles, severe surface cracking, and ceiling damage because of leaks is usually the basis for deciding to reshingle.

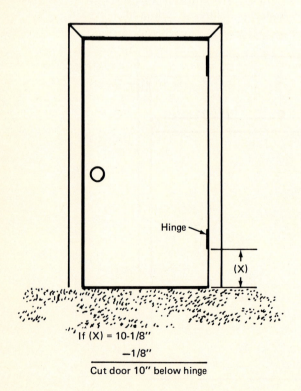

Hinge

(X)

If (X) = 10-1/8''

−1/8''

Cut door 10'' below hinge

FIGURE 26:14

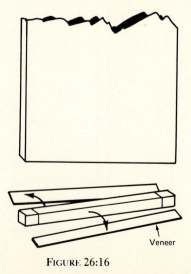

Veneer

FIGURE 26:16

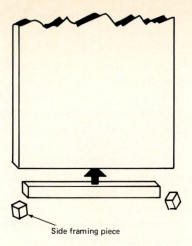

Side framing piece

FIGURE 26:17

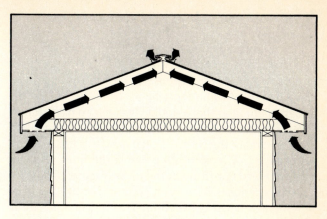

FIGURE 26:18

It is possible to cover one layer of existing shingles with another layer. Longer nails must be used to make certain that they penetrate both layers and the sheathing. If the home has already been reshingled once, it will be necessary to remove all the shingles and apply the new ones directly to the roof sheathing.

For climate control, it is advisable to install a ridge vent when the roof is reshingled. The installation involves cutting a small portion of the roof sheathing on both sides of the ridge beam to ensure adequate air flow through the vent (Figure 26:18). The roof's peak is not capped with shingles when a roof vent is installed.

When applying seal tab shingles, make certain that you invert the first course. Use $1\frac{1}{2}$" galvanized roofing nails when reshingling over another layer of shingles. Because the joints must be staggered with each row of shingles, the end piece on one row must be cut. The amount cut off can usually be used as the other end piece (Figure 26:19).

OPENINGS

Attic Stairway

Most building codes only require that a small opening be installed, usually in a bedroom closet ceiling, to allow access to the attic area. Anyone who has tried to gain entrance to the attic through this opening to hook up a TV or stereo antenna or store household items, has found it to be less than satisfactory.

The solution to this problem is to install a disappearing stairway in the ceiling of the hallway. The first item to check before starting this project is whether or not there are electrical lines in the way of the proposed opening, and if so, whether or not they can be moved out of the way.

To properly frame the opening for the stairway, several ceiling joists must be cut. Begin by marking out the rough opening on the ceiling from the attic side. To simplify the project, it is advisable to use one of the existing joists as part of the frame. Remember that the

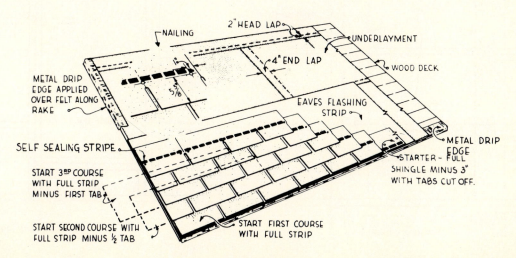

2" HEAD LAP

NAILING

4" END LAP

UNDERLAYMENT

WOOD DECK

METAL DRIP EDGE APPLIED OVER FELT ALONG RAKE

$\frac{5}{8}$

EAVES FLASHING STRIP

SELF SEALING STRIPE

METAL DRIP EDGE

STARTER - FULL SHINGLE MINUS 3" WITH TABS CUT OFF.

START 3ᴿᴰ COURSE WITH FULL STRIP MINUS FIRST TAB

START SECOND COURSE WITH FULL STRIP MINUS ½ TAB

START FIRST COURSE WITH FULL STRIP

FIGURE 26:19

frame must be doubled on all four sides in order to be structurally sound. The distance between the two top plates outlining the two hallway partitions can be used to center the frame in the hallway.

Mark both sides of the joists that must be cut and removed. The cut line must be 3″ from the line of the rough opening so that a double header can be installed. Use a chalkline to snap the cut lines and a combination square to mark the width of each joist; this helps obtain square cuts. A reciprocating saw is the most efficient tool to use to cut through the joists (Figure 26:20).

After the joists have been cut and removed, a measurement is taken between the two remaining stationary joists. From this measurement, the four joists used to frame the length of the opening are cut. The first one, on each side of the opening, is installed against the butt ends of the shortened joists and is then nailed into place. Nails are also driven through the two stationary joists into these frame members. Make certain that they are square with the stationary joists before you completely nail them in place.

Each of the second long joists can then be nailed to its adjacent partner and to the stationary joists as well.

Nail a second trimmer joist to the stationary reference joist. It should be long enough so that its ends rest on the two top plates outlining the hallway.

The fourth side of the frame can now be constructed. Two joists are cut to fit between the four joists outlining the length. Tail joists are used to connect these short joists to a stationary joist. They are attached by nailing through both the stationary joist and the first short joist installed. The second short joist is then nailed to the first one (Figure 26:21).

After the frame has been completed and checked for squareness, nails are carefully driven down through the ceiling material to outline the opening from below; removal of ceiling material is best accomplished from the hallway. It is advisable to cover the entire floor thoroughly and renail the ceiling material to its new frame before you cut out and remove the ceiling material.

The installation of the stairway is best done from the attic. You may need help to lift the unit into the attic and to nail the unit into place.

Nail two 1 × 3s across the base of the frame next to the ceiling. They are used to support the unit while it is being nailed and shimmed to its frame (Figure 26:22).

The project is complete when the appropriate trim molding has been installed around the door's frame.

Skylights

The frame for a skylight is constructed just like all other openings. Its doubled frame is built from the inside before the opening in the roof is made. Nails driven completely through the roofing materials at each of the four corners of the frame will help you determine how

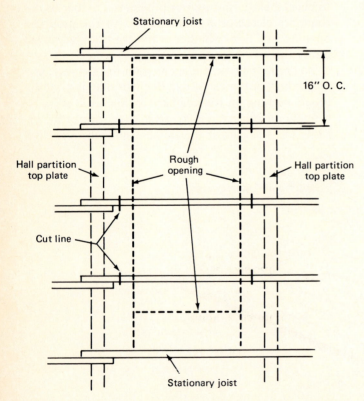

FIGURE 26:20

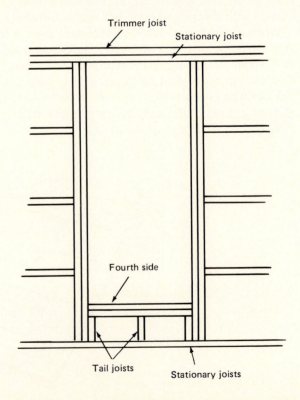

FIGURE 26:21

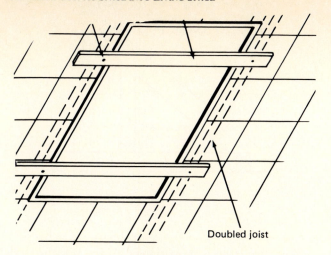

FIGURE 26:22

many roof shingles to remove with a ripping chisel. Then the roof opening can be cut out and removed.

Make certain to replace the shingles below the frame before you place the unit in its opening so that the unit's flashing will cover these shingles. On the other three sides, however, the roof shingles are installed over the flashing, leaving a gap of $\frac{3}{4}''$ next to the sides of the unit for drainage.

After the unit is installed, it is necessary to frame out the attic area as well as to determine how large to make the ceiling opening so that it, also, can be framed. Figure 26:23 shows how the area can be framed to achieve a much larger ceiling opening.

CONVERTING ATTIC SPACE INTO LIVING SPACE

The first concern in this type of alteration is how to gain access to the attic. In many instances, a portion of the present dwelling must be sacrificed to provide the space needed to install an adequate stairway. Sometimes, when an existing cellar stairway is located in the center of the house, the attic stairway can be positioned in the same stairwell so that the loss of existing living space is avoided.

The second concern is how the new area will be serviced with heat, electricity, and plumbing, and whether or not the present systems can provide adequate service to the new living space.

Many homeowners do not want to incur the additional expense necessary to construct a dormer or to add a skylight to make the area larger and brighter. They also realize that their taxes would increase with the addition of either one. They are primarily interested in simply adding one or more bedrooms in the attic area of their house.

There must be sufficient head room (7' 6" is probably the minimum) in the peak of the roof so that it can be properly framed to provide the ventilation required of all attic living areas (Figure 26:24).

Short lengths of 2 × 4 are nailed to each pair of rafters. They must be positioned parallel to each other in the same horizontal plane and be at least 12" below the roof's peak. They are used to define the highest flat portion of the ceiling; as support for the wallboard ceiling and the batt insulation; and also to guarantee the desired air flow through the gable vents (Figure 26:25).

Knee walls are usually installed at the point where the distance from the floor to the rafter is 30" to 36" (Figure 26:26). They can be constructed in one of two ways:

1. With a top and bottom plate, requiring the top of each stud to be cut at the same angle as the slope of the roof rafter (Figure 26:27)

2. with only a bottom plate, with the top edge of each upright stud nailed to the side of a rafter (Figure 26:28).

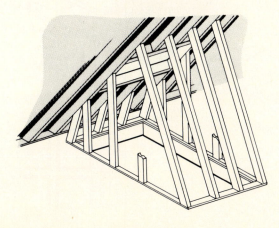

FIGURE 26:23

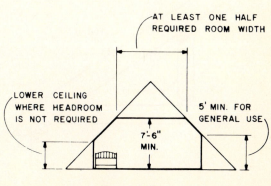

AT LEAST ONE HALF REQUIRED ROOM WIDTH

LOWER CEILING WHERE HEADROOM IS NOT REQUIRED

5' MIN. FOR GENERAL USE

7'-6" MIN.

FIGURE 26:24

FIGURE 26:25

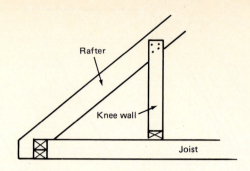

FIGURE 26:28

Batt insulation is then stapled to the upright studs comprising the knee wall and to the rafters, making certain that the foil surface always faces the room's interior. R-11 ($3\frac{1}{2}$") batt insulation is the maximum that can be used between the rafters to ensure the desired flow of air (Figure 26:29).

Figure 26:30 shows the order of locations for the wallboard. Apply the wallboard to the small flat ceiling area (1), first; then to the sloping rafter area (2); and finally to the knee wall area (3). Apply joint compound and tape to all the wallboard joints.

It may be necessary to furr around the windows in the gable ends to provide a nailing base before the gable end walls can be covered with wallboard (Figure 26:31).

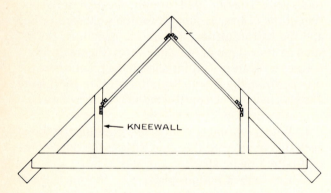

FIGURE 26:26

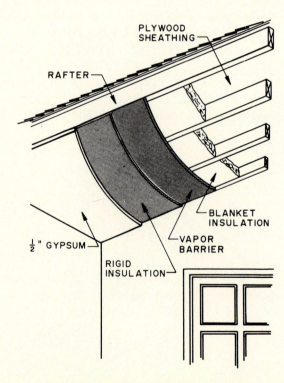

FIGURE 26:29

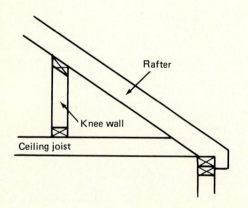

FIGURE 26:27

FIGURE 26:30

SOLAR RETROFIT

Before an *active* solar system can be installed, a home must be well insulated and reasonably air tight. This system also needs electrical energy to operate the pumps and fans used within the system. The basic components of this system are the collectors, a storage facility, and a distribution system (Figure 26:32).

The *collectors*, flat panels that are placed on the roof or on the ground beside the house, absorb the sun's rays and convert this solar energy into heat energy (Figure 26:33). The collectors should face south and have a tilt approximating the home's geographical latitude (Figure

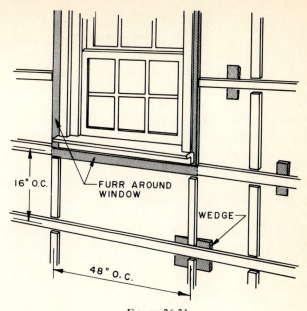

FIGURE 26:31
Furring is placed around windows to provide a nailing base.

26:34). Each collector is supplied with either air or a water-antifreeze mixture, which must be heated by the sun's rays before it can be conveyed to the storage tank (Figure 26:35).

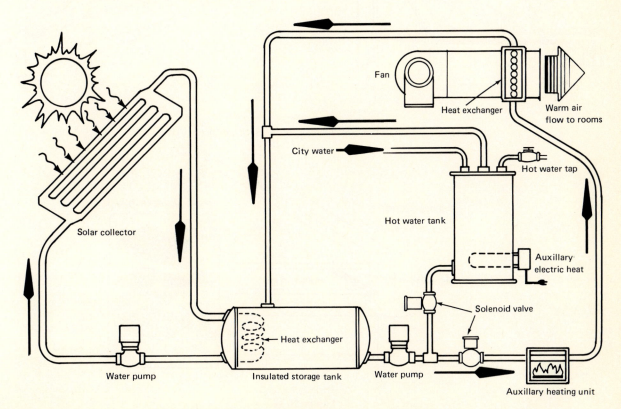

FIGURE 26:32

A comparison of the characteristics of the two systems, air and liquid, is presented in Table 26:1. For optimum efficiency, it is advisable to select the collector that provides the most heat per dollar invested.

Figure 26:36 is a detailed presentation of how the collectors are mounted on the roof.

An insulated water storage tank is used as the source of heat for the home's hot water heating system (Figure 26:37). Its size, in gallons, is determined by calculating the total square foot area of the collectors and multiplying this value by 1.5. A representative schematic for a liquid collector space heating system is presented in Figure 26:38.

The size of an air storage tank that is designed to blend with an existing forced air system should be two

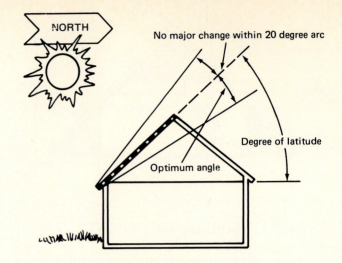

FIGURE 26:34

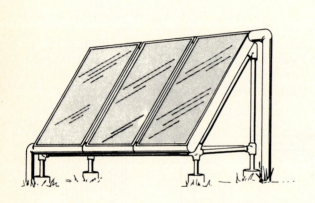

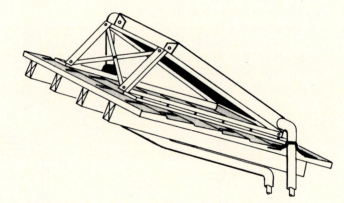

FIGURE 26:33
Rack and frame mounting. (Source: Northeast Solar Energy Center)

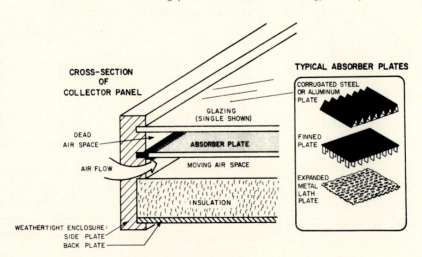

FIGURE 26:35
Air collectors. (Adapted from *Solar Energy for Buildings Handbook*)

TABLE 26:1
Characteristics of Air and Liquid Systems

Air systems	Liquid systems
Air is heat transfer medium.	Various liquids are heat transfer media.
Won't freeze, cause corrosion, or contaminate potable water supply	Measures must be taken to prevent corrosion, freezing, or contamination of potable water supply.
Leaks are harmless though difficult to detect.	Leaks can cause damage or system failure but are relatively easy to locate.
Require larger collector area	Require smaller collector area
Operate at lower temperatures	Liquids are more effective heat transfer media with higher heat capacity
Can be used for direct space heating or with a forced air system	Better for DHW, can be used with forced air space heating systems
Storage and ductwork requires much more space.	Piping and storage are smaller, easier to retrofit.
	More systems available, better selection
	Passive systems using liqquid flat plate collectors in DHW systems available.

Note: In liquid systems, water-anti-freeze solutions, silicon or mineral oils, and refrigerants are all used and have differing characteristics. The above descriptions are intended to introduce the basic differences between two broad systems types; specific descriptions of particular systems vary considerably.

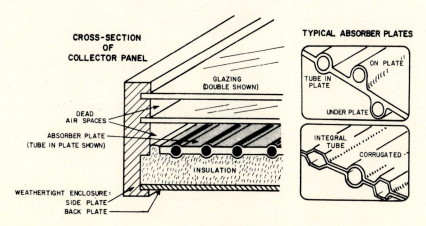

FIGURE 26:35
Liquid collectors. (Adapted from *Solar Energy for Buildings Handbook*)

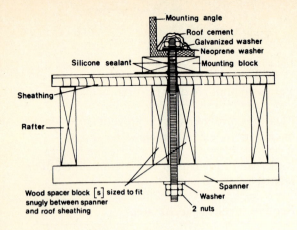

SPACER BLOCK WITH SPANNER MOUNT

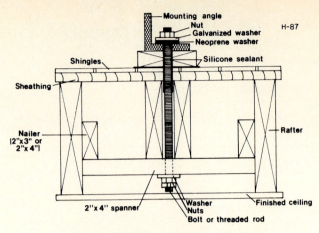

SPANNER MOUNTING IN CONCEALED SPACE

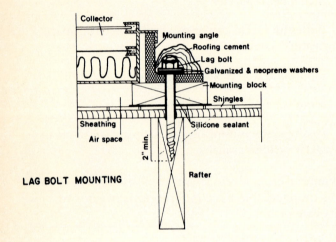

LAG BOLT MOUNTING

FIGURE 26:36
(A) Spacer block with spanner mount. (B) Spanner mounting in concealed space. (C) Lag bolt mounting. (Source: *Installation Guidelines for Solar DHW Systems*)

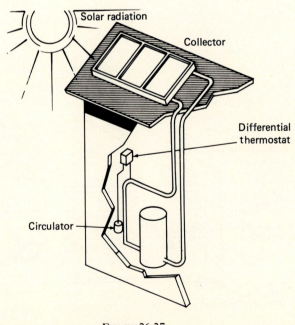

FIGURE 26:37
Solar hot water system.

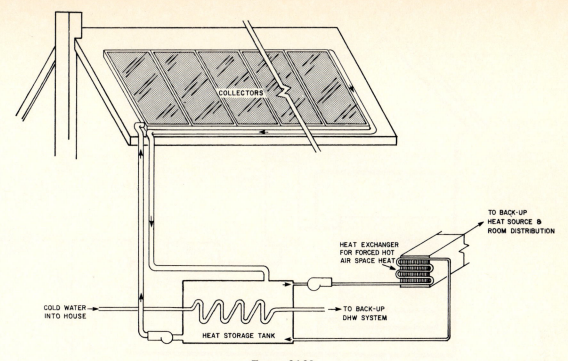

FIGURE 26:38
Liquid collector space heating system. (Source: Northeast Solar
Energy Center)

and a half to three times greater than a comparable water tank because of the presence of gravel or stones inside the tank. Figure 26:39 shows an example of an active space (air collector) heating system.

With either system, a solar storage tank must have a large enough capacity to keep the house warm during periods of cloudy weather. When cloudy weather persists, the normal backup system must carry the load.

Before you install any active system, to simply heat the domestic hot water or to heat the entire house, you must consider the expected durability and longevity of the system; the anticipated frequency and cost of repairs; how to protect the system against leakage, corrosion, and freezing-thawing conditions; the location and installation of each component; and, finally, the amount of actual physical maintenance required of the owner to keep the system operating at peak efficiency.

A comparison between the various solar domestic hot water systems is presented in Table 26:2.

Schematic drawings for the various domestic hot water systems are presented in Figures 26:40 through 26:44.

It is important to realize that solar retrofit systems may not be practical for all homes in all locations. The overall cost of installation may far exceed the expected gains from reduced fuel consumption, and therefore may not be worthwhile.

ADDITIONAL LIVING SPACE

The problems involved in making any addition to an existing home are varied and solutions should be determined by the carpenter before any construction begins. This is true whether the addition is a deck, rooms to be built in an attic area, or an extension of the present living space beyond the exterior walls.

This section will discuss the general problems inherent in most additions. Where warranted, reference will be made to a specific type of addition.

Access to the new area from the present living area must be well thought out. Attempting to add rooms in an attic area without an existing stairway requires that it be installed before the rooms are built. A proposed deck attached to the rear of a home must have a doorway connecting the house with the deck. When absent, a window can be removed and replaced with an exterior door.

Every attempt should be made to *match the two floor levels*. With decks or screened-in porches, a step down may be necessary. For an addition, however, the new floor level should match perfectly with the old floor, especially when the new area will be an expansion of the old area (e.g., a kitchen enlargement). Extending the upstairs hallway to reach two new bedrooms constructed

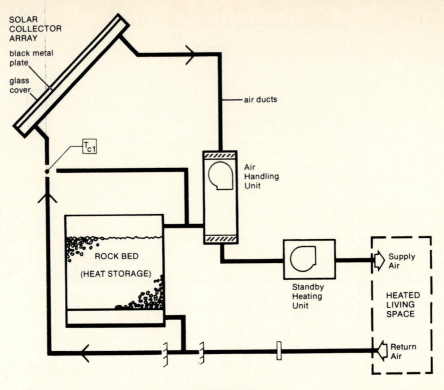

FIGURE 26:39
Active space heating system.

above a garage will appear like "a tack-on" if the new floor is not level with the old one.

How to handle the present *exterior wall surface* depends on the type of addition constructed. When the new area will be interior living space, the surface is removed so that the studs can be covered with the desired interior wall covering. Also, every attempt should be made to apply the same exterior wall surface to the new area so that the new area is not so obvious.

The only way that the bulk of an *exterior bearing wall can be removed* so that a new expansion area blends spaciously with the existing living space is to install a full-length double header or beam that is well supported at both ends. Its presence can be concealed by covering it and the balance of the ceiling with false styrofoam beams or by installing a suspended ceiling at a lower height.

It is possible to support an addition to the first floor without a *foundation*. Instead it can be supported by beams resting on steel lally columns that are buried in concrete. The size of the support members employed must conform to those specified by the local building code. The floor area must be well insulated before the base of all the floor joists are covered with exterior sheathing to ensure that the new area is energy effi-

cient. Also, the addition's house plate or header must be connected to the existing header with machine bolts, washers, and nuts.

For many additions, the *roof* presents a variety of problems. These include how and where to attach the new roof to the existing structure, what pitch must be used to attain the desired ceiling height, what changes will be needed in the roof's drainage system, and what type of roofing material is the most appropriate to use. Extending a gable roof design over a new addition using continuous lines may be more complicated than a step-down version, but its appearance is worth the extra effort.

Building a full length dormer on the second floor not only involves constructing a new roof but also removing half the existing roof. This is an elaborate undertaking which must be accomplished quickly to protect the interior from rain damage.

Since most interior additions will require both *heat and electricity*, and possibly some plumbing, the first concern is to determine whether or not the present systems are large enough to carry the extra load. The other concern is to determine the best possible means of routing the lines from their source through the existing house to the new area.

TABLE 26:2
Solar Domestic Hot Water Systems

Type	Main Features	Advantages	Disadvantages
Closed loop system	Flat plate collectors Closed loop of piping from collectors to storage tank Uses external energy (pumps, etc.) Uses non-freezing collector fluid Widely available	Won't freeze More and better established competition Small consumption of external energy	More expensive Fluid maintenance required More components required
Drainback system	Flat plate collectors Water is collector fluid Potable water goes through heat exchanger in storage tank (not through collectors) Unpressurized storage tank Large heat exchanger	No anti-freeze used Simplest of active flat plate systems (no valves)	Larger pump System must drain thoroughly
Draindown system	Flat plate collectors Potable water circulated through collectors Line pressure feeds collectors Has automatic drainage valves	No heat exchanger or extra storage tank needed High efficiency possibly due to elimination of heat exchanger	Larger pump System must drain thoroughly No corrosion inhibitor possible Freeze danger with valve failure
Air collector DHW systems	Flat plate (air) collectors Air-to-water heat exchanger Ductwork and blower (rather) than pipes and pump) Larger collector area than liquid systems	Won't freeze Leaks won't cause damage	Very hard to detect leaks More space required for ducts
Thermosiphon systems	Flat plate collector Uses a closed loop of piping but no pump or other external power (a passive DHW system) Sometimes use refrigerant as collector fluid Systems using water are rare in the Northeast Storage tank higher than collector	No external power Probably most efficient May be more difficult to retrofit	Uncommon in Northeast because of freezing (if water is collector fluid) Needs structural support for high storage tank

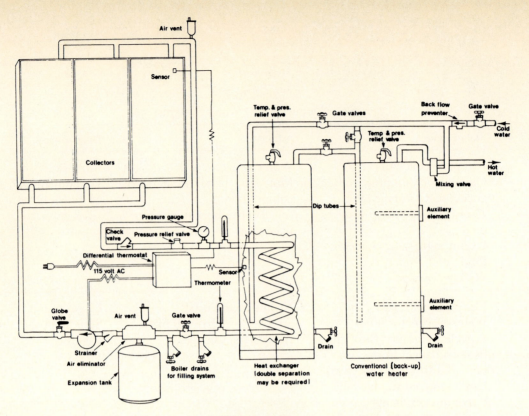

FIGURE 26:40
Closed loop DHW system schematic. (Source: *Installation Guidelines
for Solar DHW Systems*)

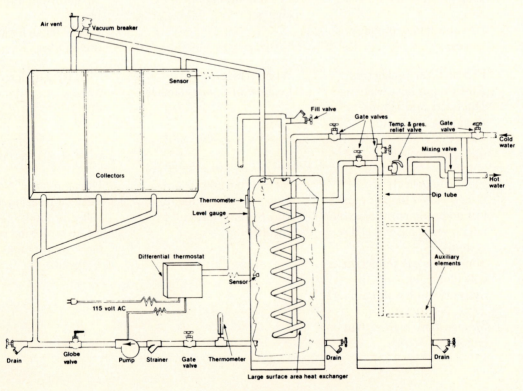

FIGURE 26:41
Drainback system schematic. (Source: *Installation Guidelines for
Solar DHW Systems*)

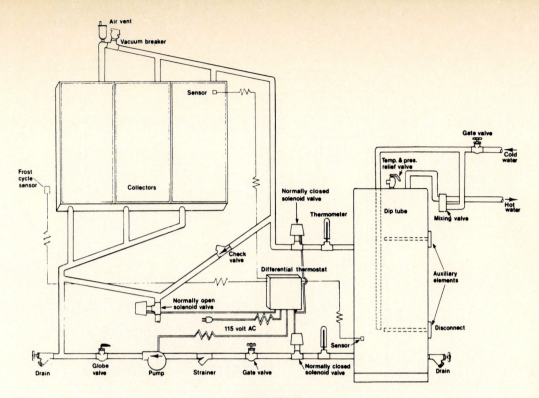

FIGURE 26:42
Draindown system schematic. (Source: *Installation Guidelines for Solar DHW Systems*)

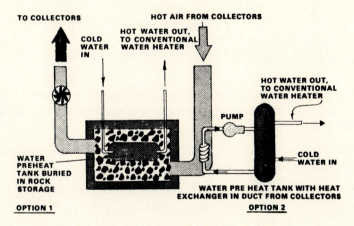

FIGURE 26:43
Air collector DHW system. (Source: *Solar Energy for Buildings Handbook*)

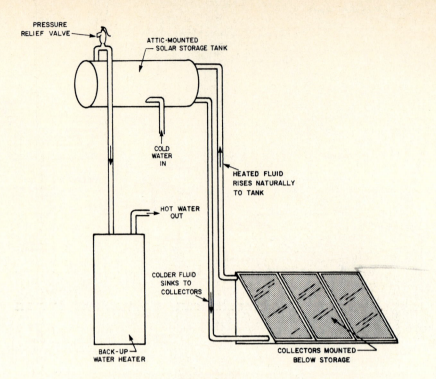

FIGURE 26:44
Thermosiphon DHW system. (Source: Northeast Solar
Energy Center)

QUESTIONS

1. Discuss the procedures that are involved in removing a window from its opening.

2. Explain the steps necessary to prepare an exterior wall surface for the installation of a sliding door. Assume that the wall area in question is a solid one with no opening.

3. What is the most appropriate means of collapsing a bearing partition without losing its structural effectiveness?

4. What are the structural concerns that must be examined before a closet can be constructed within the confines of a room?

5. Explain how to determine the amount to cut off a door's base after carpeting has been installed.

6. List three (3) criteria used to ascertain when a roof needs reshingling.

7. Tell how you would frame a hall ceiling area for the installation of a disappearing attic stairway.

8. List four (4) problems that are common to most additions.

27

HOUSE DESIGN ESTIMATION OF COSTS

A simple one-floor, three-bedroom ranch style home has been selected as the example to supply you with the following information:

1. A complete listing of the materials required for its construction
2. How to approximate these materials by using the proper standardized formulas
3. The proper sequence of assembling the various materials
4. An approximation of the cost of all the non-carpentry items (e.g., subcontracting)

5. A price comparison between the cost of building a house at the site with one of comparative size that has been constructed at the factory (i.e., modular)

Figure 27:1 shows the sample house that will be used to determine the estimations. The exterior dimensions of the house are 28′ × 44′. The house has a full basement beneath the entire living area and a one-car garage.

A portion of the basement must be partitioned off to house the heating system. To minimize the distance which the heat pipes must carry the heat away from the

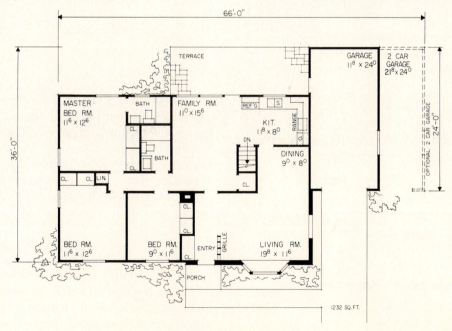

FIGURE 27:1
Courtesy of Home Planners, Inc.

355

furnace, the chimney should be near the center of the house. The possibilities for the balance of the basement area will vary with the desires of the owner but could include the following: recreation area or playroom, laundry area, storage, workshop, hobby area, and utility room.

In estimating, it is important to realize that some calculations are based on one dimension (length or a linear measurement), others on two dimensions (length times width, or an area measurement), while concrete, sand, and top soil involve volume measurements, or three dimensions (length, width, and height or depth).

The area calculations for most homes can be reduced to simple rectangles or triangles. The area for a *rectangle* is determined by multiplying its length by its width (Figure 27:2). The area for a *triangle* is determined by multiplying the base of the triangle by one half its height (Figure 27:3).

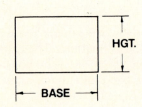

Rectangles: Base x Height = Area

FIGURE 27:2
Rectangles: base × height = area.

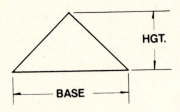

Triangle: Base x ½ Height = Area

FIGURE 27:3
Triangle: base × $\frac{1}{2}$ height = area.

FOUNDATION

To estimate the amount of concrete that will be required to fill the forms for foundation footings, multiply the width of the form (in feet) times the thickness (in feet) times the total length (in feet). This value, expressed in cubic feet, must then be divided by 27, the number of cubic feet in one cubic yard, to obtain the number of cubic yards required.

The footings for 8″ concrete blocks should be at least 8″ deep ($\frac{2}{3}$ foot) and 16″ wide ($\frac{4}{3}$ foot) and should be both wider and longer than the actual house dimensions.

CALCULATIONS

House $\frac{4}{3}' \times \frac{2}{3}' \times 46' \times 2$ sides $= \frac{8}{9}' \times 92 =$
  81.77 cubic feet of concrete

  $\frac{4}{3}' \times \frac{2}{3}' \times 29' \times 2$ sides $= \frac{8}{9}' \times 58 =$
  51.55

Garage
  $\frac{4}{3}' \times \frac{2}{3}' \times (13 + 13 + 26 + 8) = \frac{8}{9}' \times 60 =$
  53.33

  Total   186.66 cubic feet of concrete

  186.66 cubic feet $\times \dfrac{1 \text{ cubic yard}}{27 \text{ cubic feet}} =$
  6.9 yards concrete

6.9 yards of concrete + 10% waste factor =
  7.6 yards of concrete

Four concrete piers will be needed for the steel lally columns to rest on. They will support the main beam or girder and should be poured at the same time as the footings. Each pier should be 2′ square and 3′ deep. Collectively, the four will require $\frac{1}{2}$ yard of concrete.

Footings plus piers will require 8 yards of concrete. Three hundred (300) linear feet of $\frac{3}{8}$″ diameter steel reinforcing rod will be needed to reinforce both the piers and the footings.

The number of 8″ × 8″ × 16″ concrete blocks required for the 8′ tall house foundation plus the 40″ garage wall is obtained by determining the square foot area of each wall. This is done by multiplying the height (in feet) by the length (in feet) to obtain the area of the wall. This value, in square feet, is then multiplied by $\frac{9}{8}$ (1.125 on your calculator) to determine the exact number of blocks.

CALCULATIONS

House 8′ × 44′ = 352 × 2 sides =   704 sq. ft.
        8′ × 28′ = 224 × 2 sides =   448
Garage 3⅓′ × 56′   =   187
                      Total   1,339 sq. ft.
1,339 sq. ft. × 1.125 = 1506 blocks +
          10% waste (150) = 1,656 total blocks

Every 100 sq. ft. of concrete block wall surface requires 2.6 cubic feet of mortar.

CALCULATIONS

1,339 sq. ft. ÷ 100 = 13.39 × 2.6 =
34.8 cu. ft. mortar + 10% = 38 cu. ft. mortar
required

The materials needed to make this quantity of mortar are

1. 14 cubic yards of mason sand
2. 75 bags mortar
3. 21 bags Portland cement

To waterproof the exterior surface of the foundation will require five five-gallon (25 gallons total) pails of asphalt foundation coating.

Anchor bolts should be placed no more than 5′ apart in the top course of the concrete block wall and be left high enough (2″ to 3″) above the wall to perform their designated function. At least 40 will be needed for this foundation.

Sill sealer (sold in 50′ rolls), with a thickness of ¼″ and a width of 5½″, is then placed on top of the wall before the *sill plates* (2 × 6) are positioned over the anchor bolts and subsequently anchored to the foundation. The total linear length for both sealer and plates must be at least 188′ (Four rolls will be required.).

The four lally columns, spaced in the center of the basement area at 8½′ intervals, will rest on the concrete piers, previously poured, and will support either a 44′ steel I-beam or a triple built-up beam. The latter is constructed from six 16′ 2 × 8s and three 12′ 2 × 8s using the following sequence: 16-16-12
                                           12-16-16
                                           16-16-12

Both the basement floor (4″ thick) and the garage floor (6″ thick) can be poured at any time after the foun-

dation has been completed. Some builders prefer to do it immediately, while others wait until the house has been completely framed.

CALCULATIONS

Basement ⅓′ × 28′ × 44′ = 410.66 cu. ft. ÷ 27
                = 15.2 + 10% waste = 16.7 cu. yards
Garage ½′ × 12′ × 24′ = 144 cu. ft. ÷ 27
                = 5.3 + 10% waste = 5.8 cu. yards
The total number of yards of concrete + waste is 22.5 for both.

Before the concrete is poured in either area, 6 × 6 # 8/8 welded wire reinforcing mesh should be placed in the area and raised off the ground with small concrete blocks so that it will be effective at reinforcing the concrete. The combined square foot area for the two areas is *1,520.*

METHODS USED TO CALCULATE FRAMING MEMBERS

There are two methods that can be used by builders to calculate the number of framing members required, when the spacing between members is 16″ o.c. These two methods can be used to determine the number of floor joists, studs, and rafters.

The first, which will be referred to as Plan A throughout this chapter, uses the following formula:

¾ (wall length, in feet) + 1 + extras

The second, referred to as Plan B, counts *one* framing member for every foot of linear length, plus an optional 10 percent for waste.

METHODS USED TO CALCULATE SHEET MATERIALS

This method, referred to as Plan C, determines the number of 4′ × 8′ sheets needed to cover a specific area (floor frame, roof frame, and exterior walls) by using the following formula:

Area to be covered (sq. ft.) − area of openings = Net area (sq. ft.)
Net area ÷ 32 sq. ft. = Number of sheets required + 10% waste

FLOOR FRAME

The *floor joists,* spaced 16″ apart (16″ o.c.) rest on the

sill plate at one end and overlap the main beam or girder at the other end. They are connected to two 22' long 2 × 8 *joist headers* that are positioned along each of the two longer foundation walls (88 linear feet needed).

CALCULATIONS

Plan A $\frac{3}{4}$ (44') + 1 + extras = 33 + 1 + 5 (steps, partitions) × 2 sides = 78 (2" × 8" × 14")
Plan B 44' × 2 sides = 88 (2 × 8 × 14')

The joists are most rapidly covered with 4' × 8' sheets of plywood having a thickness of $\frac{1}{2}$" or $\frac{5}{8}$", depending upon the local building code requirements. The number of sheets required to cover the entire floor frame is determined as follows:

Plan C Area (28' × 44') − stairwell opening (3' × 8') = 1,232 sq. ft. − 24 sq. ft. = 1,208 sq. ft. 1,208 sq. ft. ÷ 32 sq. ft/sheet = 38 sheets + 4 (10% waste) = 42 sheets

WALL FRAME

Both the exterior walls and the interior partitions require a *single bottom plate*, which is nailed to the subfloor, and a *double top plate*, which is placed on top of the upright studs. The total linear length required for the plates is determined as follows:

1. Sum the length of all outside walls (144') and interior partitions (144'). The plans should give the dimensions of most walls.
2. Multiply this summed length in feet by three for the plates involved—bottom sole plate and doubled top plate.
3. Increase this final tripled length by 10 percent to allow for any waste incurred, for fire blocking, and for backing for vertial siding whenever necessary.
4. Use the sum of the tripled length plus the waste allowance to determine the number of pieces of plate material required.

CALCULATIONS

House exterior walls 88' + 56'   = 144 linear feet
     interior partitions   = 144 linear feet
Garage rear, side partition 12' + 24'
                     = 36 linear feet
  front, house side 2' + 6'= 8 linear feet

Total   = 332 linear feet
        × 3 = 996 linear feet
996 + 100 (10% waste) = 1,100 linear feet required

Studs are used to frame the walls of the house and garage, the window and door openings, the gable ends of the roof, the closed cornice area and the interior partitions. They are usually 2 × 4s, 8' in length.

CALCULATIONS

Plan A $\frac{3}{4}$ (332') + 2 × (number of corners, openings, and intersections = 40)
  249 + 80 = 329 studs (for both house and garage)
Plan B 332 studs + 10% waste (33) = 365 studs

Table 27:1 presents a comparison between the two plans for number of studs.

TABLE 27:1

	Plan A	Plan B
Walls and partitions (house, garage)	329	365
Gable ends (house, garage)	20	22
Closed cornice area	10	10
Totals	359	397

Even though there is a slight difference in the total number of studs from the two methods, it would be advisable to order the larger quantity (round off the total to 400 studs).

The *double headers*, positioned against the top plate above every exterior door and window opening, should be constructed of 2 × 12s. The linear length required is 170'.

A triple built-up beam will be needed to support the ceiling joists in the garage area. Three 12' long 2 × 8s are sufficient.

Plan C is used to determine the number of 4' × 8' sheets of $\frac{3}{8}$" thick sheathing that will be needed to cover the wall frame of the house and garage and the gable ends.

CALCULATIONS

House   128′ × 8′ = 1,024   sq.   ft.   *Garage*
44′ × 8′ = 352 *Gable ends* 6″ × 8′ × 60 = 240 sq. ft.
Total area (1,616 sq. ft.) − 100 sq. ft (window and door openings) = 1,516 + 10% (151) = 1,667 sq. ft.
1667 sq. ft. ÷ 32 sq. ft. = 52 sheets are needed

Five rolls of 15 lb. felt (one roll covers 400 sq. ft.) are needed to cover the exterior wall sheathing. One or two large boxes of *staples* will be needed to attach the felt to the sheathing.

The two methods used earlier to determine the quantity of floor joists needed are used to calculate the number of *ceiling joists*.

CALCULATIONS

Plan A $\frac{3}{4}$ (56′) = 42 + 1 + 5 extras = 48 × 2 sides = 96 (2″ × 6″ × 14″)
Plan B 56′ (house and garage) × 2 sides = 112 (2″ × 6″ × 14′)

ROOF

The *roof rafters*, spaced 16″ o.c., are also determined by using Plans A or B. The length of the house rafters is 16′, while those on the garage are only 14′.

CALCULATIONS

Plan A $\frac{3}{4}$ (44′) = 33 + 1 + 2 extra = 36 × 2
sides = 72 rafters (2″ × 6″ × 16′)
$\frac{3}{4}$ (12′) = 9 + 1 + 2 extra = 12 × 2
sides = 24 rafters (2″ × 6″ × 14′)
Plan B house 44′ × 2 sides = 88 rafters
garage 12 × 2 sides = 24 rafters

At their lower ends, the rafters rest on the double top plate and are nailed to companion ceiling joists. A 1 × 8 *ridge beam*, 46′ long for the house and 14′ long for the garage, are used to connect the rafter pairs at their top edges.

Preengineered roof trusses can be used as a means of reducing the number of personnel-hours that would be needed to cut and install the 2 × 6 rafters and ceiling joists. They must be 28′ 0″ in length and 23 are needed to cover the area above the house, when given their normal spacing of 24″ o.c. Seven trusses, 24′ 0″ in length, are needed to cover the garage area.

To determine the number of sheets of plywood sheathing ($\frac{1}{2}$″ × 4′ × 8′) that are required to cover the rafters or trusses, Plan C is used.

CALCULATIONS

House 16′ (with overhang) × 44′ = 702 sq. ft. ×
2 sides = 1408 sq. ft. ÷ 32 = 44 sheets
Garage 14′ × 24′ = 336 sq. ft. × 2 sides = 672
sq. ft. ÷ 32 = 21 sheets
44 + 21 = 65 + 7 (waste) = 72 sheets total

Before the roof is shingled, it must be covered with *15 lb. builder's felt* (one roll covers 400 sq. ft.) and have a metal *drip edge* installed around the entire perimeter of the roof (210 linear feet). Five rolls of felt will be needed, and 21 10′ lengths of drip edge should be sufficient.

Before you can determine the number of squares of roof shingles that will be required to cover the roof area, you must realize that one square of roofing shingles is designed to cover 100 sq. ft. of roof area. The total roof area (in square feet) is divided by 100 to obtain the number of squares (There are three bundles of roof shingles per square.). The total roof area, plus 10 to 15 percent extra for the roof's slope and for capping the ridge, will give the approximate number of squares.

CALCULATIONS

Total roof area = 2,080 sq. ft. ÷ 100 =
21 squares + 2 = 23 squares (69 bundles)

Ten 1 × 6 × 8′ pieces are used as *collar beams* to connect every fifth pair of rafters together. Trusses will require an elaborate gridwork of 2 × 4 braces to make them structurally sound.

EXTERIOR TRIM

The lower cut edges of the rafters or trusses are trimmed with a 1 × 8 *fascia board*. The end rafters, on each gable end, are first covered with a 1 × 8, which is then capped with a 1 × 4 board. The quantities required are 210 linear feet of 1 × 8 and 100 linear feet of 1 × 4.

The *closed cornice* requires ten pieces of 2″ × 4″ × 10′ to serve as *ledger boards,* and ten pieces of 2″ × 4″ × 8′ that will be cut into 1′ long pieces to serve as *lookouts*. Four sheets of $\frac{1}{4}$" × 4' × 8' A/C plywood must be ripped lengthwise to serve as the *soffit*.

ROOF DRAINAGE

The *roof drainage system* is usually installed directly after the roof shingles have been nailed into place. Table 27:2 lists the items and their quantities that are required to properly install the system. The individual items are presented in Figure 27:4.

TABLE 27:2

Item	Quantity
Leader, 10' long	4
Gutter, 21' long	5
Gutter end cap	4 (2 rights, 2 lefts)
Leader band	8
Connector	6
Ferrule/spike	75 sets
Outlet tube	4
Elbow	12
Gutter sealant	4 tubes

EXTERIOR COVERINGS

Since both the size and quality of the *exterior doors* and *windows* can vary greatly from one house or geographic location to another, it is difficult to specify exact sizes or state estimated prices. The plan shows a total of eleven windows and three exterior doors.

Likewise, the decision of which *exterior covering* to use is based on cost, appearance, ease of application, energy-saving efficiency, as well as availability, and therefore will be left up to you. To cover the exterior, 1,700 sq. ft. or 17 squares of material will be needed. Table 27:3 is the same as Table 15:2, repeated here so that the various exterior coverings per square foot of coverage can be compared with respect to cost. A notation of any waste factor associated with each material is also included.

Before any exterior covering is applied to the house in the area where the garage roof adjoins the gable portion of the house, *metal step flashing* must be installed. Thirty pieces should be sufficient to provide adequate protection. At the same time, the tops of the windows should be covered with *window head flashing* (43 linear feet will be sufficient).

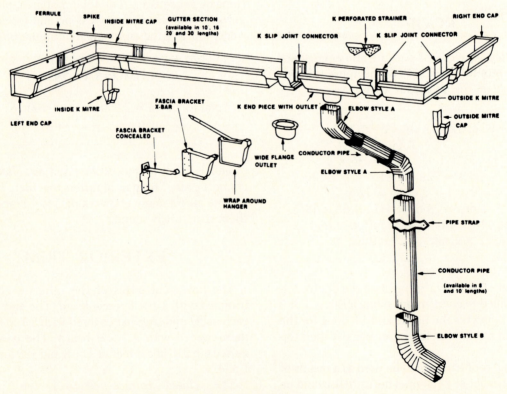

FIGURE 27:3
Components of a roof drainage system.

TABLE 27:3
Comparison of the Various Exterior Coverings

Item	How sold, size	Cost/sq.ft.	Waste	Lap amount	Comments
Sheathing					
Nailable fiberboard	$\frac{1}{2}'' \times 4' \times 8'$	$0.286	None	None	
Plywood C/D	$\frac{5}{8}'' \times 4' \times 8'$	0.294	None	None	
	$\frac{1}{2}'' \times 4' \times 8'$	0.366	None	None	
Finished exterior covering					
Texture 111	$\frac{5}{8}'' \times 4' \times 8'$	0.617		None	
	$\times 4' \times 9'$	0.631		None	
	$\times 4' \times 10'$	0.659		None	
Shingles, #1 grade	1 box = 1 square	0.563	5%	4", double course	
Barn shakes (16")	4 bundles = 1 square	0.586	5%	$8\frac{1}{2}''$, single course	
Undercourse shingles	2 bundles/square	0.12		4" lap	
Asbestos shingles	3 bundles = 1 square	0.575	8%	$1\frac{1}{2}''$ lap	12" × 24" standard size
Bevel siding (cedar)	$\frac{1}{2}'' \times 6''$	1.26	33%	1" lap	
	$\frac{1}{2}'' \times 8''$	1.14	33%	$1\frac{1}{4}''$ lap	
	$\frac{3}{4}'' \times 10''$	1.32	29%	$1\frac{1}{2}''$ lap	
Cedar T/G	$1'' \times 6''$	1.07	16%	None	
	$1'' \times 8''$	1.00	14%	None	
Pine #2 T/G	$1'' \times 6''$	1.13	16%	None	
	$1'' \times 8''$	1.12	14%	None	
Shiplap	$1'' \times 8''$	0.752	16%	$\frac{3}{8}''$ lap	
	$1'' \times 10''$	0.733	14%	$\frac{3}{8}''$ lap	
Masonite (tempered)	$\frac{1}{8}'' \times 4' \times 8'$	0.240		None when used vertically, battens cover seams	
	$\frac{1}{4}'' \times 4' \times 8'$	0.286		$1''-1\frac{1}{2}''$ lap when used horizontally	
Aluminum siding	double 4" or 8"	0.800		1" lap	
Vinyl siding	double 4" or 8"	0.690		$1\frac{1}{2}''$ lap	pieces cut $\frac{1}{4}''$ short for expansion
Brick	$2\frac{1}{4}'' \times 3\frac{3}{4}'' \times 8''$ used	1.00			
	new	2.00			

INSULATION

The net area for the exterior walls of the house is 1,152 sq. ft. Batt insulation, with the foil facing the room's interior, is placed between the upright studs and is then stapled to them (R-11 is used for 2 × 4 walls and R-19 for 2 × 6 walls or ceilings). The net area for the ceilings in the house area is 1,232 sq. ft.

WALLBOARD

To determine the number of sheets of gypsum wallboard ($\frac{1}{2}$″ × 4′ × 8′) to cover all the upright studs and ceiling joists in both the house and garage, Plan C is used. The sum of all the exterior walls and interior partitions times the height (8′) minus the door and window openings gives the net area in square feet that must be covered (Remember to count both sides of the interior partitions). This net value divided by 32 gives the number of sheets required.

CALCULATIONS

Walls 432′ × 8′ = 3,456 sq. ft. − 399 sq. ft.
                (openings) = 3,057 sq. ft.
Ceilings 1,232 sq. ft. (house) + 288 sq. ft. (garage)
                          = 1,520 sq. ft.
            Total   = 4,577 sq. ft.
4,577 sq. ft. ÷ 32 = 143 sheets

To cover the wallboard joints, eleven 200′ rolls of wallboard tape and five five-gallon containers of joint compound will be needed, plus 20 8′ lengths of metal corner-bead to protect all the exposed outside corners.

INTERIOR TRIM

The house requires the installation of 14 interior door units. Bifold doors will enclose the four closets, while flush doors will be used in the remaining doorways.

Even though prehung door units may be slightly higher in price than purchasing the door, its jambs, the trim molding, hinges, and lock hardware separately, the savings in installation time far outweigh the few dollars saved.

Approximately 360 linear feet of *baseboard molding*, 260 linear feet of *window (casement) trim molding*, and 60 linear feet of *window stool molding* will be needed.

FINISHED FLOOR COVERINGS

If you know in advance of the actual construction what type of finished floor covering will be used in each room, you can properly prepare the subfloor areas. Whenever wall-to-wall carpeting will be installed throughout the home and linoleum will be placed in the kitchen and bathroom areas, a layer of particleboard is nailed directly to the subfloor.

When exposed hardwood oak floors are desired, they are nailed directly to the subfloor. Many builders prefer to hire a subcontractor who specializes in laying finished hardwood floors than to do the job themselves.

NAIL ESTIMATIONS

A carpenter must not only know which nails to use for fastening the various structural members of a house, but also be able to estimate the quantities of nails for each phase of the construction.

Since many nail estimations are determined by using an *estimation factor* that is based on a specific number of pounds of nails per 1,000 board feet (bd. ft.), it may be necessary to refer to Table 27:4, which lists the number of board feet in various lengths of timber. This value multiplied by the number of pieces will determine the total number of board feet.

The calculations used to estimate the quantities of nails for each structural item are presented in Table 27:5. Where applicable, the 16″ o.c. spacing between adjacent members has been used. Wider spacings would require fewer nails.

Listed in Table 27:6 are the quantities and types of nails needed to construct the house. Every builder realizes the advantages associated with the purchase of nail types in 50-lb. boxes They are less expensive than when purchased by the pound; they are well labeled and packaged so that the balance left over from one project can easily be stored until they are needed for the next project; many of the boxes are also waterproof. Thus, any nail type that exceeds 20 lb. in weight should be purchased by the 50-lb. box.

SUBCONTRACTING

The following list includes those components of a completed house which most carpenters prefer to subcontract to specialists:

1. Electrical wiring
2. Plumbing installation
3. Heating system

TABLE 27:4
Board-feet Measure for Various Timbers (Lumber Scale)

Timber Size	Length of Timber								
	8	10	12	14	16	18	20	22	24
1 × 2	$1\frac{1}{3}$	$1\frac{2}{3}$	2	$2\frac{1}{3}$	$2\frac{2}{3}$	3	$3\frac{1}{3}$	$3\frac{2}{3}$	4
1 × 3	2	$2\frac{1}{2}$	3	$3\frac{1}{2}$	4	$4\frac{1}{2}$	5	$5\frac{1}{2}$	6
1 × 4	$2\frac{2}{3}$	$3\frac{1}{3}$	4	$4\frac{2}{3}$	$5\frac{1}{3}$	6	$6\frac{2}{3}$	$7\frac{1}{3}$	8
1 × 6	4	5	6	7	8	9	10	11	12
1 × 8	$5\frac{1}{3}$	$6\frac{2}{3}$	8	$9\frac{1}{3}$	$10\frac{2}{3}$	12	$13\frac{1}{3}$	$14\frac{2}{3}$	16
1 × 10	$6\frac{2}{3}$	$8\frac{1}{3}$	10	$11\frac{2}{3}$	$13\frac{1}{3}$	15	$16\frac{2}{3}$	$18\frac{1}{3}$	20
1 × 12	8	10	12	14	16	18	20	22	24
2 × 4	$5\frac{1}{3}$	$6\frac{2}{3}$	8	$9\frac{1}{3}$	$10\frac{2}{3}$	12	$13\frac{1}{3}$	$14\frac{2}{3}$	16
2 × 6	8	10	12	14	16	18	20	22	24
2 × 8	$10\frac{2}{3}$	$13\frac{1}{3}$	16	$18\frac{2}{3}$	$21\frac{1}{3}$	24	$26\frac{2}{3}$	$29\frac{1}{3}$	32
2 × 10	$13\frac{1}{3}$	$16\frac{2}{3}$	20	$23\frac{1}{3}$	$26\frac{2}{3}$	30	$33\frac{1}{3}$	$36\frac{2}{3}$	40
2 × 12	16	20	24	28	32	36	40	44	48
2 × 14	$18\frac{2}{3}$	$23\frac{1}{3}$	28	$32\frac{2}{3}$	$37\frac{1}{3}$	42	$46\frac{2}{3}$	$51\frac{1}{3}$	56
2 × 16	$21\frac{1}{3}$	$26\frac{2}{3}$	32	$37\frac{1}{2}$	$42\frac{2}{3}$	48	$53\frac{1}{3}$	$58\frac{2}{3}$	64
3 × 6	12	15	18	21	24	27	30	33	36
3 × 8	16	20	24	28	32	36	40	44	48
3 × 10	20	25	30	35	40	45	50	55	60
3 × 12	24	30	36	42	48	54	60	66	72
3 × 14	28	35	42	49	56	63	70	77	84
3 × 16	32	40	48	56	64	72	80	88	96
4 × 4	$10\frac{2}{3}$	$13\frac{1}{3}$	16	$18\frac{2}{3}$	$21\frac{1}{3}$	24	$26\frac{2}{3}$	$29\frac{1}{3}$	32
4 × 6	16	20	24	28	32	36	40	44	48
4 × 8	$21\frac{1}{3}$	$26\frac{2}{3}$	32	$37\frac{1}{3}$	$42\frac{2}{3}$	48	$53\frac{1}{3}$	$58\frac{2}{3}$	64
4 × 10	$26\frac{2}{3}$	$33\frac{1}{3}$	40	$46\frac{2}{3}$	$53\frac{1}{3}$	60	$66\frac{2}{3}$	$73\frac{1}{3}$	80
4 × 12	32	40	48	56	64	72	80	88	96
4 × 14	$37\frac{1}{3}$	$46\frac{2}{3}$	56	$65\frac{1}{3}$	$74\frac{2}{3}$	84	$93\frac{1}{3}$	$102\frac{2}{3}$	112
6 × 6	24	30	36	42	48	54	60	66	72
6 × 8	32	40	48	56	64	72	80	88	96
6 × 10	40	50	60	70	80	90	100	110	120
6 × 12	48	60	72	84	96	108	120	132	144

4. Water and sewer hookup
5. Excavation and landscaping
6. The foundation and all other masonry projects (e.g., entrance steps, chimney)

As a general rule, the combined cost of the first five components in the list above is roughly equal to the cost of all the other house materials. It should be noted that

1. The estimated price stated by electricians is based on the number of outlets required, plus the number of appliances to be wired, and does not include the electrical fixtures.

2. The plumber's estimated price includes both the fixtures and their installation.

3. The installers of the heating system support the manufacturer's guarantee that the system will be completely energy efficient.

A summary of the cost of materials for each structural component of the house, based on prices quoted in New York state, is given in Table 27:7. Note that there is no mention of the cost of the land nor of the sales tax assessed, because these two items may vary greatly in different localities.

TABLE 27:5
Estimation of Nails for a Specific Structural Item

Item	Size	Estimation factor (lbs/1,000 bd.ft.)	Calculations	Nail size/Quantity
Floor joist	2" × 8" × 14'	8 lb.*	$18\frac{2}{3}$ × 80 pieces = 1,500 bd.ft	16d/12 lb.
Ceiling joist	2" × 6" × 14'	13 lb.	14 × 100 pieces = 1,400 bd.ft	16d/18 lb.
Triple built-up girder	6" × 8"	32 lb.	4 bd.ft/1 linear ft. × 44' = 188 bd.ft	16d/6 lb.
Studs	2" × 4" × 8'	22 lb.	$5\frac{1}{3}$ × 400 pieces = 2,133 bd.ft	8d/47 lb.
Plates	2" × 4"	22 lb.	1,100 linear ft. × $\frac{2}{3}$ = 733.3 bd.ft	16d/16 lb.
Rafters	2" × 6" × 16' × 14'	12 lb.	16 × 72 pieces = 1,152 bd.ft. (house) 14 × 24 pieces = 336 bd.ft (garage) Total = 1,488 bd.ft.	16d/18 lb.
Gypsum wallboard		6 lb./1,000 sq.ft.	4,577 sq.ft.	$1\frac{1}{4}$" wallboard/28 lb.
Roofing nails	(3 tab Sq.Butt)	336 nails/square	23 squares	$1\frac{1}{4}$" roofing/28 lb.
Subflooring		40 nails/sheet	1,232 sq.ft. = 42 sheets	8d/18 lb.
Roof sheathing			2,080 sq.ft. = 72 sheets	8d/30 lb.
Exterior sheathing			1,667 sq.ft = 52 sheets	8d/22 lb.

* $18\frac{2}{3}$ board feet (From Table 27:4) 2" × 8" × 14'

Nail Summary

Nail/fastener type	Weight estimated	Purchase weight
16d common	70 lb.	100 lb.
8d common	117 lb.	150 lb.
$1\frac{1}{4}''$ drywall	28 lb.	50 lb.
$1\frac{1}{4}''$ galvanized roofing	28 lb.	50 lb.
8d finishing	30 lb.	50 lb.
6d finishing	25 lb.	50 lb.
4d finishing	2 lb.	2 lb.
8d casing	35 lb.	50 lb.
6d casing	25 lb.	50 lb.

Note: The quantity and style needed to apply the exterior covering is dependent upon the choice of covering selected. Staples 3 large boxes (used to apply 15 lb. felt over sheathing and batt insulation)

TABLE 27:7
Cost of Materials, Labor, and Subcontracting

Materials

Foundation	$ 4,500
Floor frame	1,650
Wall frame and exterior sheathing	1,750
Ceiling frame	550
Roof frame	1,650
Exterior trim, drainage system	200
Nails, 15 lb. felt, staples	300
(*) Exterior doors, windows	3,500
(*) Exterior covering	1,200
Insulation	550
Wallboard and supplies	750
Interior doors, molding	1,100
Garage door	300
Total	$18,000

Labor for above (materials cost + 20%) = $21,600

Subcontractors

(*) Electrician	$ 3,500
(*) Plumber	3,500
(*) Heating system	3,500
(*) Excavation, grading	2,000
(*) Water and sewer	2,500
(*) Finished floors	1,000
(*) Kitchen cabinets, countertops	2,000
Total	$18,000

Summary of Costs:

Materials	$18,000
Subcontractors	18,000
Labor	21,600
Total	$57,600

*Estimated price

Note also that the prices for all the subcontracting, the exterior doors, the windows, and the exterior covering are estimates that may be either higher or lower depending upon the choice of materials and/or geographical location.

FRAMED VERSUS MODULAR HOMES

This same house, constructed as a factory-built modular home and erected on the foundation, would be less expensive than the cost to construct a site-built house. Table 27:8 lists the expected prices for a modular home. Note that the total cost for subcontracting is less than half that estimated for the framed house, since most modulars arrive with those items completely installed. The heating, electrical, and plumbing systems,

however, still require the services of subcontractors to connect them properly. The cost of excavation, landscaping, and water and sewer hookup is the same for both houses.

TABLE 27:8

House, transportation, erection	$16,000
Foundation materials (4500) labor (4200)	8,700
Subcontractors	8,300
Extra labor and materials to complete	5,000
Total	$38,000

Because of the lower price and the speed with which they can be assembled and readied for sale, modular homes are becoming more appealing; especially to the noncarpenter investor who can afford to subcontract every item required for completion.

QUESTIONS

1. Give the mathematical expressions used to determine the area of a rectangle and a triangle.

2. Determine how many $8 \times 8 \times 16$ concrete blocks will be needed to build a wall 8' high and 40' long.

3. How many floor joists will be needed to cover a house having a length of 50' when given a spacing of 16 o.c.?

4. Determine the number of exterior studs needed to frame the first floor living area of a building 24' by 50'.

5. How many 4×8 sheets of plywood would be needed to cover the wall frame in Question 4?

6. Determine the number of roof rafters needed to build a roof over a wall frame measuring 24' by 50'.

7. How many 4×8 sheets of wallboard would be required to cover 3850 square feet of interior wall space?

8. Explain why the total cost for subcontracting is less for a modular home than for a framed one.

Appendix

STAIRWAY SPECIFICS

TERMINOLOGY
(Consult Figure A–1)

Riser   the vertical member of the step. Stairs lacking risers are described as having open risers (Figure A–2).

Tread   the horizontal member of a step.

Unit rise   the vertical distance from the top of one tread to the top of the next.

Total rise   it is equal to the sum of all the risers.

Unit run   the horizontal distance from riser to riser.

Total run   the total horizontal distance that a flight of stairs and its landings covers.

Headroom   the clearance space between the stair's tread and the floor level directly above.

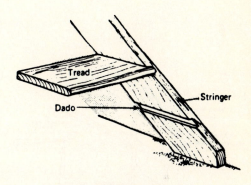

FIGURE A–2
Steps with open (no) risers.

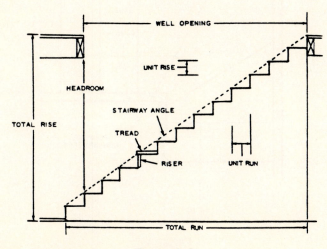

FIGURE A–1
Stair layout terms.

Landing: any area within a flight of stairs that interrupts its flight or where a change of direction is made (Figure A–3).

Stringer   an inclined member to which the treads and risers are joined. It runs parallel to the slope of the stairs. It made be constructed in one of three ways;

Housed   the stringer is routed to receive the ends of the treads and risers (Figure A–4).

Open   a stringer having its top edge notched to follow the riser and the tread (Figure A–5).

Closed   one having parallel edges, with the top edge extending above the ends of the stairs (Figure A–6).

Handrail   needed whenever there are more than three risers in the stairs. It should be placed 30 inches above the steps and 34–36″ above landings.

Winder   a step having a tread narrower at one end than at the other (Figure A–7).

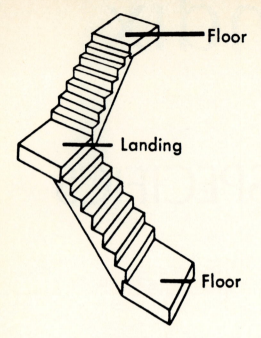

FIGURE A-3
Stairway specifics.

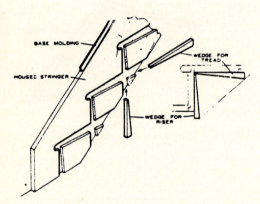

FIGURE A-4
Housed stringer.

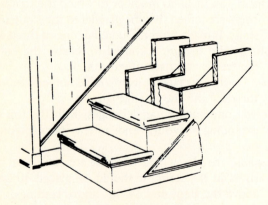

FIGURE A-5
open (Plain) stringer.

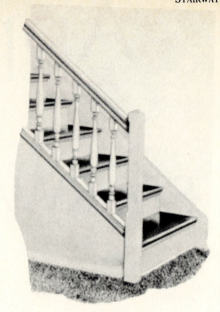

FIGURE A-6
Closed string stairway.

Stair Riser-Tread Formulas
  a) twice the riser height plus the tread width equals 24–25″
  b) the riser height plus the tread width equals 17 to 18″
  c) the riser height times the tread width equals 72 to 75″
A *stair's best or most ideal*
  dimensions are: riser $7\frac{1}{2}$″; tread 10″
  width: main stairway (minimum 32″, more desirable 36–42″)
  utility stairs 30″ minimum

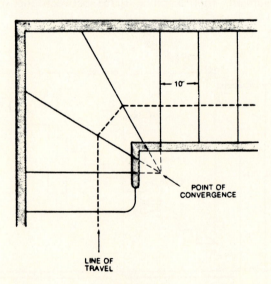

FIGURE A-7
Winder details. Point of convergence should be away from corner of stairs.

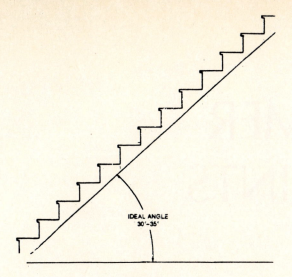

FIGURE A–8
The ideal incline angle for stairs is between 30 and 35.

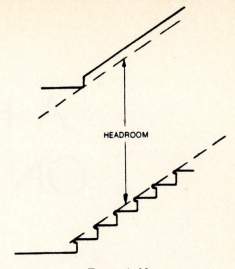

FIGURE A–10
Minimum headroom requirements: main stairs 6′–8″; basement stairs 6′–6″.

slope: 30 to 60° with the horizontal (Figure A–8)

handrail height: 30″ above steps, 34–36″ above landings (Figure A–9)

Headroom height: 6′8″ (80″) or greater for main stairways (Figure A–10)

6′4″ (76″) or greater for utility stairs

Note: 14 risers is common number used between the first and second floors (105″ is total height)

Additional information

1) The number of treads is always one less than the number of risers (Figure A611)

2) A landing is needed whenever the flight is more than 15 steps

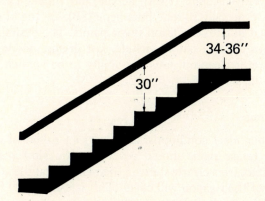

FIGURE A–9
Preferred hand rail heights.

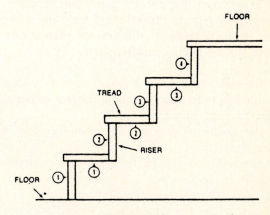

FIGURE A–11
The number of treads in a stairway is always one less than the number of risers.

A PRIMER
ON PAINTS

All paints have been formulated to cover the surface of a solid object with a thin protective film or coating. For a paint to be successful, it must be applied to the surface as a liquid and then be quickly transformed to a hard, dry surface.

Modern technology, coupled with years of trial and error experiences, has enabled paint manufacturers to know exactly how to ideally blend the various ingredients in each paint in the correct proportions to achieve the following characteristics:

   a. the ability to flow or be spread properly (viscosity)
   b. the ability to make adequate contact with a solid surface (wetting capability)
   c. the ability to dry fast and then become hard (drying time)
   d. the ability to cling to the solid's surface (adhesive strength)

All paints have the same basic composition. They all contain a *pigment*, a fine powder that provides color and hides the surface it covers, and a *vehicle*, consisting of two liquid components. The purpose of the non-volatile portion of the vehicle, known as the *binder* or *resin*, is to hold the pigment to the solid surface; while the purpose of the *volatile* or *solvent portion* is to aid both the application and the drying process.

The shine or glossiness of a paint is related to its *pigment to vehicle ratio*. A high gloss surface is the result of a low ratio while a flat or low luster finish is the result of a high pigment to vehicle ratio. A clear finish results whenever the vehicle is the only ingredient applied.

Paints are classified as being either *solvent-reduced* or *latex emulsion* types. The applied surface of a solvent-reduced paint becomes hard when its organic solvent evaporates and the resin is left to either oxidize or polymerize. The resin can be either natural (linseed oil, tung oil, or soybean oil) for "oil base" paints or synthetic (vinyl, epoxy, or polyester) for "alkyd base" paints.

Polyvinyl acetate (PVA), acrylic and rubber based resins, especially styrene butadiene (SB), are the binders used in latex emulsion tyype paints. The desirable features of latex paints include: fast drying, odorless, easy water clean-up, plus the ability to be applied to damp surfaces.

Coverage

Under ideal conditions, an American gallon of exterior house paint should cover 500 ft^2. Most paint manufacturers clearly state on their label what the contents of their gallon will cover. This is usually 400 ft^2.

The procedures that can be used to estimate the number of gallons required are:

1. allow one gallon for each interior room plus two additional gallons to cover the gable ends,
2. multiply the "average height" for the house times the house perimeter to obtain the total surface area in square feet. Divide this amount by the manufacturer's stated coverage to determine the number of gallons needed for one coat coverage.

The "average height" is obtained in one of two ways:
   a. for a flat roof—foundation to eaves
   b. for a pitched roof—foundation to eaves plus *two* additional feet

3. trim paint is determined as follows; one gallon per average sized house or one gallon for every five gallons of exterior paint applied.

Preparation of New Wood

This should follow a definite series of steps. They are:

1. remove all dust and dirt from the surface,
2. apply shellack to all exposed knot areas,
3. cover the wood with the proper primer,
4. use caulk to seal all open joints and mitered cuts,
5. apply the finish coat(s).

Problems Associated with Painted Surfaces

Problem	Cause(s)	Solution
Alligating*	too many paint layers applied or incompatibility between paints	scrape to bare wood, sand edges smooth, apply primer and then repaint
Bleeding*	wood's natural oils or oils from an undercoat rising to the surface	apply primer over the area and then repaint
Blistering* Flaking Peeling*	exterior surface being bathed by moisture coming from the house	correct the moisture problem first, and then scrape the entire area smooth, apply primer, and then repaint
Chalking*	a natural phenomena of painted surfaces	apply non-chalking paint if a real problem
Checking* (Wrinkling)	surface coat applied before undercoat was thoroughly dry	sand or scrape before repainting
Stains a) Mildew*	damp, heavily shaded area	scrub with 1:8 bleach solution, prune trees
b) Nailheads*	exposed, non-galvanized	set below the surface, cover head with putty and then paint
Sagging	too many applications of paint or paint applied over glossy surface	either scrape to bare wood or sand the surface smooth before repainting

(*) Consult Figure: for a pictorial representation of the problem

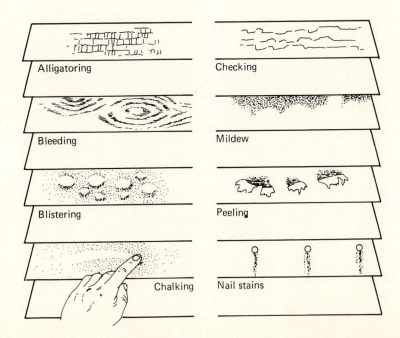

Circular Saw Blade Types

	Type	# Teeth	Type of finish	Used for
	Rip	36	medium smooth	framing, cutting with the grain
	Combination	44	medium smooth	for thick hard/soft woods both with and across the grain
	Chisel tooth	20	rough	crosscuts, rips and mitres those materials that tend to dull blades rapidly
	Plywood/Paneling	132–150*	smooth, splinter free	plywood, paneling, Cellotex and thin plastics
	Cross Cut	90–100*	smooth	across grain hard/soft woods, fiberboard, hardboard and veneer
	Hollowground/Plywood	130–200*	super smooth, splinter free	plywood, veneer, and paneling
	Hollowground/Planer	50– 70*	smooth, splinter free	hardwoods, for cabinets and fine furniture
	Novelty combination	50– 60*	Smooth	all purpose blade
	Special purpose blades Nail cutting (Carbide)	12	rough	embedded nails
	40 Tooth (Carbide)	40	almost smooth, splinter free	hard/soft woods, particle board, fiberboard, plywood, paneling, and non ferrous metals
	60 Tooth (Carbide)	60	extremely fine and smooth	cabinet work and prefinished materials
	Metal cutting			ferrous and nonferrous metal tubing, sheets and bar stock
	Masonry cutting			tile, patio block, slate, and concrete block

*Range based on diameter of saw blade

GLOSSARY

Acoustical materials   those capable of absorbing sound waves.

Actual dimension   the true size of a piece of lumber after it has been milled and dried.

Adhesive   any material used to connect or fasten two pieces together.

Air dried lumber   lumber stacked in yards having a minimum moisture content of 12–15%.

Airway   spacing between roof insulation and roof sheathing to ensure air movement.

Alligatoring   cracks in a paint surface exposing an older coat beneath.

Anchor bolts   steel bolts used to anchor a wooden sill plate to the underlying masonry foundation or floor.

Apron   trim molding placed beneath the window stool.

Architect   an individual who designs buildings and draws their respective plans.

Asphalt   petroleum material used to waterproof roofs, foundation walls and flooring tiles.

Attic ventilators   screened openings used to ventilate an attic area.

Backfill   returning earth to an excavated area, usually around foundation walls.

Balusters   vertical members (spindles) connecting the top and bottom rails.

Balustrade   railing composed of toprail, balusters, and sometimes a bottom rail.

Baseboard   any board nailed to the wall at the floor line.

Basement   the area below the ground.

Base molding   trim molding placed on top of the baseboard.

Base shoe   molding nailed to baseboard where it meets the floor (called a carpet strip).

Batten   narrow wooden strips used to conceal joints in vertical siding.

Batter board   used to establish and maintain the outside dimensions of the foundation walls.

Bay window   a window that projects beyond the building's walls.

Beam   a horizontal structural member used to support floor joists.

Bearing partition (wall)   one that supports any structural load member above it (roof).

Bed molding   molding used at an angle between a vertical and horizontal surface.

Bird's mouth   a notch in a rafter enabling it to rest solidly on the wall's top plate.

Blind-nailing   driving nails into the tongue of wood flooring so that they are invisible.

Blind stop   window molding used as a stop for screens and to resist air infiltration.

Board   nominal one inch lumber dressed to nominal widths ranging from 2 to 12 inches.

Board and batten   vertical boards whose joints are covered with battens.

Board foot   equal to a board 12″ long, 12″ wide and 1″ thick (nominal measurements).

Brace   any board placed at an angle to stiffen a stud wall.

Bridging   pieces of wood or metal placed midway in the span between adjacent floor and ceiling joists to stiffen them.

Building code   specifications used to guarantee a structurally sound building.

Built-up roof   consists of several asphalt layers covered with stone, used on flat roofs.

Butt joint   formed when two pieces of wood meet at right angles.

Canopy   a small roof overhang supported by pillars.

Cantilever   a projection having no support at one end.

Cant strip   a triangular shaped piece connecting wall and deck surface.

Cap   any top member of a column, door cornice, pilaster or molding.

Casement window   those hinged on one side to swing open along its entire length.

Casing   door or window trim used to cover jambs or walls.

Cats   horizontal 2 × 4's nailed between studs to serve as a nailing surface.

Caulking   any pliable material used to seal cracks, and repel water.

Ceiling   any surface overhead.

Cement (Portland)   one of the ingredients of concrete.

Center to Center   the distance from the center of one framing member to the next one (usually 16"O.C. or 24" O.C.).

Chalk line   a dry chalk saturated string used to mark a straight line between two points.

Checking   cracks in a painted surface.

Check rails   double hung window components designed to be air tight when the window is closed, located at the base of the top sash and the top of the bottom sash.

Chimney   used to remove smoke and gas from a furnace or fireplace.

Clapboard   beveled siding that is overlapped horizontally.

Cleat   wood strip used for support and for fastening.

Clinch   to bend a protruding nail tip over on an angle.

Collar beam   nominal 1" or 2" thick members used to connect two opposite common rafters.

Column   a vertical support member, a post, pillar or pier (made of concrete).

Combination doors and windows   provide both summer and winter protection.

Concrete   a mixture of Portland cement, sand and gravel or crushed stone.

Condensation   water droplets form when the air reaches its dew point.

Conduit   tube used to carry and encase electrical wires.

Cope   a molding cut to butt against another piece at an inside corner.

Corner   where two perpendicular walls meet.

Corner bead   metal strip placed on an outside corner.

Cornerpost   a stud that allows both interior and exterior nailing at corners.

Cornice   an open or boxed structure at the roof's eave connecting soffit and fascia.

Counterbore   drilling a hole to set a screw below the wood's surface.

Countersink   drilling a hole to have the head of the screw flush with the wood's surface.

Crawl space   area between ground and first floor with a height less than 6 feet.

Cricket   a drainage diverting roof structure above the chimney.

Cripple stud   short studs between sole plate and window sill, between header and top plate.

Cross bridging   diagonal braces between the floor joists.

Crown molding   molding used to cover an interior angle.

Dado   a groove cut across a board.

Decay   wood decay due to the action of fungi.

Dewpoint   temperature at which water vapor becomes a liquid.

Dimension lumber   lumber 2"–5" thick and at least 2" wide.

Door   any building unit, except windows, that is designed to close an opening.

Door frame   finished frame which encloses a door opening.

Door jamb   it encases a door opening, has two side jambs and one head jamb.

Dormer   an opening in a sloping roof, usually with one or more windows.

Double hung window   has two sashes, top and bottom, that pass each other when they are either raised or closed.

Downspout   a metal pipe used to carry rainwater from a roof gutter.

Dressed lumber   lumber planed down from its rough nominal size to its dressed size.

Drip   projection at roof edge designed to have rainwater drip over the edge instead of down the face of the exterior wall.

Drip cap   wood or metal strip positioned above windows and doors to cause water to drip beyond their frames.

Drip edge   metal strip nailed to edge of roof to serve as a drip.

Dry wall   a finished gypsum wallboard or paneling wall surface.

Dry well   a covered hole in the ground used to accept rainwater from the gutters or soapy water from the washer.

Ducts   metal pipes used to distribute the heat from a forced air heating system.

Eave   roof edge running along the lowest part of the roof.

Ell   a building extension at right angles to the main structure.

Excavation   any hole in the ground.

Expansion joint   asphalt fiber strips placed between concrete segments, *e.g.* walkway, to prevent the concrete from cracking when it expands.

Exterior finish   any building material suitable to be used as the home's exterior.

Face nailing   driving a nail through the face of the top piece into a second underlying piece.

Fascia (Facia)   a horizontal flat panel of a cornice, used to conceal the ends of roof rafters.

Felt   asphalt saturated paper used beneath roof shingles, exterior siding and finished floors.

Fieldstone   natural stone used to face a fireplace and for retaining walls.

Fire-retardant chemical   material designed to suppress a flame.

Fire stop   2 × 4's placed horizontally between studs to retard fire in a wall cavity.

Fish plate   wood or plywood pieces used to connect beams or rafters where their ends butt together.

Flagstone   flat, cut stones used for exterior floors, sidewalks and patios.

Flashing   metal used to weatherproof the joint where the roof meets the vertical wall.

Flue   space in a chimney through which smoke and exhaust gases ascend.

Fly rafter   the end rafter of a gable roof, supported by lookouts and roof sheathing.

Footing   wide concrete base which the foundation rests on.

Forms   panels made of 2 × 4's and plywood that are used to contain the poured concrete until it thoroughly sets up.

Foundation   poured concrete or block wall, below grade, used to support the wooden members of the first floor.

Framing-balloon   where 2 × 4 studs extend from the foundation's sill plate to the roof line.

Framing platform   where 2 × 4 walls rest on a floor platform.

Frieze   horizontal trim connecting the soffit with the top of the exterior siding.

Frost line   depth to which the frost normally penetrates the soil.

Fungi   microscopic plants that cause wood to decay.

Fungicide   any chemical designed to destroy fungi.

Furring   strips of wood used as a nailing base for the finish material.

Gable   the roof portion above the eaves of a double-sloped roof forming a triangular shaped vertical wall above the eaves.

Girder   the main support for the interior ends of the floor joists.

Glazing compound   material used to waterproof glass panes in any wooden frame.

Grade   the existing ground surface.

Grain-edge (quartersawn)   lumber sawed at right angles to the growth rings.

Grain flat   lumber sawed tangent to the growth rings.

Groove   a notch, on the edge of a board, that runs along its entire length.

Grounds   wood strips used as surface guides for plastering.

Grout   mortar used to fill the joints between wall and floor tile.

Gusset (Fishplate)   used to connect components of wooden roof trusses.

Gutter (Eavestrough)   collects rainwater from the roof and drains it into downspouts.

Gypsum board (Sheet rock)   dry wall plasterboard.

Gypsum plaster   plaster base coat with sand, top coat without sand.

Hanger   "U" shaped bracket which supports the end of a beam at the foundation wall.

Hardboard   manufactured material sold in 4 × 8 sheets, peg-board is an example.

Header   beam placed perpendicular to the floor joists, and above door and window openings.

Head jamb   the top member of a door or window frame.

Hearth   fireplace floor made of brick, tile or stone.

Hip   angle formed where two sloping sides of a roof meet.

Hip roof   one that slopes from all four sides of the building.

Humidifier   unit designed to release additional water vapor into the home.

I-beam   steel beam used to support floor joists.

Insulation   any material designed to reduce heat loss or sound transmission through walls.

Interior finish   materials used to cover interior ceilings and walls.

Jack rafter   one that spans from the top plate to a hip, or a valley to the ridge.

Jamb   top and side frames of door and window openings.

Joint   any space between two adjacent surfaces.

Joint compound   material used to cover nailheads and seams in gypsum wallboard.

Joist   structural member designed to support a floor or ceiling.

Joist hanger   metal fastener used to support the end of a joist.

Kerf   void left in wood after a saw blade cuts through it.

Kiln dried   controlled oven drying reducing the moisture content of the wood to 6–12%.

Knee wall   short vertical wall connecting rafters to attic floor.

Knot   an imperfection in a wood's surface representing a former tree limb or branch.

Lally column   concete filled steel post used to support a beam.

Landing   platform between stair flights, or flat surface at top of stairs.

Lap joint   joint formed when the top member of the double top plate overlaps the lower member.

Lath   material used as a plaster base.

Lattice   a framework of crossed wooden or metal strips.

Ledger   a strip of lumber, nailed to a girder or header, that is used to support floor joists.

Level   condition existing whenever any surface is at a true horizontal.

Lineal foot   actual length of a piece of wood.

Lintel   horizontal support member placed above a door or window opening.

Lookout   short horizontal bracket connecting the overhanging fly rafter to a common rafter.

Louver   screened opening used to ventilate attic and crawl spaces.

Lumber   classified as boards, dimension, dressed, matched, timbers and yard lumber.

Mantel   shelf above a fireplace.

Masonry   brick, stone, tile, concrete blocks joined together with mortar.

Matched lumber   tongue and grooved lumber.

Metal lath   used to form or shape openings and as a base for plaster.

Millwork   lumber shaped or molded in a millwork shop.

Miter joint   where two connecting pieces have their ends cut at a 45 degree angle.

Moisture content   percentage of water present in oven dried wood.

Molding   any strip of wood used for decorative purposes.

Mortar   material used to join masonry units together.

Mortise   a slot cut in a piece of wood that is used to accept the tenon of another piece.

Mullion   a vertical wood divider strip between two window or door openings.

Muntin   the wooden dividing member between glass panes in a window sash.

Natural finish   material placed on a wood surface that does not alter its original color or grain.

Newel   any post used to attach a railing or balustrade.

Nominal dimension   the stated size of a piece of lumber, its actual size is smaller.

Non bearing wall   one that only supports its own weight.

Nosing   a molding whose edge projects beyond the underlying piece, *e.g.* stair tread over a riser.

Notch   a groove cut in the end of a board.

On center (o.c.)   standard spacing between joists, studs and rafters, it is measured from the center line of one member to the center of the next.

Panel   a thin piece of wood placed between the stiles and rails of a door, also 4 × 8 sheets used as wall coverings (paneling).

Paper (Felt, Sheathing paper, Building paper) used beneath floors, siding, and roofing.

Particle board (Flakeboard) manufactured sheet material composed of wood chips, shavings and used in cabinet making and as a floor underlayment.

Parting stop or strip molding strip used in side and head jambs to separate the two sash in a double hung window.

Partition a wall used to subdivide the available space into rooms.

Penny (d) designation given to nail length, originally referred to the price per hundred (English system).

Pier a short support column.

Pilaster built in projecting column designed to reinforce a foundation wall.

Pile a timber driven into the earth to support a structure, it replaces a normal foundation.

Pilot hole a small hole drilled into a piece of wood to prevent it from splitting when it receives a screw or nail.

Pitch the roof's slope which is expressed as a ratio between its rise and span.

Pitch pocket an opening in the wood's surface that oozes pitch.

Plan the top view of a building.

Plaster a mixture of sand, lime and gypsum troweled on ceilings and walls.

Plastic petroleum products used as building materials.

Plate (sole, sill, top) horizontal members connecting wall studs.

Plough a lengthwise grooved-cut in a board.

Plumb exactly vertical or perpendicular to the horizontal.

Ply denotes the number of layers of roofing, veneers in plywood.

Plywood sheet material made of an odd number of veneers, minimum of three, that are joined with glue, and have the grain of each ply at right angles to the one below.

Pores wood vessel openings visible on the wood's surface.

Post and beam type of construction using heavy beams and posts for support and decorative effects.

Preservative chemcial that suppresses wood decay and insect attack.

Pressure treated wood wood impregnated with a solution making it impervious to weather, insects and decay.

Primer first coat of paint applied to bare wood.

Purlin horizontal support timber used to help support roof rafters and sheathing.

Putty material used to seal glass in sash, and as a wood filler.

Quarter round molding whose surface equals ¼ of a circle.

Rabbet a groove, running across the grain, at the end of a board.

Rafter the structural member of a roof.

Rafter hip it forms the hip of a roof (external roof angle).

Rafter valley it forms the valley of a roof (internal roof angle).

Rail horizontal cross member of a window or paneled door.

Rake trim used to finish off a roof edge that runs parallel to its slope.

Reinforcing (Rebar) steel rods or wire embedded in concrete to strengthen it.

Ridge top horizontal member of the roof frame connecting rafter pairs together.

Ripping to saw a board parallel to its grain.

Rise vertical distance from top plate to roof peak.

Riser a lumber piece enclosing the vertical distance between stair treads.

Roll roofing asphalt impregnated material 36″ wide covering 108 square feet.

Roughing in term used to describe the entire house framing process.

Rough opening opening in frame for a window, door, stairwell, or chimney.

Run-roofs horizontal distance that a rafter covers.

Run-stairs horizontal distance covered by a flight of stairs.

R-value measure of a material's resistance to the flow of heat.

Saddle (cricket) used between the rear side of a chimney and the sloping roof.

Sapwood outer portion of wood next to the bark.

Sash single window frame encasing one or more panes of glass.

Sash balance counterbalancing device used in double hung windows.

Scaffold temporary platform used during construction.

Scratch coat first coat of plaster.

Screed board used to level concrete, board used as guide to maintain plaster thickness.

Scribing method used to fit molding into irregular areas.

Sealer chemical applied to bare wood to seal its surface.

Seasoning moisture removal from green wood.

Shake thick hand-split wood single.

Sheathing material used to cover the home's structural components.

Shims wedge shaped wooden pieces, usually singles, used to fill gaps between two components or to level building components.

Shingles materials used as the finished exterior for roofs and exterior siding.

Shiplap lumber designed to make a lapped joint.

Shoring timbers used to brace a structure to prevent it from sagging.

Shutter decorative or protective frame placed at the side of each window and door.

Siding finished exterior house covering.

Sill lowest component of the structure's frame.

Sleeper 2″ thick board embedded in a concrete floor serving as a nailing base for a wood floor.

Slope rise of a roof expressed in inches per foot of run.

Soffit the underneath section of a cornice or boxed eave.

Soil stack vertical vent pipe for the house plumbing.

Solid bridging solid wood pieces nailed between joists in the center of their span.

Span the distance between two supporting points.

Splash block masonry feature used to divert rainwater from the downspouts.

Square unit measurement of one hundred square feet of roofing or siding.

Stair a set of steps.

Stair horse a stair stringer with the tread and riser sections cut into it.

Stile   the upright framing member in a panel door.

Stool   flat molding placed horizontally over the window sill and between its jambs.

Storm sash   additional window enclosing the existing one.

Story   the living area between the floor and ceiling.

Story-pole   wooden rod used to measure heights and locate rows of siding.

Stringer   diagonal stair support, also structural support member for floor and ceiling openings.

Strip flooring   narrow tongue and grooved wood strips.

Stucco   exterior masonry finish using Portland cement as its base.

Stud   Vertical member of a framed wall.

Subfloor   wood material nailed directly to the floor joists.

Suspended ceiling   one hung by wires and brackets from the joists in a gridwork pattern.

Tail beam   shortest structural component of a floor or ceiling opening.

Termite shield   metal shield placed on a foundation wall to keep termites away.

Threshold   tapered wood piece placed between door and its sill.

Timber   wood whose smallest dimension is 5″ or more.

Toe-nailing   driving nails at an angle through two wood pieces that are perpendicular to each other.

Tongued and grooved   matched lumber where the tongue of one piece slips into the groove of the adjacent piece.

Tread   horizontal stair component that is stepped on.

Trim   wood material used to finish the exterior and interior of the home.

Trimmer   joist to which a header is nailed in any framed opening.

Truss   prebuilt rigid framework used to support loads over a span (floors, roofs).

Undercoat   first coat of paint applied to bare wood before the finish coat is applied.

Underlayment   material placed beneath a finished surface, *e.g.* flooring to ensure a smooth surface.

Valley   the angle formed where two sides of a sloping roof meet.

Vapor barrier   material used to suppress the movement of water vapor into the exterior walls.

Veneer   a thin sheet of wood used to decorate the surface of a more common wood.

Vent   ductwork used to carry air throughout the home.

Vermiculite   insulating material used to fill concrete blocks.

Wainscoting   thin, tongue and groove, panel used to cover walls and ceilings.

Wane   presence of bark or absence of wood on the corner or edge of a piece of wood.

Warp   variety of lumber defects resulting from uneven shrinkage of the wood cells.

Weatherstrip   material placed around exterior surfaces of doors and windows to suppress the passage of air.

Weephole   a hole left in brick or masonry veneer walls to enable trapped moisture to escape to the building's exterior.

Wood rays   storage and transport cells that run radially through the wood.

INDEX